PHYSICS OF ELECTRONIC CERAMICS

PART B

CERAMICS AND GLASS: SCIENCE AND TECHNOLOGY

A Series of Monographs

Edited by

JOHN B. WACHTMAN, JR.
Inorganic Materials Division
National Bureau of Standards
Washington, D.C.

Volume 1
Radome Engineering Handbook: Design and Principles
edited by J. D. Walton, Jr.

Volume 2
Physics of Electronic Ceramics *(in two parts)*
edited by L. L. Hench and D. B. Dove

Volume 3
Characterization of Ceramics
edited by L. L. Hench and R. W. Gould

IN PREPARATION
Electrical Conductivity of Ceramics
edited by N. M. Tallan

Ceramic Processing
edited by P. J. Jorgensen

Mechanical Properties of Ceramics
edited by J. B. Wachtman, Jr.

PHYSICS OF ELECTRONIC CERAMICS

(in two parts)

PART B

Edited by L. L. Hench and D. B. Dove

College of Engineering
University of Florida
Gainesville, Florida

MARCEL DEKKER, INC. New York 1972

MARCEL DEKKER, INC.
95 Madison Avenue, New York, New York 10016

LIBRARY OF CONGRESS CATALOG CARD NUMBER 74-142893

ISBN NO. 0-8247-1314-1

PRINTED IN THE UNITED STATES OF AMERICA

CERAMICS AND GLASS: SCIENCE AND TECHNOLOGY

FOREWORD

Ceramics and glass are ancient materials which continue to be used in the old ways but which have also been refined and developed by modern science and technology into many new and useful types of material. Ceramics and glass can be manufactured with unique chemical, thermal, electrical, magnetic, optical, mechanical, and mass transport properties. The list of their applications is much too long to give here in detail but some examples will serve to illustrate the great diversity and importance of these applications. Special chemical and thermal properties are exploited in furnaces and crucibles which are essential to the refining, alloying, and casting of metals and to other types of high-temperature processing. Special electrical properties form the basis for applications as insulators, dielectrics, piezoelectric transducers, and an ever-increasing family of semiconductor devices. Magnetic ceramics include cores for computer memories and high-coercive-force permanent magnets for electric motors. Special optical properties of some ceramics make them suitable for laser hosts, modulators, infrared windows, etc. The high elastic moduli and hardness of certain ceramics make them suitable for special applications requiring mechanical stability such as gyroscope mounts and wear surfaces while the extremely low thermal expansion achievable is appropriate to telescope mirrors and to applications involving thermal shock. The intrinsic strength of many ceramics has never been fully exploited because of their brittleness and accompanying sensitivity to flaws, but great improvements in the strength of bulk commercial ceramics and the technique of surface compression combine to offer much promise in this area while the recent development of high-strength, high-temperature ceramic fibers opens new possibilities for composite materials. The high speed of sound in certain ceramics has led to their successful use in armor and the low damping characteristic of many ceramics makes them useful as delay lines. The great range of rates of mass transport achievable in ceramics makes them useful as protective coatings at the extreme of low diffusion and makes them promising as fuel cell components at the other extreme of high diffusion rates. The thermal stability of certain nuclear ceramics has led to their adoption as fuel elements for most power reactors.

Additional examples could be quoted to illustrate proven uses of ceramics in modern technology and the promise of other uses to come, but the continuing improvement of traditional ceramics with large volume use should also be noted. Glass, electrical and sanitary whitewares, structural clay products, enamels, and chinaware are still benefiting from science and technology through improved properties and reliability.

Ceramics materials are of great interest to scientists quite apart from their many practical applications. Certain phenomena occurring in ceramics lie at the frontier of science and stimulate scientific work of the highest sophistication and quality. These phenomena include the detailed behavior of superconducting polycrystalline oxides which are semiconductors at room temperature. The voltage and time dependent behavior of semiconducting glasses is not well understood and the question of the structure of glasses and its relation to their properties remains the basis of an active and challenging field of science. The discovery of the widespread occurrence of phase separation in glass and the use of controlled nucleation and controlled crystallization in glass to produce crystalline ceramics has raised stimulating new scientific questions as well as enriching technology. The range of behavior from glass to crystalline ceramics is so nearly continuous that the term "ceramic" is today frequently used to include the whole range of these materials. The chemistry of crystalline ceramics continues to provide scientific challenge as well as the discovery of useful materials. Layer structures, stacking-block structures, magnetic structures, the extent and character of solid solution, the type and interaction of point defects, and the structure of nonstoichiometric solids all need to be systematized and quantitively understood. The chemistry of surfaces, including grain boundaries, is receiving increasing study. Mechanical properties have been extensively studied but the details of mechanical behavior in specific materials offer many opportunities for research and development. The list of scientific challenges could be extended; ceramic materials are the seat of many phenomena worthy of study.

Some of the above applications are appropriate to a specific type of engineering and some of the scientific problems can be viewed simply as a part of solid state physics and chemistry. It is, however, usually advantageous to view both applications of ceramics and relevant aspects of science together from a materials science viewpoint in which the underlying chemistry, the kinetics of processing to produce a material which is often in a metastable state, and the physics of properties are all taken into account. Understanding the properties of ceramics depends upon understanding how microstructure and small changes of composition affect these properties. Experimental knowledge of the factors controlling properties depends critically upon measurement techniques, particularly upon techniques for measuring properties at high temperature and upon characterization techniques. The series "Ceramics and Glass: Science and Technology" is intended to present a survey of current knowledge of processing, characterization, and physical properties of ceramics taking into account the underlying chemistry; it is made up of two parts: (1) textbooks, monographs, and handbooks, and (2) selected conference proceedings. The first component in the series is made up of volumes each of which gives a complete treatment of its subject area rather that just a report of progress. Sufficient knowledge of the basic physics, chemistry and mathematics is assumed to treat each subject on the level of a first year graduate or advanced undergraduate course in materials science. This level is deliberately chosen to make the textbooks, monographs, and handbooks broadly useful to scientists and engineers in industry and in research and development laboratories as well as to students. The second component of the series is made up of conference proceedings selected for their subject, quality, and timeliness. The growth of technical society meetings in size and diversity of subject matter has caused increasing emphasis to be placed on small, separate topical conferences. Such conferences typically cover the selected topics with a series of invited papers plus a limited number of contributed papers. When successful, the proceedings of such a conference give a state-of-the-art presentation useful for several years until the new information can be assimilated in an organized manner into textbooks and handbooks. Good conferences and their proceedings have an important intermediate place between individual research papers and textbooks. The series "Ceramics and Glass: Science and Technology" accordingly includes selected conference proceedings but recognizes and maintains the distinction between these and the textbook, monograph, and handbook series component.

The present volume, "Physics of Electronic Ceramics," edited by L. L. Hench and D. B. Dove, is the proceedings of a conference held November 10-14, 1969, at the University of Florida, Gainesville, Florida. This conference was structured to deal first with general underlying electronic behavior (semiconductor and dielectric behavior), then with aspects associated with particular forms of material (bulk polycrystalline ceramics, glasses, glass ceramics, thin film, and surfaces) and finally with types of behavior of high technological importance (ferroelectrics, magnetic ceramics, and optical materials). Invited papers provided a survey and introduction to many of these areas covered by the conference so that the book can be profitably used by the general materials scientist or engineer as well as by the expert in one of these areas. A detailed treatment of the underlying science of electrical conduction is presented in textbook style by another volume in the series, "Electrical Conductivity in Ceramics and Glass" edited by Norman M. Tallan. The two books thus compliment each other despite some overlap in the subject matter. The present book presents aspects of the technology of electronic ceramics in terms of the underlying science upon which this technology depends -- a major goal of the series. Electronic and technical ceramics constitute the most rapidly growing component of the ceramic industry. It is hoped that this report of a conference designed to summarize current knowledge in this area will be a standard reference for users of electronic ceramics as well as for workers and teachers in this field.

John B. Wachtman, Jr.
Chief, Inorganic Materials Division
National Bureau of Standards

PREFACE

This book comprises the written proceedings of the Electronics Phenomena in Ceramics Conference held at the University of Florida, November 10-14, 1969. The objective of the Conference was to provide a review of the current status of physical theory for the field of electronic ceramics and to demonstrate applications of the physical theory to commercial electronic ceramic products where possible. The term "electronic ceramics" is taken in this volume to include such topics as semiconducting compounds, amorphous semiconductors, insulators, ferroelectrics, ferrites, and materials with laser and other optical applications. Separate sections on thin films and surface defect structure and behavior are included because of their relevance to the practical application of electronic ceramics in all phases of microcircuit technology. A Keynote lecture, "Quantum Mechanics and Ceramics," by Dr. J. C. Slater, and a Summary Address by Dr. D. G. Thomas on "Ceramics and Glasses-Some Uses in the Communications Industry," provide the theoretical-applications framework for the proceedings.

The volume has been organized to serve as a text for a graduate course in electronic ceramics. Each section begins with a review of physical theory pertinent to the section and this is followed by progressively more applied topics that represent the most recent advances in the field. The level of treatment is such that a student should have had a previous course in modern physics and preferably a solid state physics course on the level of Kittel's "Introduction to Solid State Physics."

Several factors contribute to make such a text especially timely. Research topics that classify as electronic ceramics have grown rapidly in both breadth and depth in the 1960s. For example, abstract coverage of electronic ceramics in the American Ceramic Society is well in excess of 600 items yearly. At a recent International Congress on Magnetic Materials, over 100 papers devoted to nonmetallic compounds were presented. Three international conferences on amorphous semiconductors alone were held in 1969.

The impact of such growth on the individual is that it tends to force narrow specializations and produces a barrier to communication and eventual application of research outside specific fields. Education in electronic ceramics both on the university level and the continuing education level suffers from the difficulty of having to reduce this large volume of material from many specialities into a digestible set of principles, and supporting facts. It is the purpose of this volume to aid in such a crosslinking between theorist, manufacturer, user, and teacher.

The Electronic Phenomena in Ceramics Conference is the fifth in a series of University Conferences on Ceramic Science, held previously at North Carolina State University at Raleigh, University of Notre Dame, University of California at Berkeley, and SUNY College of Ceramics, Alfred University.

Sponsoring agencies for the 1969 Electronic Phenomena in Ceramics Conference were the Electromagnetic Materials Branch of the Air Force Materials Laboratory, Wright-Patterson Air Force Base, Ohio; the Metallurgy and Ceramics Division of the U. S. Army Research Office, Durham, North Carolina; and the Department of Metallurgical and Materials Engineering, University of Florida. A special acknowledgment is extended to the personal encouragement and support of Lt. Col. D. J. Iden and Maj. C. Ehrenfried of AFML; Drs. H. M. Davis and G. Mayer of AROD; and Dr. F. N. Rhines of the department of Metallurgical and Materials Engineering. Thanks is also heartily extended to Mrs. Elizabeth Godey for her efficient secretarial management and to Mrs. Dottie Finder and Mrs. Melanie Peele for their registration assistance.

L. L. Hench

D. D. Dove

Gainesville, Florida

CONTRIBUTORS TO PART B

J. D. AXE, Brookhaven National Laboratory, Upton, New York

L. E. CROSS, Materials Research Laboratory and Department of Electrical Engineering, The Pennsylvania State University, University Park, Pennsylvania

W. K. CULBRETH, JR., Florida Tile Industries, Inc., Lakeland, Florida

R. A. EPPLER, Ceramics Group, Glideen-Durkee Division, SCM Corporation, Baltimore, Maryland

J. B. GOODENOUGH, Lincoln Laboratory, Massachusetts Institute of Technology, Lexington, Massachusetts

J. J. HAMMEL, PPG Industries, Inc., Glass Research Center, Pittsburgh, Pennsylvania

G. R. HARRISON, Sperry Microwave Electronics Division, Sperry Rand Corporation, Clearwater, Florida

L. R. HODGES, JR., Sperry Microwave Electronics Division, Sperry Rand Corporation, Clearwater, Florida

N. J. KREIDL, Department of Ceramics Engineering and Materials Research Center, University of Missouri-Rolla, Rolla, Missouri

R. PEPINSKY, Department of Physics and Department of Metallurgical and Materials Engineering, University of Florida, Gainesville, Florida

C. F. RAPP, Owens-Illinois, Inc., Toledo, Ohio

Y. R. SHEN, Department of Physics, University of California and Inorganic Materials Research Division, Lawrence Radiation Laboratory, Berkeley, California

F. J. SCHNETTLER, Bell Telephone Laboratories, Murray Hill, New Jersey

D. G. THOMAS, Bell Telephone Laboratories, Murray Hill, New Jersey

CONTENTS

CONTENTS OF PART A

PHYSICS OF
ELECTRONIC CERAMICS
PART B

VII. FERROELECTRICS

Chapter 21

SOME STRUCTURAL MECHANISMS IN FERROELECTRICITY

Ray Pepinsky
Department of Physics
and
Department of Metallurgical
and Materials Engineering
University of Florida
Gainesville, Florida

I. INTRODUCTION

When von Hippel's group at M. I. T. was investigating the dielectric properties of $BaTiO_3$ ceramics, in 1945, the investigators were astonished to discover that ceramic specimens prepared under certain conditions were piezoelectric. The anomalous dielectric properties of this material had been discovered earlier and independently by Wainer and Salomon[1] in this country and Ogawa[2] in Japan. The report of piezoelectricity in the ceramics was received with skepticism by some members of the M. I. T. group. If individual crystallites in the ceramic were indeed piezoelectric, it was at first assumed that the processing would result in a distribution of crystallite orientations which was essentially isotropic, and that a charge separation developed by stress on one crystallite would on the average be cancelled by that on another with opposite orientation.

The high dielectric peak at 120°C and the piezoelectricity eventually led to the conclusion that the material must be ferroelectric [3][4].* This was confirmed by the appearance of dielectric hysteresis. The piezoelectricity appeared because pyroelectric domains within each crystallite could be reoriented by an electric field; and when the ceramic was heated above 120°C and then cooled in the presence of an electric field, a predominance of domain orientations in the field direction remained after the field was removed.

The high dielectric constant and piezoelectric nature of the ceramic rendered the material of very great technical importance. Strong, chemically stable piezoelectric components could be fabricated in advantageous shapes for use as electromechanical transducers, particularly as sonar transmitters and detectors. Addition of proper amounts of other perovskites to $BaTiO_3$ in the compounding of the ceramics resulted in even more advantageous properties.

About this time (1946) Schubnikov[5] in the U.S.S.R. had experimented with Rochelle salt "ceramics," and had discussed their piezoelectric properties on the basis of the FE behavior.

---

*In the subsequent discussion, the symbol FE will be used to indicate the terms ferroelectricity or ferroelectric; AFE will indicate antiferroelectricity or antiferroelectric; and PE will indicate paraelectric. <u>HT</u> indicates high-temperature; <u>LT</u> implies low-temperature.

Importance of Single - Crystal Studies

Basic information concerning the physical properties of $BaTiO_3$, and particularly its domain structure and its alteration by electric fields, was advanced tremendously when good macroscopic single crystals were first prepared by Blattner, Matthias, and Merz[6]; and the development by Remeika[7] of a much-improved method for flux growth of large single crystals vastly facilitated many physical investigations and permitted single-crystal technological applications. The neutron diffraction analyses reported later in this chapter could not have been accomplished without Remeika's crystals.

Understanding of the properties of a ceramic can only be achieved through examination of the materials of which it is constituted both in single-crystal form and in the sintered polycrystalline state. Single crystals are particularly necessary for crystal-structural studies of FE's. The FE structures are always pseudo-symmetric, that is, their true structure differs only slightly from a structure of higher symmetry. This is precisely why it is possible, with a relatively low electric field component directed along the direction of spontaneous polarization of some domains, to cause domains to reorient which were not yet so polarized. All that is required, for reorientation of the electrical axis of a domain, is that some atoms move a distance which is generally only a fraction of a usual bond length. There are some cases where particular atoms or groups of atoms move by an Ångstrom or more in this redirection of the electric axis; but even in such cases other atoms move only slightly, and yet the displacement of every atom in the structure is significant.

These explanations are offered, perhaps unnecessarily, in order to justify the writer's concern not with ceramic specimens, in the following discussion, but with macroscopic single crystals. Furthermore, the illustrations are not restricted to materials such as $BaTiO_3$ which form ceramics in the usual sense of that term.

It is necessary for the writer to assume that readers will have some familiarity with the phenomenology of FE and AFE, or are willing to acquire such information. The latest of Landolt-Börnstein's "Numerical Data and Functional Relationships in Science and Technology" volumes is a magnificent outline of FE and AFE materials and their properties[8]; it was edited by T. Mitsui with the collaboration of 13 other Japanese investigators. Jona and Shirane's treatise[9], although prepared about a decade ago, is in the writer's view still the outstanding general discussion of this field; but readers will find the volumes by Martin[10] and by Fatuzzo and Merz[11] useful. Burfoot's book[12] is more theoretical and less helpful to the uninitiated.

II. PREMONITORY INDICATIONS OF CRYSTAL TRANSITIONS AND THE IMPORTANCE OF CRYSTAL CHEMISTRY

In the paper just previous to this, John Axe discussed with great clarity one of the important premonitory indications of an impending crystal transition. There are other such indicative phenomena in the PE phase. The most notable and perhaps the most significant of these is the usual Curie-Weiss-like variation of the PE dielectric "constant" (in one or more crystallographic directions) with temperature, as the transition point is approached.* There may also be temperature variations of elastic constants, changes in optical properties, etc. It must be realized: (a) that variations of any or all of these properties with temperature, and/or the appearance of a soft transverse optical mode, do not require that an FE or AFE transition is impending; (b) that the absence of such variations as a gradual function of temperature does not require that a transition is not about to occur; and (c) all of these variations in properties, if and when they appear, require an explanation on crystal-structural grounds just as much as do the appearance or nonappearance of a transition and the nature of the transition if it does occur. Conclusion (c) is the most important one, and provides the justification as well as the motivation for structural studies of FE and AFE materials.

A. Examples of Dielectric Anomalies Not Accompanied by FE or AFE

Figure 1 illustrates a dielectric anomaly in the clathrate crystal $(\text{hydroquinone})_3$ MeOH[13]. The structure of this crystal consists of a three-dimensional network of hydroquinone molecules, linked by H-bonds, with the methyl alcohol molecules enclosed in cavities in the network: one MeOH per cavity. No FE activity could be detected, nor was it possible to affect the dielectric behavior by means of a strong electric field in a manner which might suggest AFE behavior. The temperature of the transition was inconvenient for other measurements in our laboratory at the time of our interest in

*Similar behavior may appear in a FE phase, to signal the occurrence of a further transition.

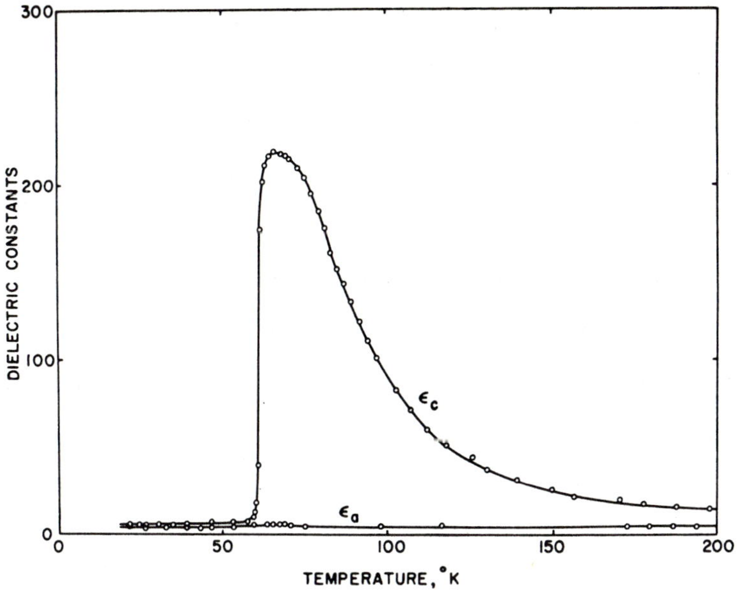

FIG. 1
$3C_6H_4(OH)_2 \cdot CH_3OH$: dielectric constant versus temperature curves.

the crystal; but no doubt other physical properties vary with temperature in much the same way as does the dielectric constant.

Similar dielectric anomalies are encountered in a good number of polar-molecular crystalline solids, with no expectation that the crystals become either FE or AFE. Soft transverse optical modes, changes in elastic constants, and other property changes, no doubt accompany the dielectric anomalies, nevertheless.

B. "Well-behaved" FE Transitions

KH_2PO_4, which transforms from PE above 122°K to FE below[14], shows in the former phase premonitory indications of the coming transition over a large temperature range. As the temperature is lowered, the dielectric constant ε_c along the PE tetragonal axis begins to follow a Curie-Weiss law at a temperature about 50° above the transition point, and reaches a value of almost 10^5 just before the transition occurs[15]. Figure 2 is the well-known illustration of this behavior[15]. The symmetry of the PE phase is $I\bar{4}2d$, and thus piezoelectricity is permitted. The piezoelectric strain constant d_{36} (which relates the shear stress X_Y to the polarization P_3 on the c face of the crystal) follows a similar Curie-Weiss curve, and reaches an extremely high value ($\approx 1500 \times 10^{-12}$ coulombs/newton) near the transition point[16]. The electrooptical modulus r_{63} shows a very similar behavior[17].

The Curie-Weiss behavior of the above-indicated properties characterizes what is here designated as "well-behaved." We shall now discuss some less-disciplined FE's.

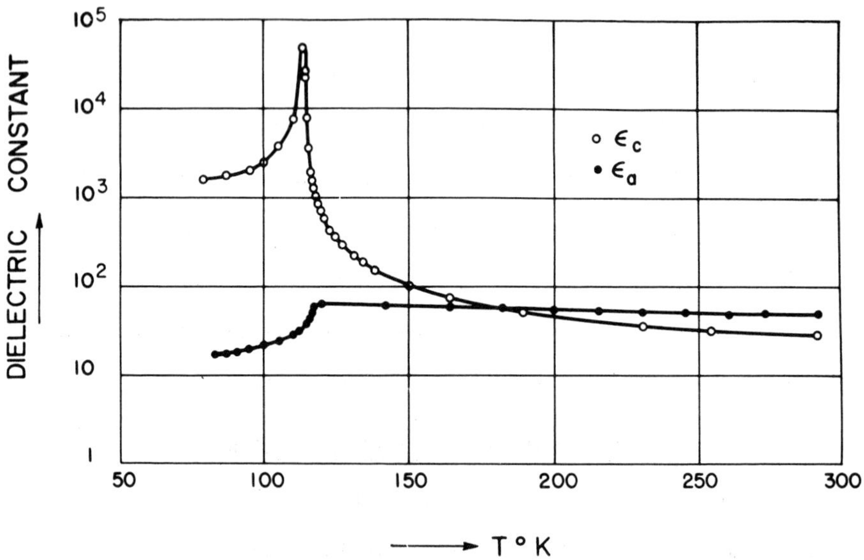

FIG. 2
Dielectric constant of KH_2PO_4 as a function of temperature (after Busch).

C. FE's with No Transitions

Evidence in the PE phase of an impending FE transition does not always appear, at temperatures some distance from the transition point. The first badly behaved FE was discovered by Matthias in 1955; this was (guanidine) $Al(SO_4)_2 \cdot 6H_2O$ and isomorphous crystals with trivalent Cr, Ga, or V in place of Al and SeO_4 instead of SO_4[18]. These materials had no PE phase at all; they remained in the same FE phase throughout the entire temperature range of their stability in the solid state (at normal pressure)—although the coercive field climbed to intolerably high values at low temperature. There are now four other such FE's known:

a. $LiH_3(SeO_3)_2$[19]

b. $Li(N_2H_5)SO_4$[20]

c. (glycine)$_2 \cdot MnCl_2 \cdot 2H_2O$[21]

d. $[N(CH_3)_4]HgCl_3$, where Cl can be replaced by Br or I and in some cases $[N(CH_3)_4]$ by $[P(CH_3)_4]$[22].

One cannot make direct use of variation of properties with temperature in the search for FE's of this type. A logical search strategy will be based upon chemical and crystal-chemical relationships to known FE's, or similar relationships to known materials with properties close to FE or AFE behavior. All five of the transitionless room-temperature FE's were discovered by use of certain rational criteria.

D. PE-FE Transitions Without Premonitory Indications

There is another group of FE's in which no premonitory evidence of an FE transition appears in the PE phase, at least if this evidence is sought in the usual manner. Three materials displaying this type of anomalous behavior are $(NH_4)_2SO_4$[23], $(NH_4)_2Cd_2(SO_4)_3$[24], and $NaNH_4SO_4 \cdot 2H_2O$[25].

The behavior of the first of these is shown in Figs. 3, 4, and 5, which compare some of the dielectric properties of $(NH_4)_2SO_4$ with pseudoisomorphous $(NH_4)_2BeF_4$[26][27][28]. The former becomes FE below -49°C, along the orthorhombic c axis; the latter becomes FE along b axis, at -97°C. The small but significant differences between the symmetries and structures of the two phases of each compound were first determined in our laboratories[29][30] and those of $(NH_4)_2SO_4$ refined later at Brookhaven[31]; some references will be made to these crystals in a later section.

As Fig. 3 indicates, neither the sulfate nor the fluoberyllate is well behaved in the PE phase just above the transition point, but the sulfate is particularly anomalous; neither shows Curie-Weiss behavior even if the measurements are made at a very slow time rate of temperature change. The maximum value of ε_c for the sulfate is ≃165, although this cannot be shown on the figure; ε_c decreases slowly as the temperature of the sulfate is lowered in the FE phase, while the corresponding rate of change of ε_b in the FE fluoberyllate is very much faster. Figure 4 compares the spontaneous polarizations of the two materials, of course in the FE phase; and Fig. 5 the coercive fields for both.

Figure 6[32] illustrates the similarity in behavior of $(NH_4)_2Cd_2(SO_4)_3$ to that of $(NH_4)_2SO_4$; and Fig. 7 shows the dielectric constant ε_c for $NaNH_4SO_4 \cdot 2H_2O$[33]. Measurements are made at a rate of about ~0.1°/min. in all of these observations. Takagi and

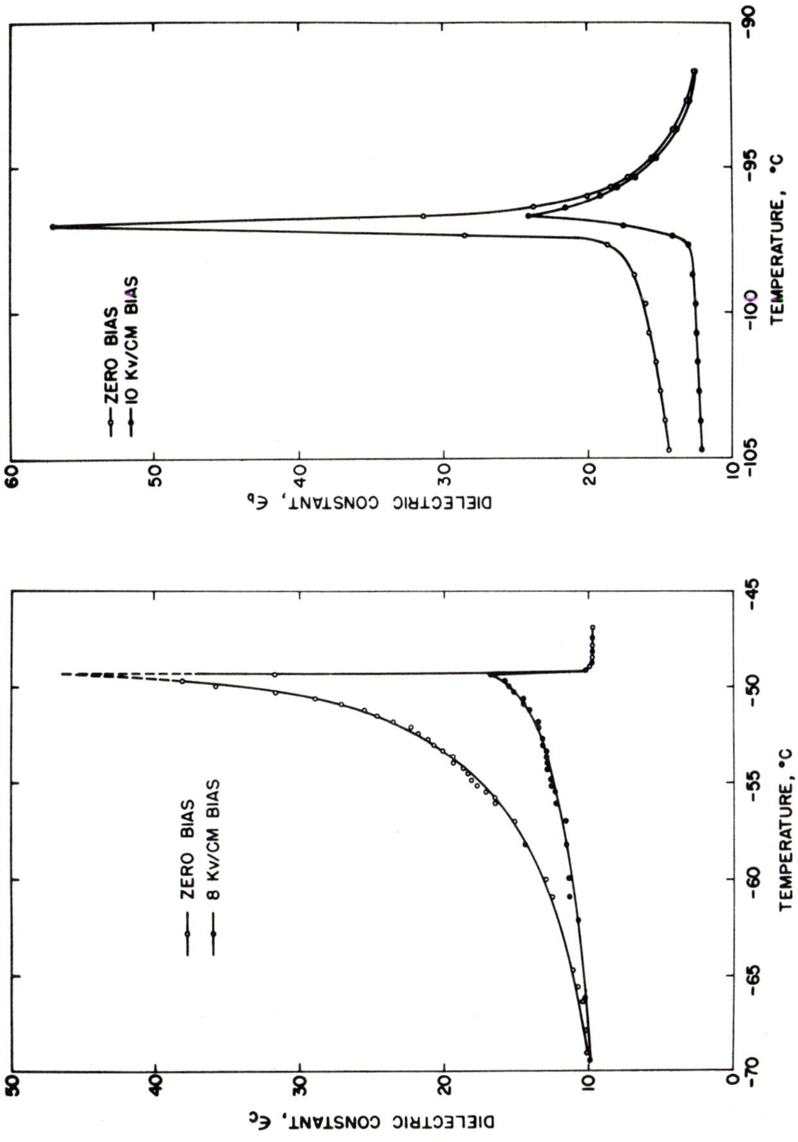

FIG. 3

(a) $(NH_4)_2SO_4$: effect of dc biasing field on dielectric constant ε_c; (b) $(NH_4)_2BeF_4$: effect of dc biasing field on dielectric constant ε_b.

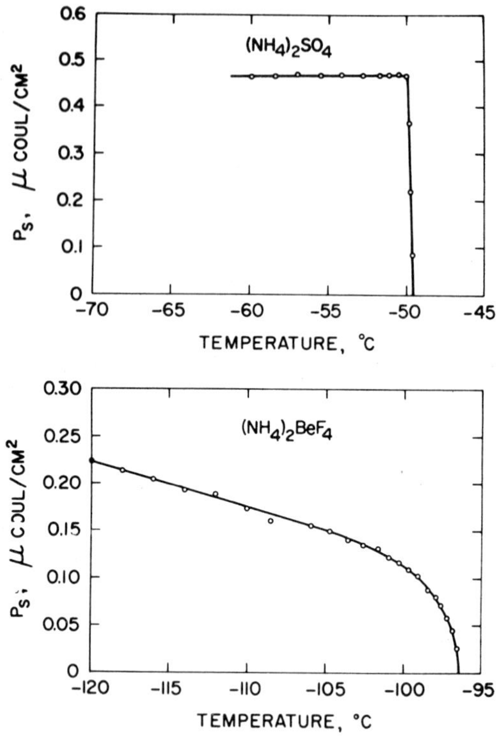

FIG. 4
Spontaneous polarization of $(NH_4)_2SO_4$ and $(NH_4)_2BeF_4$ in the neighborhood of the Curie point.

and Makita[34] have indicated that if measurements are made at an even slower rate, the discontinuity in $(NH_4)_2SO_4$ is less sharp. This suggests some type of very slow relaxation process near the transition point; and perhaps the same sort of process applies in the other crystals as well. If so, one would still not have indication of an impending transition in a usual exploration of dielectric behavior. The shape of the specific heat anomaly (λ-type) in $(NH_4)_2SO_4$ does indeed appear to be dependent upon rate of measurement, particularly for rates below a few degrees per hour[28].

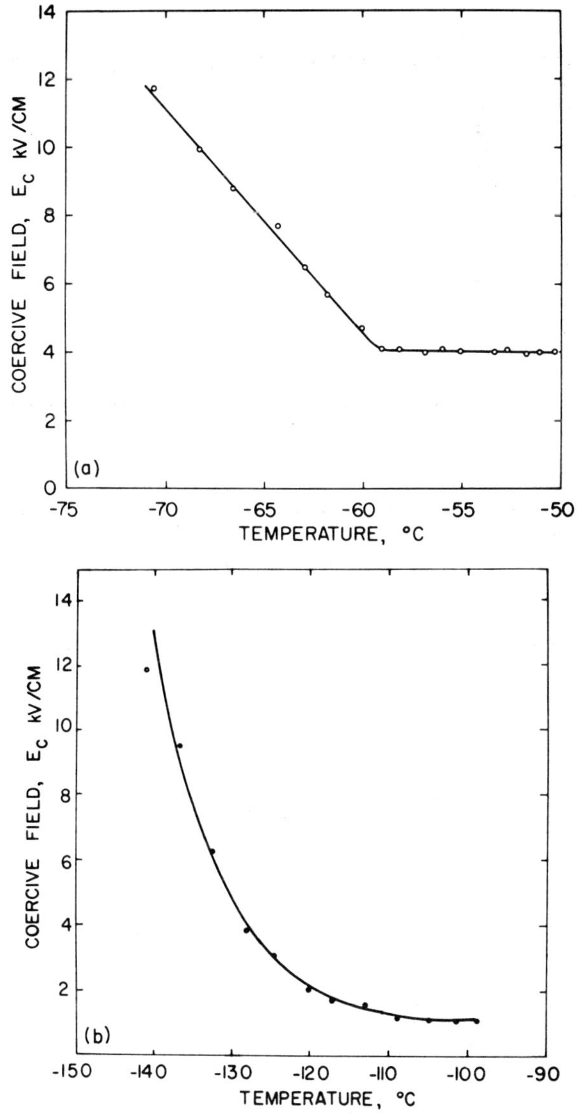

FIG. 5
(a) $(NH_4)_2SO_4$: coercive field versus temperature; (b) $(NH_4)_2BeF_4$: coercive field versus temperature.

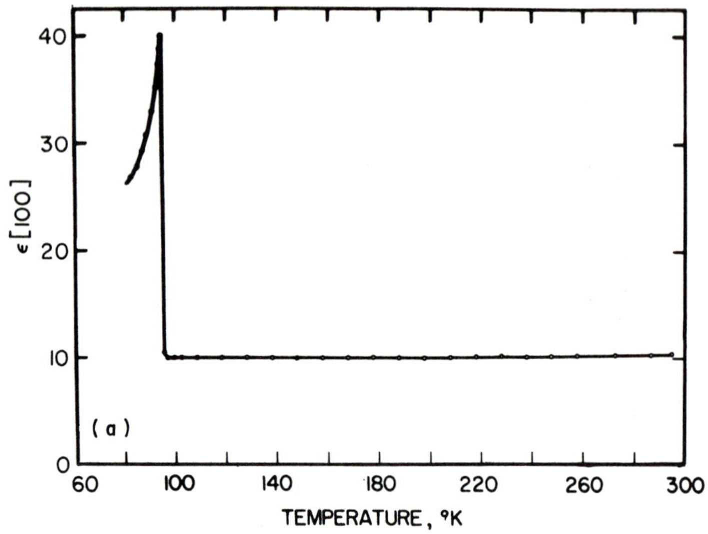

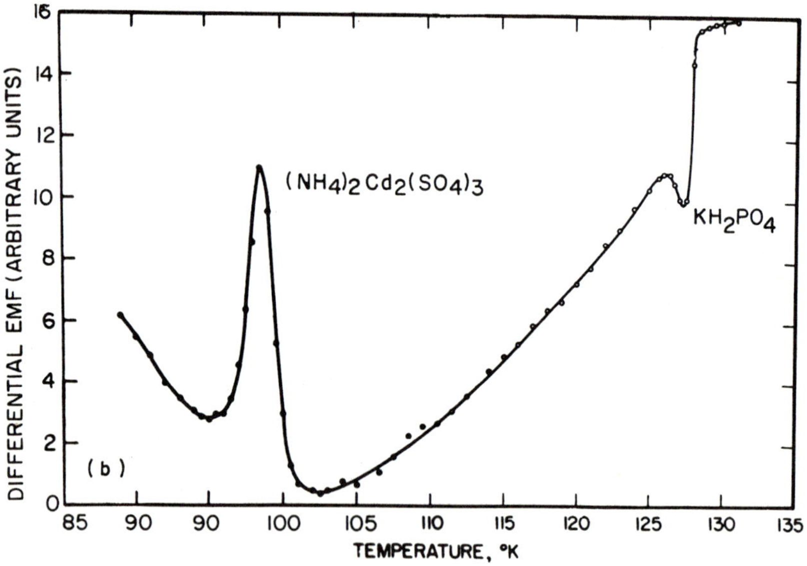

FIG. 6

(a) Dielectric constant ε[100] of $(NH_4)_2Cd_2(SO_4)_3$ as a function of temperature; (b) differential thermal analysis: $(NH_4)_2Cd_2(SO_4)_3$; (c) $(NH_4)_2Cd_2(SO_4)_3$: spontaneous polarizations P_s[100], P_s [100], and P_s[100] as a function of temperature.

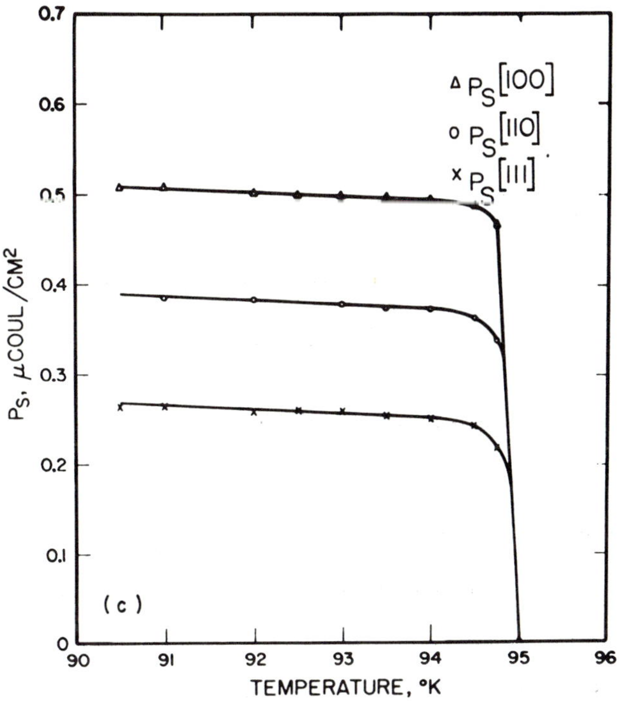

FIG. 6 (cont.)

A significant difference between the $(NH_4)_2SO_4$ and $(NH_4)_2BeF_4$ transitions still remains to be explained, however. Merz[35] first showed (for $BaTiO_3$) that application of a strong field along the direction of the FE axis when the crystal is just above the transition temperature results in a shifting of the transition temperature.

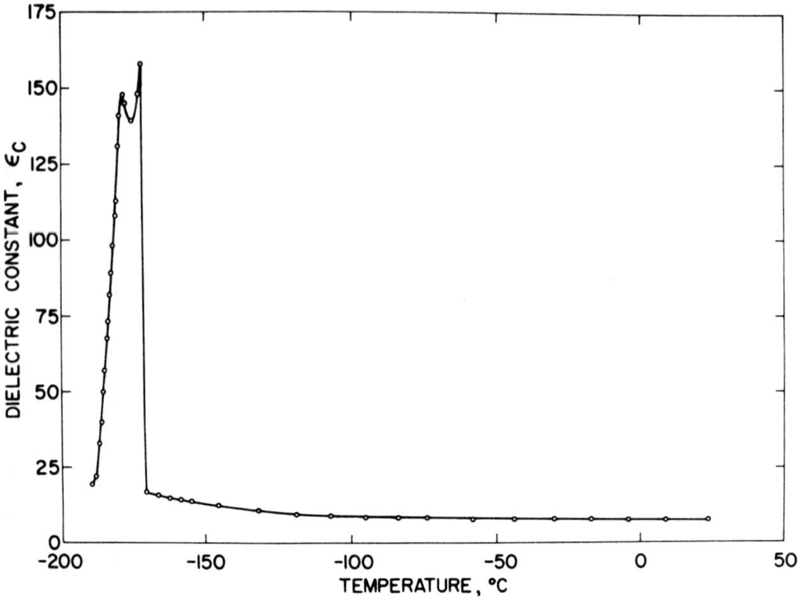

FIG. 7
$NaNH_4SO_4 \cdot 2H_2O$: dielectric constant versus temperature.

Also if the field is alternating in direction, double loops appear as shown in Fig. 9, which is to be compared to the normal FE hysteresis loop shown in Fig. 8. Double loops of the type in Fig. 9 can be observed in $(NH_4)_2BeF_4$, just above the transition temperature; but none can be found in $(NH_4)_2SO_4$ even with an applied field of 30kV/cm[28]. These double loops also cannot be found in $(NH_4)_2Cd_2 \times (SO_4)_3$, even with a field of 32 kV/cm[32].

The $(NH_4)_2Cd_2(SO_4)_3$ is remarkable in that the sharp rise in ε shown in Fig. 6 was not ameliorated when the dielectric measurements were accomplished at extremely slow rates; and yet double hysteresis loops as in Fig. 9 were observable just above the transition point[32].

The differences between $(NH_4)_2SO_4$ and $(NH_4)_2BeF_4$ are surely due, ultimately, to the greater strength of N-H. . .F bonds than

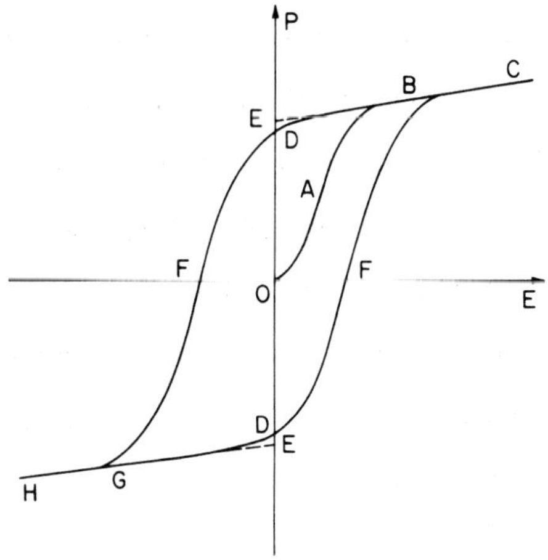

FIG. 8
Ferroelectric hysteresis loop (schematic): OE = spontaneous polarization P_s, OF = coercive field strength E_c.

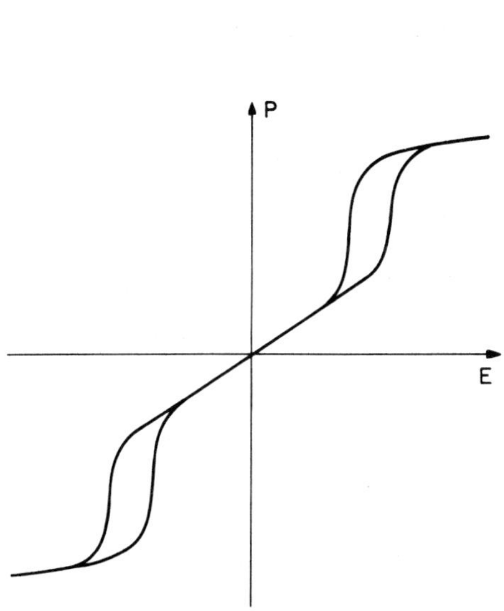

FIG. 9
A double hysteresis loop of $BaTiO_3$ observed just above the Curie point (schematic). The induced transition is shown by the sudden jump in polarization.

N-H•••O bonds; and the great sensitivity, of the dielectric anomalies and nature of the transitions to these bond differences, is not surprising although we cannot yet account for them in detail. The complete structure analysis of the $(NH_4)_2BeF_4$ transition is not yet available; and when it is, details of the superstructures found in both materials must be clarified. Finally, we must explain the remarkable phase diagram for the $(NH_4)_2(SO_4, BeF_4)$system, which results in quite a new symmetry for crystals $(NH_4)_2(SO_4)_x(BF_4)_{1-x}$ where $0.20 < x < 0.70$[28].

Deuteration of $(NH_4)_2SO_4$ has very little effect, if any, upon either dielectric or thermal properties; but $(ND_4)_2BeF_4$ has its transition point about 3° higher than that of $(NH_4)_2BeF_4$, and the transition energy and entropy change are also larger for the deuterated fluoberyllate[28]. This again is reasonable in terms of the difference in H-bond strengths.

E. The AFE Problem

The definition of AFE involves not only the presence in a crystal of dipoles arranged in antiparallel arrays, but also the occurrence of a reversible transition accompanied by a dielectric anomaly. The definition is necessarily so restrictive because the first condition of the two in the definition is satisfied widely; AFE further implies that at some temperature in another phase an instability has developed in the structure such that its dielectric polarizability alters; with further temperature change this leads to the transition to the new phase. If this involves an antiparallel disposition of dipoles, it is indeed AFE.

A foremost investigator in FE has proposed (privately) that the occurrence of a reversible dielectric anomaly must imply either an FE or AFE transition. Since dielectric anomalies do abound, this is a tempting idea; but it seems to exclude some structurally and thermodynamically acceptable alternatives. Two classes of crystals which can serve as counter-examples to the proposal are now discussed.

F. Dielectric Anomalies Not Necessarily Implying FE or AFE

The first class is comprised of molecular crystals, the structures of which represent a balance between intermolecular bonding which is chiefly Van der Waals, packing which is controlled by molecular shape, and thermal motion of the molecules as a whole and perhaps of rotatable intramolecular groups. As the temperature of such a crystal is raised toward its melting point, one or more transitions can occur, accompanied by specific heat anomalies; these are due to stepwise disordering of molecular orientation. The latent heat of melting of such crystals is low, since in a sense premelting has occurred at the earlier (lower) transition points.

If the globular molecules are polar, dielectric anomalies should appear at the "premelting" transitions; and indeed they do. Morgan[36] has provided an excellent review of such phenomena: in camphor and borneol and various derivatives, in cycloparaffins and derivatives, in substituted ethanes, and in substituted benzenes. Many more examples are known, and very many more are predictable. Anomalous dispersion of the dielectric behavior is observed, as one would also predict.

When the separation between the dipoles is great compared to that in usual FE's or AFE's, because of molecular size and/or shape, the coercive fields required for FE are likely to be so high as to be unachievable. Weak AFE is a possibility, but by no means necessary since molecular orientations are chiefly determined by packing conditions.

Nevertheless, one ferroelectric is known in this class: thiourea[37]. This was discovered by accident, during an unsuccessful attempt to incorporate protonated thiourea into aluminum and chromium alums, as the writer had been successful in doing with protonated urea and mono-, di-, and tri-methylamine[38][39]. The N-H•••S and N-H•••N bondings are so weak that they do not seriously restrict molecular motion in the crystal; and a series of transitions occur, at least two of which are FE and others perhaps AFE. Thiourea is a unique FE and details of the mechanisms of its behavior are difficult to extract. More such small-molecular FE's will surely be discovered.

A second class of crystals in which dielectric anomalies appear but in which FE or AFE need not occur is exemplified by the quinol • MeOH clathrate whose dielectric behavior was shown in Fig. 1. As earlier described, the methanol molecules, which are of course polar, are trapped as "guests" in the "host" hydroquinone cage structure[40][41][42]. There are many types of such inclusion compounds, as the latter two references indicate. Some are of the cage type, some have two-dimensional networks of channels in which polar molecules can be cocrystallized, and some have unidirectional channels in which the "guests" reside. The size of the cages and channels are determined by the chemical groups which interlink to form them, and the guest molecules, which must of course be of limited dimensions, have considerable freedom of thermal motion within these spaces: orientational and perhaps to a lesser degree translational freedom. Bonding between the host network and a guest is weak and in most cases only Van der Waals.

If the methanol molecules are restricted in orientation, once they have cocrystallized within the cages, they are certainly free to occur with the OH group statistically head-up or head-down with respect to the trigonal host-structure axis. At some elevated temperature they may be free to reverse their orientation completely; this is uncertain and of no immediate concern to us. At low temperature their thermal motions and that of atoms of the host structure are reduced; and eventually a temperature is reached at which their disorder is sufficiently reduced for them to couple somewhat uniformly to an impressed alternating electric field. There may be, at such a temperature, some OH•••O bonding between the guest and cage atoms; and in any case there is a definite reduction in mobility within the cages.

Relaxation effects are observed[43][44], as one expects; and low-frequency dielectric peaks have been observed in a number of other clathrates, including some of the urea and thiourea channel types[45]. In none of these has any evidence of FE or AFE been discovered. Nevertheless, some very intriguing possibilities for actual design of clathrate FE's or AFE's suggest themselves[45].

G. AFE Transitions

In view of the requirement of evidence of a dielectric instability in some temperature region, there can be no AFE materials analogous to the FE's which have no phase change over their entire temperature range of chemical stability at normal pressure. The appearance of AFE as a function of pressure has not been explored, and it would be interesting to do so.

When AFE transitions occur, they are as interesting as FE transitions; furthermore, they may give a clue as to where, in closely related materials, FE behavior might be found. As winter to spring in the last line of Shelley's Ode, where AFE appears, can FE be far behind?

There are two general types of AFE transitions: (a) those showing a more-or-less "well-behaved" dielectric peak as a function of temperature; and (b) those showing only a sudden drop in dielectric constant as the temperature is lowered. Examples of class (a) are found in $NH_4H_2PO_4$ or the corresponding arsenate[15], $PbZrO_3$[46][47] and many other simple or mixed perovskites and related oxygen-octahedra structures, $RbNO_3$[48], certain of the organic-amine alums[49], $CsH_3(SeO_3)_2$[50], Cu(formate)$_2$ · $4H_2O$[51], the highest-temperature (V to IV) transition in thiourea[52], $(NH_4)_3H(SO_4)$[53] β-$LiNH_4SO_4$[54], NH_4PF_6 · NH_4F[55], and NH_4I with excess iodine[56]. An illustration appears as Fig. 10 of the dielectric behavior of NH_4PF_6 · NH_4F, which has two AFE transitions: at 172°K and 228°K[57].

Class (b) is illustrated by Fig. 11, for the case of the fluosilicate salt of the di-protonated cage compound tri-ethylene-diamine [properly designated as 1,4-diazabicyclo(2,2,2)octane][58].

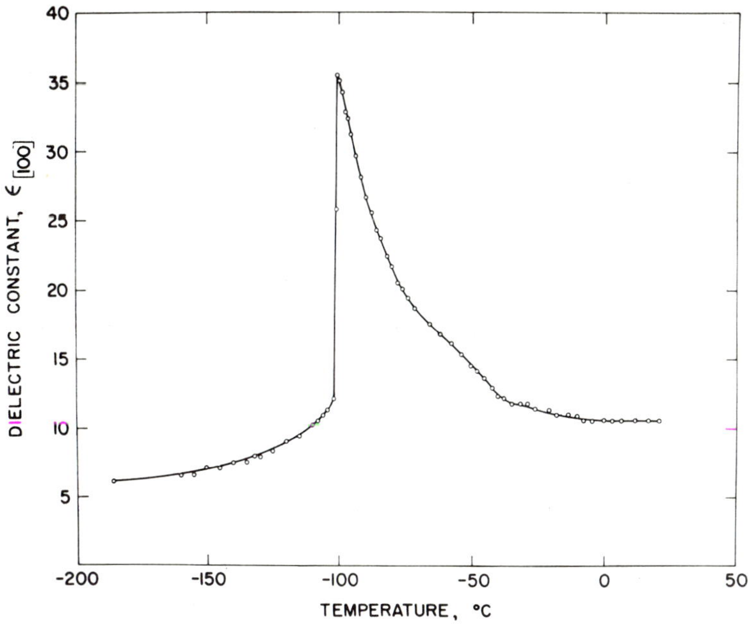

FIG. 10
$NH_4PF_6 \cdot NH_4F$: dielectric constant versus temperature.

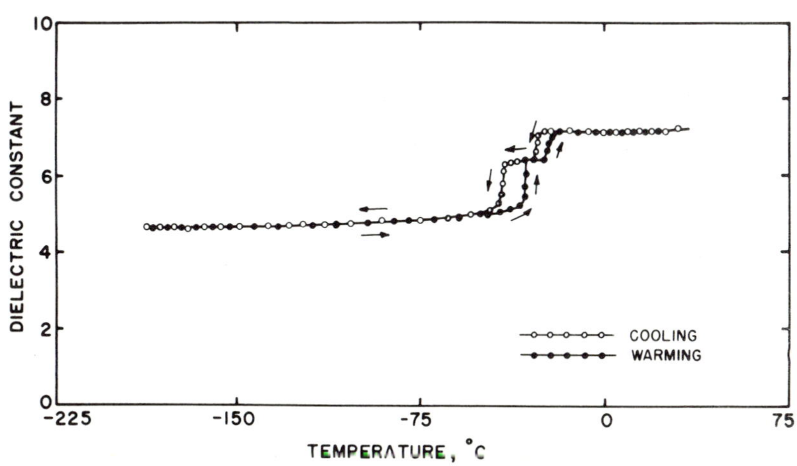

FIG. 11
$C_6H_{14}N_2SiF_6$: dielectric constant versus temperature curves.

The first-known representatives of this class were $Ag_2H_3IO_6$ and $(NH_4)_2H_3IO_6$[59]. A probable example is $KLiSO_4$[60]. We found a very large number of such step-down dielectric anomalies in our laboratory during the 1958-1962 period, but have not analyzed them to determine whether AFE actually occurred[61].

The significance of the difference between class (c) and class (b) AFE's will not be considered here. It should be mentioned that one can expect similar step-down anomalies in dielectric constants for some ferroelectric transitions.

H. Evidence Other than Dielectric of FE Behavior

One generally accepted definition of an AFE is that given in Jona and Shirane[9]: an AFE crystal is an antipolar crystal whose free energy is comparable to that of a polar crystal. This is concommitant with the possibility of driving an AFE by an electric field into an FE state, in the neighborhood of the temperature of the dielectric anomaly; and this forced transition is indeed evident in a number of AFE's: in which case one again obtains double hysteresis loops as in Fig. 9 if the applied field is alternating. (Transition should be first-order, if this occurs.)

There are borderline cases of AFE just as there are borderline FE cases (e.g., $NaNH_4$ · tartrate $\cdot 4H_2O$[62]); and definitions are likely to be strained by these. It is commonly stated that dielectric observations are essential to establishment of both FE and AFE states. This is not true, certainly, for the FE state because (a) anomalous dispersion X-ray diffraction analyses can establish the absolute configuration of a structure to which an unidirectional electric field has been applied and (b) the study of just a few pairs of

X-ray reflections (or even a single pair) from a known structure can permit determination of the state of polarization of the structure as a function of field. In the case of FE's such as Rochelle salt[63] LiTltartrate · H_2O[64], Ca_2Sr(propionate)$_6$[65], and $NaNH_4SO_4 \cdot 2H_2O$[33], for which the PE phases have optically active symmetries, establishment of the absolute configuration of the PE phase by anomalous dispersion then permits X-ray determination, in the FE phase and without the use of anomalous dispersion, of the absolute directions of small atomic shifts when an electric field is applied in a specified direction. One can, therefore, by means of X-ray diffraction measurements of a few reflections only, determine the state of polarization of the FE structure.

Furthermore, there are crystals such as $LiH_3(SeO_3)_2$ which are optically active and in which reversal of the polarization direction reverses the direction of optical rotation[66][69]; and there are other FE's in which incremental changes in optical activity accompany polarization reversal (e.g., Ca_2Sr(propionate)$_6$[70]). The state of polarization can be determined in these without recourse to static or low-frequency dielectric measurements; in the case of $LiH_3(SeO_3)_2$ one can easily trace out the FE hysteresis curve by these optical rotatory measurements only (see Fig. 12)[68]. Aizu and his colleagues[71][72] later discussed these and other even more interesting optical properties in FE's.

These comments and illustrations are not intended as deprecation of Jona and Shirane's excellent general discussion and definitions, in the first chapter of their treatise[9]. They do tend to remove the justification for a once very pertinent remark

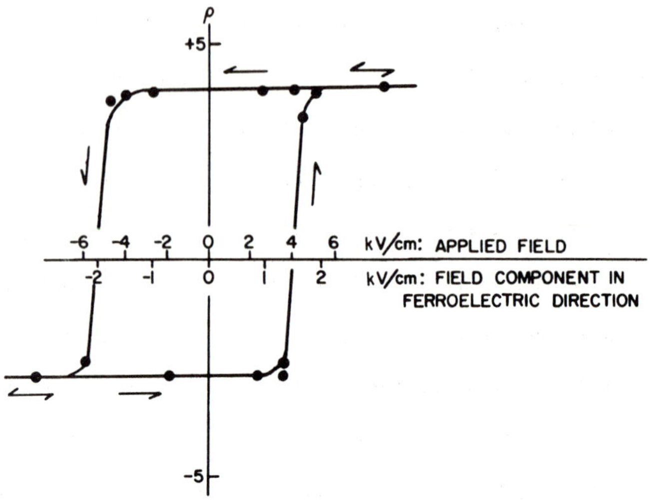

FIG. 12
Hysteresis of rotatory power of $LiH_3(SeO_3)_2$ crystal: static field applied in direction of optic axis O_I. The ρ = rotatory power in degrees/mm.

by Kaenzig: "Every year several new FE's and AFE's are discovered, and the field rapidly becomes broader, but not much deeper"[73].

I. Need for More Structural Information

The real source of both the depth and breadth of FE and AFE arises from the origin of the phenomena in the crystal chemistry of so wide a variety of such complex materials. Even $NaNO_2$[74], which is perhaps the simplest of FE's, needs more understanding; and it represents an isolated type of FE-PE behavior. $BaTiO_3$, and its immediate and more distant relatives, once looked so comparatively simple and now seems to evade complete understanding with increasing dexterity. What large understanding we have of the KH_2PO_4 and $NH_4H_2PO_4$ families is still incomplete, and does not help us very much with an understanding of tris-glycine sulfate[75] or Rochelle salt, for example.

The writer has often emphasized the viewpoint of FE and AFE as crystal-chemical accidents. This statement is intended to suggest that the origins of these phenomena arise from delicate balances of interatomic forces—alterable by temperature, electrical and mechanical stresses—which for any particular crystalline compound are specific to that compound and its immediate relatives. The generalities which are possible in ferromagnetism and its associated phenomena do not carry over well to FE, because details of atomic arrangements in crystals play a much larger role in dielectric phenomena than they do in magnetic. Thus crystal structure analyses, of all phases which appear in a material showing dielectric anomalies, are absolutely basic to understanding of the behavior.

Before roughly illustrating the structural details of a few FE mechanisms, reference will be made to magnetic resonance and relaxation phenomena as tools for elucidation of these mechanisms. The reason for this ordering of the discussion will be apparent eventually.

J. Magnetic Resonance Methods

NMR methods have been widely applied to FE's and their transitions. The measurements can be accomplished fairly rapidly; but their interpretation is often more difficult; particularly if the NMR practitioner is not fully familiar with available crystallographic and dielectric information on the crystal concerned.

For examples of careful observations and very penetrating interpretations of NMR data, readers are referred to the proton relaxation and deuteron quadrupole resonance studies carried out by O'Reilly and Tsang[75] on $(NH_4)_2SO_4$ and $(ND_4)_2SO_4$. These investigators had carefully studied the anomalous dielectric behavior of the PE and FE crystals, their thermal properties, and the X-ray and neutron structure analyses; and they correctly challenged the bearing of the latter on the transition and the character of the PE phase.

The NQR technique is capable of following the premonitory stages of a transition, on either side of a transition point and directly through the transition. This continuous monitoring of the structure could be accomplished by diffraction methods only with the expenditure of impossibly great effort. No such attempt has ever been made, because of the tremendous amounts of data which would have to be collected, to continuously explore any kind of transition by means of coherent scattering measurements: to the end that changes in Debye-Waller thermal oscillation parameters as well as atomic coordinate changes are established as a function of temperature—over the entire temperature range in which these parameters must be undergoing significant alterations. Herein lies one of the specific, extreme advantages of resonance techniques, properly applied and analyzed. The ammonium sulfate studies of O'Reilly and Tsang are eye-openers, and are unmatched by many other resonance studies of FE's.

Another feature of resonance methods which can be of importance in some structural studies is that the resonance observations do not damage the structure. There is evidence that incident X-ray beams produce marked changed in the dielectric properties of Rochelle salt and (glycine)$_3$ · H_2SO_4 during exposures for measurement of scattered X-ray intensities; and at least in the case of Rochelle salt this X-ray damage causes slight changes in the structure. These effects will be mentioned in later sections, in conjunction with interpretations of structure analyses. Similar X-ray damage can be anticipated in diffraction analyses of such FE's as $Li(N_2H_5)SO_4$[20] and $(glycine)_2 \cdot MnCl_2 \cdot 2H_2O$[21], which tend to decompose at room temperature and pressure; and some damage may occur as well in the $LiH_3(SeO_3)_2$ family[19] and $(guanidine)AlSO_4 \cdot 6H_2O$ and isomorphs[18]. The use of NQR for studies of damaged Rochelle salt, to determine what effects the damaging radiation has had on the bulk structure, has not occurred to the investigators who previously reported such damage phenomena. This sort of investigation will also be considered, at the end of the Rochelle salt discussion below; and resonance observations involving this feature of the Rochelle salt problem and other FE problems are planned here at the University of Florida.

Thermal neutron beams used in diffraction analyses do not produce such damage; but collection of accurate three-dimensional neutron scattering data is even more time-consuming and expensive than the gathering of analogous X-ray data, and can be accomplished in only a few laboratories where high-flux reactors are available. Thus again it becomes even more unfeasible to determine atomic coordinates and thermal parameters accurately as a continuous function of temperature over the range of a dielectric anomaly. Of course one can follow X-ray or neutron intensity changes for a limited number of reflections, as a function of temperature; but this does not provide the wealth of information which can be extracted by resonance methods with an equivalent amount of experimental effort.

There is no question but that interpretation of NMR data should be based upon well-established structural information. O'Reilly and Tsang have also examined NMR and spin-lattice relaxations in

$(NH_4)_2BeF_4$[76]; and because structural information on this crystal was not available to them, they refrained from an attempt to carry interpretation of their data beyond the point where knowledge of the structure was essential. (Such wise reticence is not always displayed by other NMR practitioners, unfortunately.) As Figs. 3, 4, and 5 indicate[28], $(NH_4)_2BeF_4$ is a "better-behaved" FE than the sulfate. Of course its FE axis is along b, while that of $(NH_4)_2SO_4$ is along the pseudoisomorphous c direction; and other differences are apparent in the figures and discussed at length in Ref. [27] and [28].

Two features of the crystallography of these materials merit mention. Superstructuring is observed in the PE phase of $(NH_4)_2BeF_4$, such that b and c axes are actually doubled in length with respect to the "basic" cell; and the a axis is doubled in the FE phase. The structures of the two phases are thus quite complex. It is unfortunate that our neutron analysis of the structures of both phases, interrupted by a change in laboratories, has not been completed[29][30]. We did offer the hypothesis that the high-temperature phase is actually AFE. A search for a still higher temperature transition to a possible truly PE phase was negative, but this does not rule out the AFE condition.

It is particularly interesting that superstructuring was sometimes found in the room-temperature phase of $(NH_4)_2SO_4$, and sometimes appeared to be missing[27]. This is significant in the light of O'Reilly and Tsang's conclusion, from the NQR study just above the transition point, that the structure of that phase must indeed be AFE (two interpenetrating and oppositely oriented FE structures). Hamilton[77] has questioned this conclusion, on the basis of the neutron structure analysis[31]; but the fact that the latter was accomplished at a much higher temperature (300°K) was considered to be a possible reason for the discrepancy between the two results[75].

Both O'Reilly and Tsang[75] and the neutron analysts[31][77] assumed the "basic" orthorhombic cell for PE and for FE $(NH_4)_2SO_4$, and at no point considered the significance of a superstructure. The NMR investigators added a small amount of $Co(NO_3)_2$ to their solution, during crystal preparation, to "improve the quality of the crystals"; and this may have affected the crystal texture.

One might wonder: is it possible that $(NH_4)_2SO_4$ has two transition points, as the specific heat curve suggests: one at -47.5°C and the second, corresponding to the Curie point, at -50°C? The specific heat curves for $(NH_4)_2SO_4$ and $(NH_4)_2BeF_4$ are shown in Figs. 13 and 14, respectively[28]. (The curves for the deuterated materials are analogous.) The thermal behaviors are extensively discussed in reference [28]. What is pertinent here is that the peak in the sulfate at -47.5°C does alter in position and shape according to the rate of temperature change. This is also true of the shape of the dielectric constant ε_c curve for the sulfate. Our conclusion[28] was: "these data indicate the existence of a relaxation mechanism with a very large relaxation time." Another conclusion might have been that a metastable intermediate phase can exist, between the nonsuperstructured PE phase and the FE phase, with the transition temperature from the PE to intermediate phase depending upon the purity and degree of perfection of the crystals.

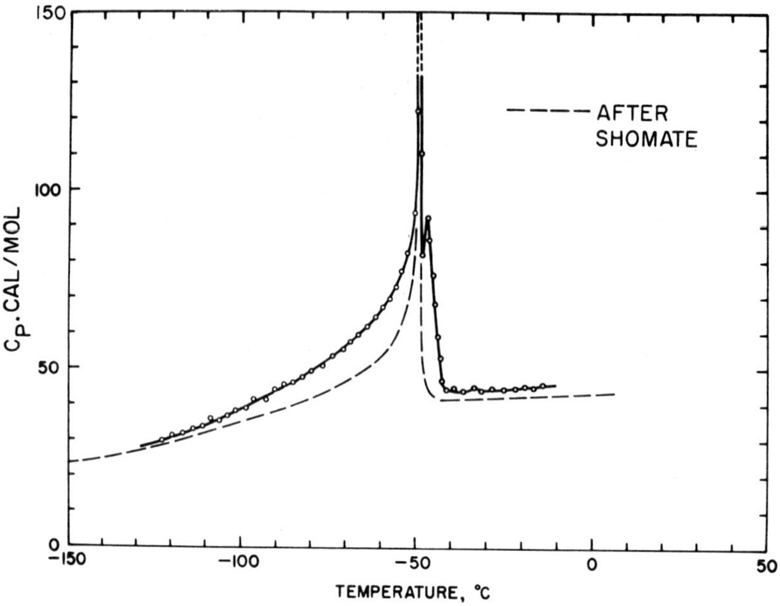

FIG. 13
$(NH_4)_2SO_4$: specific heat versus temperature.

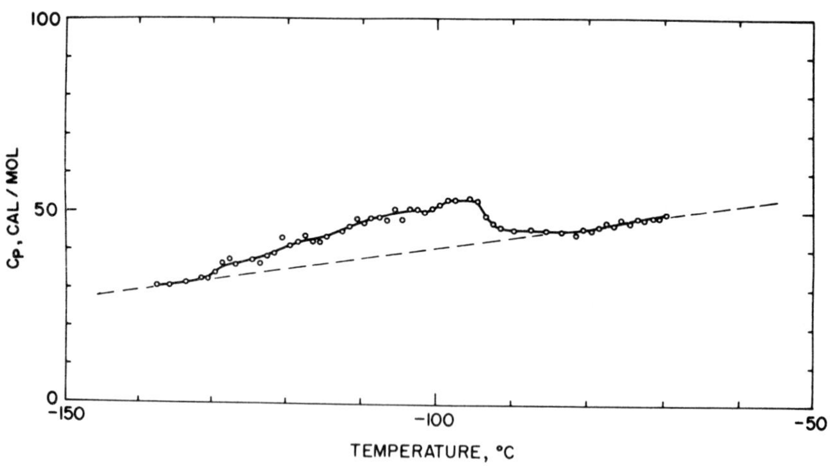

FIG. 14
$(NH_4)_2BeF_4$: specific heat versus temperature.

This sort of complication is typical throughout the field of FE. Nothing is as simple, or as adequately explained, as it first appears to be. Very many other examples, even in so seemingly simple a case as that of $NaNO_2$, could be cited. In $(NH_4)_2SO_4$ one must now look for superstructuring as a function of impurity concentrations, defects, and temperature and pressure. Once one knows what to look for, something significant generally turns up.

Even when the above problems are clarified, one is still almost at the beginning of the $NH_4 \cdots SO_4$ system. Consider the mixed-crystal system $(NH_4)_2(SO_4)_x(BeF_4)_{1-x}$, the phase diagram of which appears in Fig. 15[28]. The phase corresponding to $0.20 < x < 0.70$ is

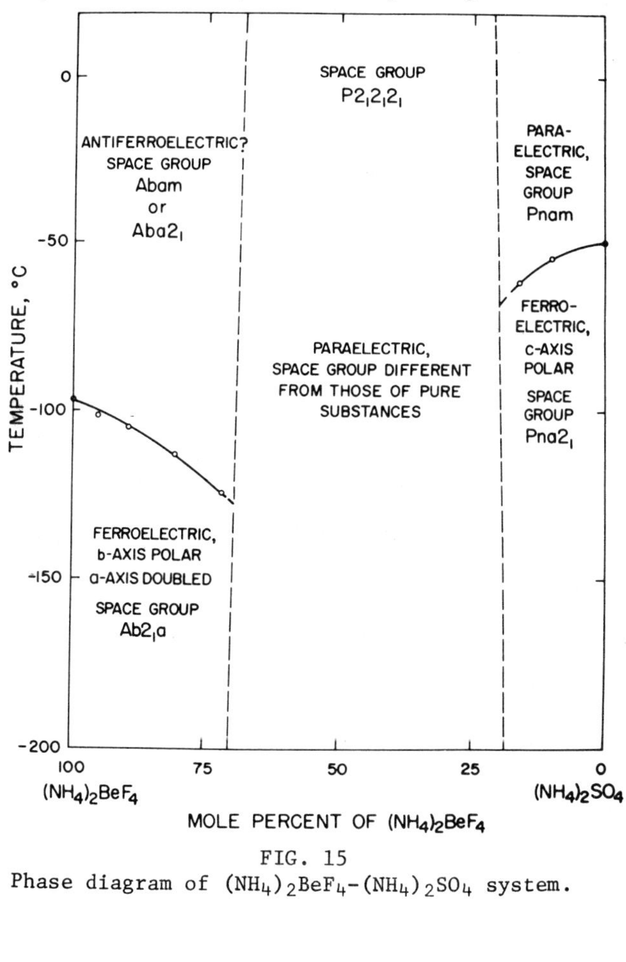

FIG. 15
Phase diagram of $(NH_4)_2BeF_4$-$(NH_4)_2SO_4$ system.

PE between +25°C and -196°C, and has a space-group symmetry differing from that of either of the end-members. The superstructuring of none of the phases is indicated on the figures; the intermediate-composition phase does show superstructure maxima resembling those in a fiber diagram[27]. The $(NH_4,Rb)_2SO_4$ crystals show disordering in the (001) plane. The writer has prepared a vast number of other simple, double and mixed sulfates, selenates, fluoberyllates, chromates, monofluophosphates, tetrachlorozincates, and so on, of monovalent cations; and so many interesting properties appear in these that lifetimes of work would be required to understand them. Resonance methods can contribute much to partial understanding; one cannot hope to follow through every one of the problems to its end point with diffraction analyses.

When one of the monovalent cations is Li+ or a proton, other systems of unsolved problems appear. Typical among these are the structures and transition mechanisms in NH_4HSO_4[78], $RbHSO_4$[79], $(NH_4)_3HSO_4$[53] and β-$LiNH_4SO_4$[54]. The crystal structures of none of these is adequately known, and thus resonance studies are hampered in their interpretation. The NH_4HSO_4 and Rochelle salt are the only two ordered crystals which show a FE phase sandwiched between two PE phases. Since one would have assumed that the presence of the NH_4^+ ion was of prime importance in the behavior of NH_4HSO_4, it was surprising to find the dielectric anomaly in $RbHSO_4$. Again, other members of the acid salt families show anomalies, and mixed crystals are interesting. The writer has found a further world of transitions, including both dielectric and conductivity anomalies, in double sulfates, etc., involving monovalent and divalent cations. FE and AFE are everywhere among these, like a virus in an epidemic.

How far we have come from the time when, for 15 years, Rochelle salt was the only known FE! The more common a property is, the more important it is in the mass and the less important is any single example. We now know so many FE materials, and enough about many of these, to realize that no general structural theory of their behavior can exist. The FE depends upon accidental details of crystal strutures, involving delicate dynamic balances of bonding forces and their alterability by temperature, mechanical stresses and electrical fields. One can develop significant generalizations in ferromagnetism and related phenomena; the predictability of magnetic behavior, in terms of chemical constitution and atomic arrangements, is very much greater than predictability of anomalous dielectric properties.

Competent structure analysts who have not immersed themselves in FE phenomena may find it an intolerable and unacceptable fact, that the results of their powerful techniques can be inadequate for the explanation of all FE phenomena; but the fact may remain: as it has for example—at least so far—in the cases of $(NH_4)_2SO_4$ and $(NH_4)_2BeF_4$. Magnetic resonance and relaxation methods have added a new dimension to FE investigations, in their abilities to explore field gradients and various dynamic effects.

Blinc[80] has recently provided an extensive review of resonance methods as applied to FE's and their transitions. Readers are referred to this for details of NQR, spin-lattice relaxation (SLR) and electron paramagnetic resonance (ESR) theory and techniques, and a survey of FE's which have been examined by these methods. One has the impression that wielders of these techniques are often still in the stage of instrumentalists looking for problems to which their methods might be applied; complete immersion in the FE problems, such as is exemplified by the work of O'Reilly and Tsang on $(NH_4)_2SO_4$, is not universal. More recently, Rigamonti[81] and O'Leary[82] have described very interesting use of resonance and relaxation methods for exploration of soft optical modes and Debye-Waller factors. Much remains to be done with both structural and resonance methods.

III. ILLUSTRATIONS OF SOME STRUCTURE ANALYSES OF FE POLARIZATION AND TRANSITION MECHANISMS

Once the existence of FE or AFE has been detected in a crystal, and the macroscopic phenomenology of the dielectric behavior and related phenomena has been established, the primary attack in any attempt at elucidation of the structural changes from which the macroscopic properties arise must be via X-ray and (if possible) neutron diffraction analyses. Methods for exploring EFG's around certain atoms in the structure are supplementary and increasingly valuable, as discussed in the previous section; but the interpretation of these is the more enhanced, the better the structural details—including, if possible, the mechanisms of transitions—are known. It is not always possible to obtain single-domain specimens in a FE phase (e.g., when the PE phase is cubic and one or more FE phase(s) not uniaxial); and then a complete determination of the FE structure may not be possible.

When single crystals of sufficient size are available, and these can be maintained as single-domain in the FE phase, the best contemporary techniques for collection of scattered X-ray and neutron intensities, coupled with proper corrections for extinction, absorption, multiple reflections, etc., and coupled with proper use of digital computing in structure determination and refinement, are capable of determination of coordinates of atoms in ordered positions to ~0.01 Å. The effects of various types of errors tend to be concentrated into the thermal oscillation parameters of the various atoms; and the accuracy of anisotropic vibration parameters is sometimes overrated as a consequence.

An ever-increasing number of studies of FE transitions and polarization mechanisms are being accomplished with the full power of modern crystallographic methods. Very few studies prior to 1960 can be considered as definitive: even the important early neutron studies of KH_2PO_4[83][87], $NH_4H_2PO_4$[88][89], $BaTiO_3$[90][91], $PbTiO_3$[92], and $PbZrO_3$[93], need restudy, since these were all done in two-dimensional projections. Other analyses have been accomplished in which the structure of a specific phase is solved but the polarization reversal or transition mechanism is not analyzed; and these will not be mentioned here. Among analyses which are in a much more satisfactory state are those of the following FE's: $NaNO_2$[94]-[101], $(NH_4)_2SO_4$[30][31], $(glycine)_3 \cdot H_2SO_4$[102][103], colemanite[104], $K_4Fe(CN)_6 \cdot 3H_2O$[105], SbSI[106], $Li(N_2H_5)SO_4$[107]-[109], $LiNbO_3$[110]-[112], $LiTaO_3$[113][114], and $Cu(formate)_2 \cdot 4H_2O$[115].

There is no up-to-date summary, by investigators familiar with both the physical behavior of FE's and modern crystallographic techniques, which does equal justice to: the phenomenology of FE and AFE; the structural and structure-dependent details which a physicist would like to know; the answers which present-day X-ray and neutron diffraction methods can and in some cases have indeed supplied; and the very important supplementary information which can be gleaned from nuclear or electron resonance studies and other techniques. There is no FE or AFE for which we know all we would like to know. The nearest-to-satisfactory situation obtains for $NaNO_2$, thanks to the extensive neutron diffraction and X-ray critical scattering studies which have been accomplished on that crystal over a wide range of temperatures: as reported in References [94]-[101].

Limitations of time and space necessarily imposed on the present discussion renders this an unsuitable place for a review of various FE structure analyses, the supplementary information available from resonance or spectroscopic data, and the explanation of physical properties in terms of these. Instead, the writer proposes to carry readers with him into the tangle of an unsolved but very living problem: that of the structural mechanism of FE in Rochelle salt. This is the oldest puzzle in the history of the field, and—as we will see—nature has contrived to render its solution extremely difficult. Most of the structural information which we will use has lain uninterpreted for many years, and largely unpublished for that reason; and what has been published has led as much to confusion as to clarity: in the theory of dielectric behavior and particularly in the interpretation of magnetic resonance and relaxation phenomena.

The choice of subject for this illustration lay originally between two FE's: Rochelle salt, and tris-glycine sulfate (TGS). These presented similar difficulties in understanding of the transition mechanisms, effects of radiation damage, and magnetic resonance and relaxation. The results of a three-dimensional X-ray analysis of TGS had been published over a decade earlier[102], but the structure of the PE phase still presented difficulties. A three-dimensional neutron analysis of TGS has just been accomplished by Professor I. Shibuya and his associates at the Research Reactor Institute of Kyoto University[103]. This latter analysis is a triumph of dedicated competence, and the details are about to be published by the Japanese investigators, who will also discuss its significance with respect to the dielectric properties.

Thus we turn to the hoary problem of Rochelle salt: not as a reviewer offering a logical discussion of a problem whose solution is at hand, but from the viewpoint of investigators seeking a path through a jungle of half-understood results.

A. Value of Earlier Structure Determinations

Some latter-day diffractionists, equipped with large scale automatic diffractometers, high-flux reactors and the best of computers and computer programs, scorn pre-1960 determinations of FE and AFE structures and structural mechanisms. This somewhat smug attitude ignores the fact that discovery of most FE materials depended upon crystal-chemical principles derived from earlier structure determinations, relatively imprecise as some of these may have been. The automatic machines, computing procedures (many early analysts had to devise and construct their own computers) and perfected radiation sources were not available earlier than perhaps a half-dozen years ago; high-flux neutron sources are available to only a handful of workers, even today; and many a good scientist still has access to none such modern equipment. The automatic machines and techniques became available because of the efforts of preautomation designers.

Earlier investigators of FE and AFE phenomena could not wait for perfection of diffraction equipment and methodology. If they had done so, there would have been little knowledge of the crystal chemistry of these materials, very few known FE's and AFE's, little if any knowledge of structural mechanisms, and little sound basis for the interpretation of such important supplementary information as that which can be provided by spectroscopic or resonance methods; and prediction of new FE's would have been even more difficult.

The X-ray determination of the PE structure of KH_2PO_4 by West[116], now over 40 years old, was a triumph of accuracy and formed the basis for the first clarification of the transition and FE polarization mechanism by Slater[117]. The now 30-year-old analysis of the PE Rochelle salt structure, by Beevers and Hughes[118], was a great tour de force of "classic" crystallography. These, and many indeed of the pre-1960 analyses, are the rocks from which our field was hewn.

Not all of the earlier attempts at structure analyses were successful. This was the case for $BaTiO_3$ and guanidinium aluminum sulfate hexahydrate, for example; and some frustrated analysts devised theories which purportedly demonstrated that parameter interaction in these pseudosymmetric FE's would not permit determination of the structures. Mention of these theories persists in some of the literature to this day, even though they are quite incorrect; successful analyses of both of the above FE structures, and their absolute configurations, have been accomplished.

Summaries of pre-1960 structural studies are excellently presented in Jona and Shirane's treatise[9].

B. Phenomena to be Explained

In each structural study of FE behavior and FE or AFE transitions, one aims to account for a variety of phenomena. Among the problems for which answers are needed are these.

(a) What structural features are responsible for the appearance, in some temperature range(s), of structural instabilities: as revealed by dielectric, optical, thermal, elastic, dilatometric, or lattice-dynamical observations?

(b) What happens, structurally, at a transition point?

(c) What are the effects of isomorphous replacement of one or another constituent atom, partially or entirely by other atoms; what atomic groups are partially or totally replaceable, and with what effects?

(d) What is the specific mechanism of polarization reversal in a FE?

(e) What structural explanations can be provided to account for the magnitude of spontaneous polarization and coercive field, and their dependence upon temperature, in a FE?

(f) Why do some transitions lead to AFE rather than FE in closely related systems involving isomorphous replacement?

(g) What are the structural effects of mechanical stresses, and of radiation?

(h) What effects do structural features have upon domain structure, and what other features affect domain structure and its alterability?

(i) Is a transition of a displacive or order-disorder type; how does this affect the thermodynamics of the transition?

(j) Why do some materials crystallize directly as FE and remain so over their temperature range of chemical stability?

(k) What is the structural basis for change of dielectric behavior and transition point(s) with hydrostatic pressure?

(l) What structural details are revealed by X-ray diffraction methods, and what by neutrons? What are the limitations of each of these techniques?

(m) What subsidiary information can we expect to obtain from methods other than diffraction (e.g., resonance and relaxation methods, infra-red absorption, angular correlation studies of radiation from decaying nuclei, etc.)?

(n) What remains to be done?

In order for discussion of these matters to be understandable, extensive knowledge of the macroscopic phenomenology of each material is necessary. Because of space and time limitations, this phenomenology cannot be presented in suitable detail in this article. Readers are referred to the Landolt-Börnstein collation[8] and Jona and Shirane's treatise[9], and of course to the original papers cited, for this information.

C. Rochelle Salt Problem

Rochelle salt, NaKtartrate · $4H_2O$, is the crystal in which FE was first discovered (by Valasek, in 1920[63]). It holds the distinction of being not only the first FE but also the most difficult to understand.

W. P. Cady, in his classic treatise on piezoelectricity[119], introduced each chapter with a correspondingly classic quotation. Chapter XXII, entitled "Rochelle Salt: Dielectric Observations," is prefaced by a sentence from E. C. Stoner: "As often happens, the phenomena which are the easiest to observe are the most difficult to interpret." But the quotation is apt only in its last phrase. It was a quarter of a century ago that Cady devoted over five chapters and well over 200 pages to summaries of this crystal's behavior; and much more has been published since. A synonym for Rochelle salt is Seignette salt; and FE is often denoted as Seignette-electricity, particularly in non-English publications. The latter term is actually a preferable one, for obvious reasons. The first treatise on the phenomenon, by I.-V. Kourtschatov[120], dealt solely with Rochelle salt (it alone was known to show this dielectric behavior, at that time). Valasek, Mueller, Mason, and others have treated the macroscopic properties extensively (see Ref. 119 for summaries; also Mason[121]). It was specifically because Rochelle salt behaved in so complex a manner that Scherrer determined to find a simpler FE material in which to seek the origin of the phenomenon[122]: a decision which led to the discovery of FE in KH_2PO_4[14].

Rochelle salt has two transition points, at +24°C and -18°C, between which it is ferroelectric. Natural tartaric acid is optically active and dextrorotatory; and since the salt is prepared from the natural acid, the crystals must be in an enantimorphic point group. The crystallography of the three phases, as determined by Ubbelohde and Woodward[123], is shown in Table I. The non-FE phases have symmetry $P2_12_12$ and the FE axis is along orthorhombic a. This suggests that the FE symmetry must by $P2_1$; this assumption could not be substantiated by X-rays, but was definitely established with neutron diffraction by Frazer, McKeown, and Pepinsky[125] (see below). The monoclinic axis would ordinarily be designated as c; but in our diffraction studies we have elected to retain the notation of a for the monoclinic and FE axis in order not to confuse comparison with the orthorhombic axes.

The low-field unclamped-crystal dielectric constants ε_a are shown in Fig. 16, as determined by Mueller[126] and Mason[127]; and Fig. 17 shows the results of measurements of the spontaneous polarization from measurements of Bradford[126] and Halblützel[128] (see also Mueller[129]). The values are very sensitive to the magnitude of the applied electric field[126].

Deuteration (of all but the two H's covalently bonded to the two central carbon atoms) raises the upper transition temperature to +35°C and drops the lower point to -22°C[130][131].

The effects of replacement of K^+ by $NH_4{}^+$ ions was first observed by Kourtchatov and Eremejev[132], and later re-examined in detail by Melmed, Jona and Pepinsky[133]. The results shown in Fig. 18 are taken from the latter reference. As little as 1% of NH_4 replacing K narrows the FE temperature range very markedly, and at 2.5% NH_4 the two dielectric peaks coalesce. A broad, low dielectric anomaly persists until the NH_4 content reaches about 18%, at which concentration a new, sharp single peak appears at about -100°C; a new FE phase then persists for rising NH_4 content up to ≃92%.

TABLE I

ROCHELLE SALT: CRYSTALLOGRAPHIC DATA FOR THE THREE PHASES*

Phase, temperature of measurement	a	b	c	Space group
High temperature, +35°C	11.87_8	14.24_6	6.21_8	$P2_12_12$
Ferroelectric, +20°C	11.86_7	14.23_6	6.21_3	$P2_1$**
Low temperature, -50°C	11.81_5	14.20_3	6.19_5	$P2_12_12$

*After Ubbelohde and Woodward[123] (in Å).

**The unique axis is along a. The monoclinic angle α is 90°03′ [124].

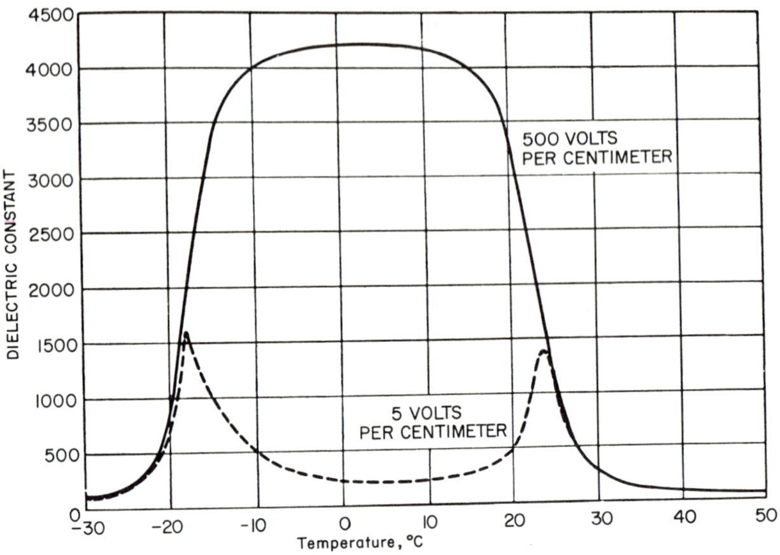

FIG. 16
Free dielectric constant for Rochelle salt for fields of 5 and 500 V/cm.

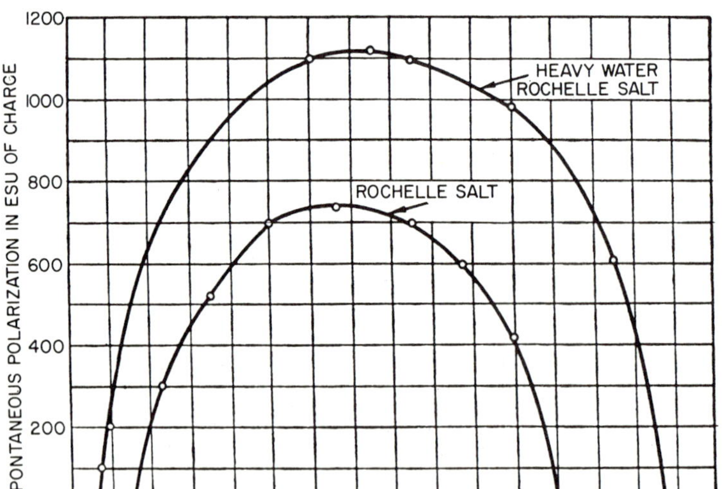

FIG. 17
Spontaneous polarization for Rochelle and heavy water Rochelle salt.

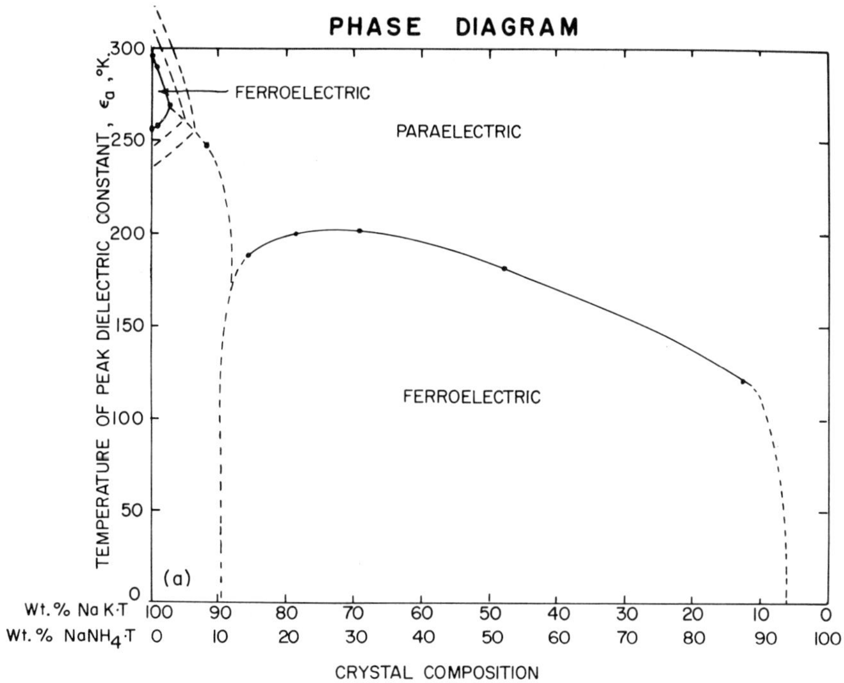

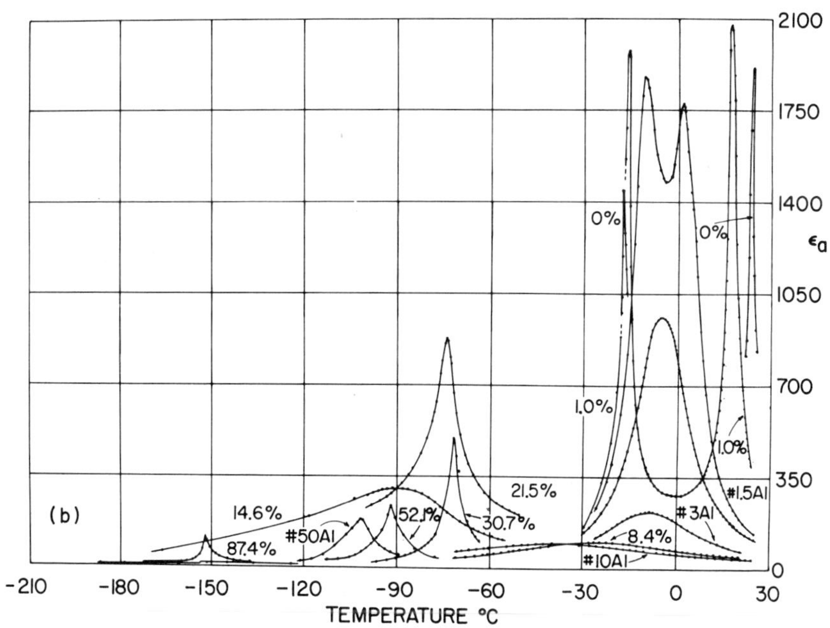

FIG. 18

(a) Behavior of ammoniated Rochelle salt: phase diagram for $Na(K,NH_4)C_4H_2O_6 \cdot 4H_2O$; (b) dielectric constant ε versus temperature for $Na(K,NH_4)C_4H_2O_6 \cdot 4H_2O$: percent of NH_4 indicated.

Cady[119] summarizes techniques for preparing and handling Rochelle salt crystals, and Mueller[124] also gives some practical advice. The crystals readily lose some of their water of crystallization, and this tendency has complicated various measurements. Dielectric properties are also strongly affected by radiation damage.

D. Rochelle Salt Structure

Tartaric acid occurs naturally in the d-form, which by convention is written as

```
  COOH
   |
 H-C-OH
   |
HO-C-H
   |
  COOH
```

The conformation of the tartrate ion is apparently stabilized by bond resonance, which extends from a carboxyl group through the α-carbon to the α-OH oxygen. Figure 19 shows the conformation of a typical ion, as found in the present analyses. It will be discussed in detail below. This tendency toward coplanarity of the carboxyl,

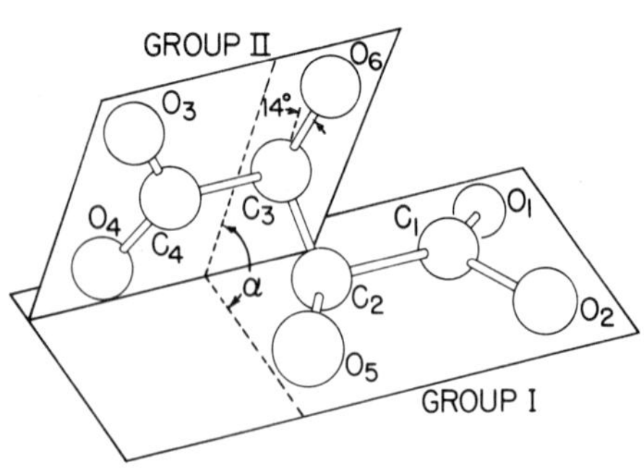

FIG. 19
Rochelle salt: perspective drawing of the tartrate ion.

α-carbon and oxygen or nitrogen of an α-OH or α-NH_2 group is found in various α-hydroxy and α-amino acids; and it is of great importance here and in the structure of (glycine)$_3$ · H_2SO_4, as we will show later. The numbering of atoms in the tartrate ion follows that originally adopted by Beevers and Hughes [118] (as will also be the numbering of water oxygens in the orthorhombic phases). The ion is thus comprised of two nearly flat groups I and II, of carboxyl, α-C and α-O atoms, twisted at an angle α with respect to one another. In Rochelle salt α is ~47°.

The hydroxyl (or carboxyl) hydrogens can exchange freely with hydrogen or deuterium atoms of the solvent, in solution, and they can of course be involved in H-bonding. Only the H's attached covalently directly to C_2 and C_3 will not exchange in solution under normal conditions, and do not participate in H-bonding.

Figure 20 shows the c-axis projection of Rochelle salt, as determined with X-rays by Beevers and Hughes[118] (B-H). The C_1-C_2-C_3-C_4 skeleton is z-shaped, and four of these are seen in the figure, related by 2_1-axes parallel to a and b and 2-axes parallel to c. The O_5 is seen to fall below or above C_4 in this projection. The positions were not determined, but some likely H-bonds are indicated: particularly O_1-O_{10}, O_9-O_{10}, and O_2-O_9. The number in parentheses beside each atom indicates its z-coordinate: i.e., its height above the projection plane. The coordination of O's about the K^+ and Na^+ ions is also indicated.

As already remarked, the B-H analysis was a great achievement in 1940, and was sufficiently accurate to serve as the starting-point for later refinements. It is not of sufficient precision, however,

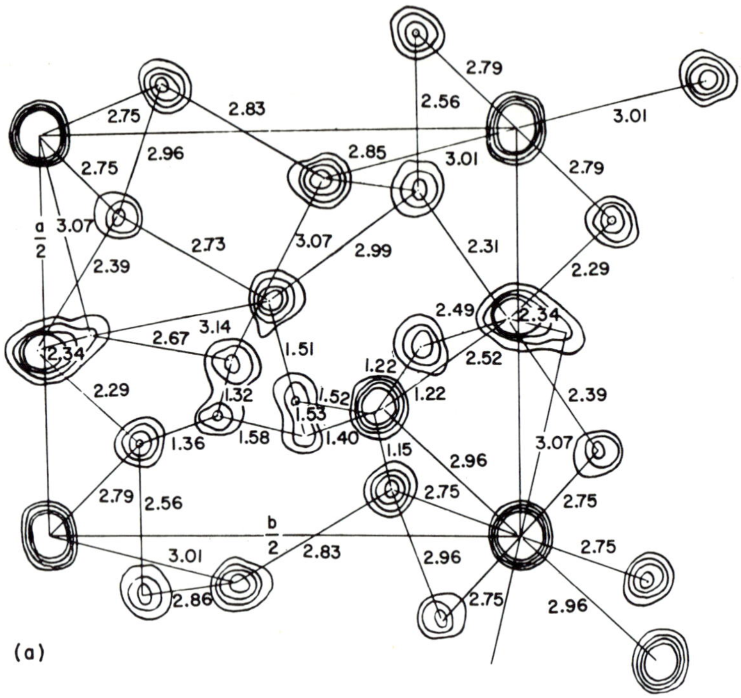

FIG. 20
Rochelle salt: c-axis projection from X-ray analysis of Beevers and Hughes: (a) Fourier projection of one-fourth of Rochelle salt cell (giving the positions chosen for the atoms and the interatomic distances in Å ; (b) schematic of bonding, full cell of Rochelle salt.

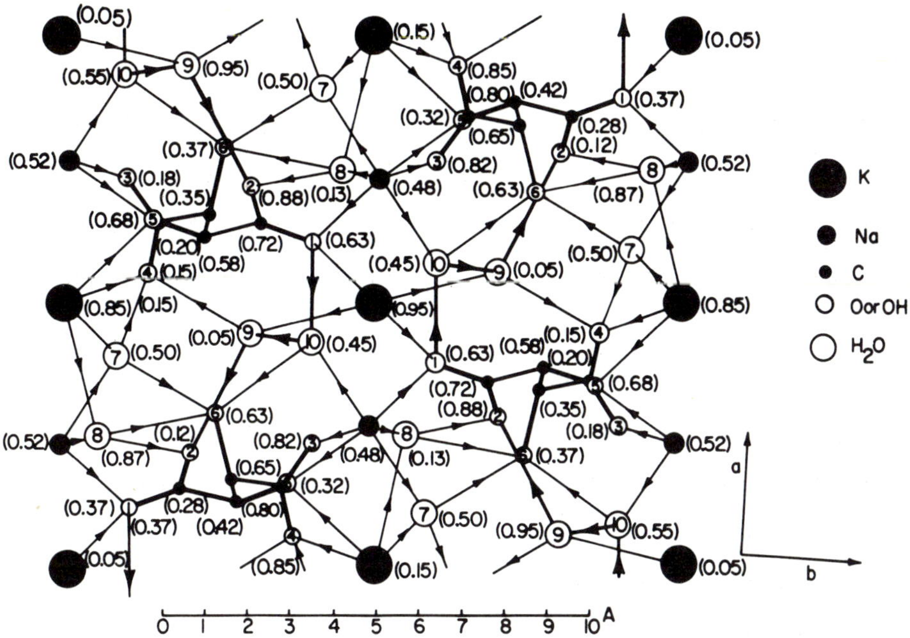

(b)

FIG. 20 (cont.)

to warrant the conclusions which were based upon it with respect to the mechanism of FE activity.

In the first place, the symmetry of the FE phase must be monoclinic, with space-group $P2_1$: as had been earlier pointed out by Jaffe[134], B-H treated the FE phase as if it had the same symmetry as the two PE phases: $P2_12_12$. There was no X-ray evidence for the disappearance of the 2_1 axis along b as the transitions occurred to the FE state; and retention of the 2_1 axes along b, with consequent retention of the accompanying 2-axis parallel to c, meant that the a axis appeared nonpolar in the deduced structure: i.e., no direct evidence of spontaneous polarization could be obtained.

Second, the accuracy of atomic coordinate and inter-atomic bond determinations was not high, even though the entire analysis was a work of genius at the time it was accomplished. No determination of H-positions could be achieved, because of the limited accuracy. And finally, no search for structural changes at the two Curie points was made.

The FE theoreticians, understandably eager to find in the B-H analysis some indication of what atoms and bonds might be involved in the FE polarization, accepted the O_1-O_{10} distance of 2.59 Å as correct; and since this bond lies parallel to the FE axis and must represent an H-bond, it was considered to be the locus of the FE activity.

Here is a quotation from Mason's treatise[121], page 237, which summarizes the deductions on which that excellent investigator's conclusions were based and expanded into a theory.

> There are three possible hydrogen bonds in the structure, between the oxygen molecule 1 and the water molecule 10,

> between the water molecule 10 and the water molecule 9, and between the water molecule 9 and the oxygen 2. The distance between the successive molecules are 1 to 10, 2.59 Å; 9 to 10 is 2.86 Å; and 9 to 2 is 3.02 Å. The bond with the shortest distance is 1 to 10 (2.59 Å) and it is the motion of the hydrogen nucleus along this bond that causes the FE effect in Rochelle salt. This bond lies nearly along the X-axis . . . which is the FE axis of the crystal. The theory discussed in the next section is based on the action of this hydrogen bond and leads to a FE effect having the right value of spontaneous polarization, two Curie temperatures, a good quantitative agreement with the measured values of the clamped dielectric constant, and agrees well with recent measurements which show that the dipole dielectric constant is relaxed at a frequency of about 5×10^8 cycles.

There follow some 17 pages of theory in which the above conclusions are apparently substantiated.

Mason's theory was very clever, but it serves as a classic illustration of the danger of extrapolation, from a single bond distance in a very approximate structure for which even the symmetry elements are not correctly assigned, to a polarization mechanism in a very complex structure. Two later series of observations, by Frazer and Pepinsky using neutron diffraction[56][125][135] and by the writer's X-ray group in full three-dimensional analyses of all three phases of Rochelle salt[137], showed that the Mason theory had depended upon an incorrect value for the O_1-W_{10} bond length, and that the proton which bonded these particular oxygens showed no change in position as the FE direction was reversed. In fact, all interatomic distances in the B-H analysis were in need of refinement.

It would be very wrong to condemn or ridicule a theory which is based upon inadequate information, and particularly to criticize a scientist who has himself contributed such a mass of careful measurement, guided by well-based theory and leading to improved theory. Many of the reliable observations we have on Rochelle salt and on many other piezoelectric materials were made by W. P. Mason; and his many volumes on crystal physics are source-books in that large area. The theory of the O_1-W_{10} bond and its significance for the dielectric anomalies in Rochelle salt was developed before more than a handful of people suspected how very complex the atomic interactions—and their temperatures and field dependence—were, and how rough the original structure analysis was.

There were other, earlier theories of the FE mechanism in this crystal. Valasek, the discoverer of FE, lamented the fact that structural information on the crystal was unavailable and indeed beyond the capabilities of X-ray diffractionists at the time he studied the material[63]. Today we know that X-ray diffraction alone cannot solve the problem of the FE mechanism.

The first structural theory, long before that of Mason, was due to I.-V. Kourtschatov and his group[120]. Their idea was that the anomalies were due to reorientation of water molecules in the hydrated structure. It was the tremendous success of Slater's theory of the ordering of H-bonded protons in KH_2PO_4[117] which led Mason and others to expect similar order-disorder effects of protons in Rochelle salt, and indeed to discard Kourtschatov's hypothesis in favor of the latter.

As our neutron analysis progressed, it became evident that a good number of H's in the structure shifted positions during the polarization reversal, and at the transition points. The results of the neutron analysis were never published in extenso, because we felt that they should have been carried further and yet we were distracted by other, newer problems. For it was just at this time that Matthias altered the character of the entire field of FE by uncovering so many entirely new types of materials, and we began our attempts to imitate him: and then we were overwhelmed by avalanches.

The Rochelle salt problem is difficult not only because of its complex structure, but because it is so easily damaged by X-rays that the very process of collecting X-ray diffraction data alters the structure to be determined. This was shown by Boutin, Frazer, and Jona in an interesting study in 1963[138].

Meanwhile a number of investigators have examined Rochelle salt with resonance techniques, and have drawn various conclusions, many of them contradictory, based on what little was available from the earlier X-ray and neutron studies. Attention seems to shift from one set of H's and H-bonds to another. If the writer had to choose a well-known quotation to head a chapter on the participation of H-bonds in the Rochelle salt dielectric behavior, he would turn to the chorus of the song the Mock Turtle sang in "Alice in Wonderland"[139]:

> Will you, won't you, will you, won't you,
>   will you join the dance?
> Will you, won't you, will you, won't you,
>   won't you join the dance?

The invitation here is to the various H-bonds; and in fact, all of them are participants in the dielectric behavior, either directly or indirectly.

E. Neutron Analysis of Frazer and Pepinsky

Our neutron study began in 1953, and continued for at least two years. The Rochelle salt structure was at that time the most complex yet attempted by neutron diffraction, and was carried out at the original graphite reactor at the Brookhaven National Laboratory, with joint support of that Laboratory and direct contractual support from the U. S. Atomic Energy Commission. Mrs. M. McKeown assisted in earlier stages of the work, and computations were accomplished on X-RAC with Office of Naval Research support. The flux density of neutrons at the reactor was such that data collection was a slow and tedious process; and a three-dimensional analysis was at that time not feasible.

The writer had earlier introduced a technique for facilitation of interpretation of neutron analyses of crystal containing H atoms which could be replaced by D's[87][135][140]; and it was used widely throughout this and various other neutron analyses at that time. The scattering factor for a proton, for thermal neutrons, is $-0.38(\times 10^{-12}\text{cm})$ and for a deuteron the corresponding factor is +0.65. Since the neutron scattering factors for most elements in the natural abundance ratios of their isotopes are positive, peaks in a neutron diffraction Patterson map which represent interactions between protons and positively scattering nuclei are negative, while interactions between two protons or between all pairs of positively scattering nuclei are positive. Thus the identification of peaks involving proton-to-positive-scatterer interactions are readily made, and proton location is simplified. The procedure can be carried further by exchanging deuterons for protons, and subtracting the resulting Patterson map from that for the protonated structure. Finally, coherent neutron scattering from all replaceable protons or deuterons in a structure can be eliminated by partial substitution of deuterons for protons; the ratio of replaceable deuterons to protons in such partial exchange must be 0.38 to 0.65: under the (reasonable) assumption that replacement is uniform for all replaceable protons or deuterons. Similar techniques can be used with other elements having both positively and negatively scattering isotopes.

This H-D replacement technique was used at the outset of the neutron analysis of Rochelle salt, to demonstrate the true monoclinic nature of the FE phase and furthermore that shifts in proton positions were chiefly involved in the transition mechanisms. The (0k0) reflections with k odd were immediately observable in the neutron spectra for the FE phase; and these disappeared below the lower and above the upper Curie points. Figure 21 shows the (050) peak for a deuterated Rochelle salt sample at 20°C. When a sample was partially deuterated, with a ratio of 38/65 for the replaceable deuterons to the replaceable protons, the odd (0k0) peaks did not appear. This deuteron-proton ratio was achieved by crystallizing the material from a D_2O-H_2O solution of proper D to H concentration. (A similar experiment was conducted on PE and FE $LiNH_4$tartrate · H_2O, with similar evidence of change from orthorhombic PE to monoclinic FE phase. This structure is closely related to that for Rochelle salt, but of course lacks 3 of the 4 waters of crystallization of the latter, has a single Curie point, and is FE in another direction with respect to orientation of the tartrate ions[64][141][142]).

Frazer subsequently examined the (0k0) reflections, for both protonated (H) and deuterated (D) crystals, and found that the ratio of a given odd (0k0) reflection for an H and a D crystal is not the same, as the value of k changes; and he concluded from this that shifts in H-positions were not alone responsible for those reflections[143]. Our earlier conclusion was that they were chiefly, not solely

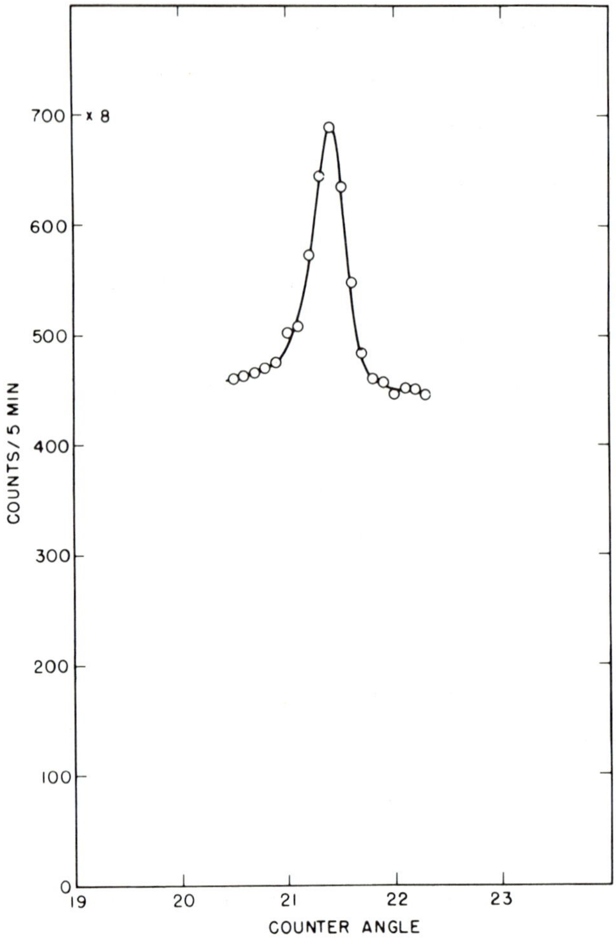

FIG. 21
"Forbidden" (050) neutron reflection from deuterated Rochelle salt in FE phase, (cylindrical sample 2 cm × 0.3 mm).

responsible; and it would be surprising if the earlier qualification did not apply. X-ray studies, reported in the next section, show clear indication that other atoms shift in position at the Curie points; and the neutron analysis reveals shifts of the K^+ and Na^+ ions, in the c-axis projection of the FE phase. These can be seen in some of the figures below. From these it does seem remarkable that the translational symmetry along the b direction has not broken down, to the extent that at least some odd (0k0) do not receive strong contributions from non-H atoms. Frazer's observations[143] indicate that these contributions are small but observable in the neutron case. Why are they not observable with X-rays?

One can expect some deviations from the systematic absences which arise from translational symmetry, if the possibility of multiple reflections is taken into account[144]. Contributions to a specific reflection (hkl) can occur through double reflection whenever two fairly strong reflections (h′k′l′) and (h″k″l″) exist for $h = h' + h''$, $k = k' + k''$, $l = l' + l''$. Thus even though odd (0k0) reflections are weak, or should be absent as a consequence of some translational symmetry, some energy may be scattered into them due to any pairs of fairly strong reflections satisfying $h' + h'' = 0$, $k' + k'' = K_{odd}$, $l' + l'' = 0$. Since the intensities of neutron reflections (h′k′l′) or (h″k″l″) may differ significantly from X-ray intensities for the same reflections, multiple reflections may occur for neutrons when they do not for X-rays, and conversely.

Frazer's observations concerning the ratios of odd (0k0) neutron reflections for H and D crystals were directed at the larger problem, of radiation damage during collection of X-ray intensities: the hypothesis being that the real cause for nonappearance of odd (0k0) X-ray reflections in the FE phase is radiation damage during X-ray exposure.

There seems to be no question, on the basis of observations of changes in odd (0k0) neutron reflections following γ-ray exposure of Rochelle salt, that radiation damage occurs[138]. The dielectric properties are significantly altered by X-ray or γ-ray dosages, but optical observation indicates that the monoclinic phase is not destroyed even though the damage may have reduced the departure from the translational symmetry along the b direction.

Do the observations of Boutin, Frazer, and Jona[138] imply that X-ray study of the FE structure is fruitless? This depends upon the extent of the damage. The fact that the monoclinic symmetry is retained even after heavy radiation dosage suggests that the effects of the damage are somehow restricted to certain atoms, the displacement of which do not upset the delicate balance of interatomic forces which lead to the stability of the monoclinic phase.

In describing the neutron analysis of the monoclinic phase, and interpreting the three-dimensional X-ray analyses of all three phases, we will bear in mind the damage problem and aim at understanding it. It will be seen that we can do so.

The orthorhombic structure is shown, in the c-projection, in Fig. 22. This first result of the neutron study is to be compared with Fig. 20 from the B-H analysis. All H-atoms are shown in Fig. 22.

Using the H-D partial replacement technique[83][135][140] or its equivalent which involves the calculation of Fourier maps with coefficients $\left(F^D_{obs} + F^H_{obs} \times |f_D|/|f_H|\right)_{hk0}$, where $\left(F^D_{obs}\right)_{hk0}$ is the cell neutron structure factor for an (hk0) reflection as observed from a crystal in which deuterons have replaced all replaceable protons, $\left(F^H_{obs}\right)_{hk0}$ is the corresponding observed structure factor for a completely protonated crystal, and $|f_D|/|f_H|$ is the ratio 0.65/0.38 of deuteron to proton scattering factor magnitudes, we obtain a

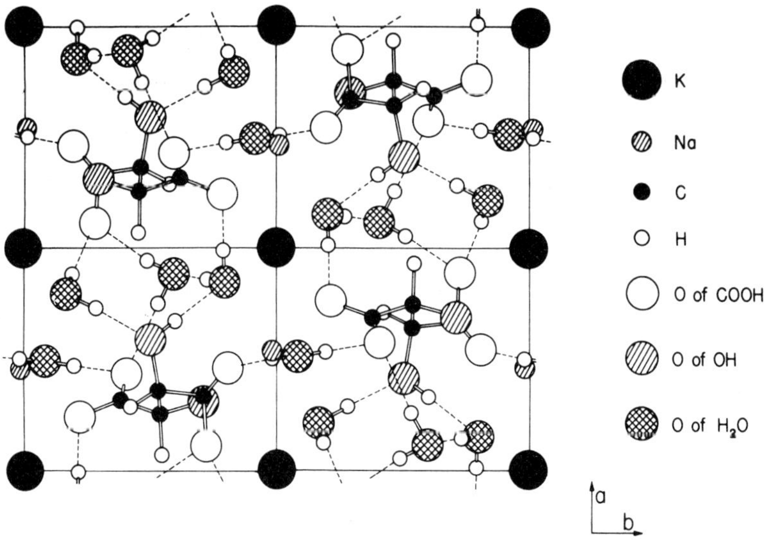

FIG. 22
The c-axis projection of Rochelle salt, from neutron data: orthorhombic symmetry.

Fourier map which indicates the positions of all nuclei except for the replaceable H or D atoms. Such a map is shown in Fig. 23. The contours for the two H atoms covalently-bonded to C_2 and C_3 appear at negative levels, and are not shown in this particular figure (although they are shown in other patterns, later). All syntheses are photographed from X-RAC.

We can thus locate all of the C, O, Na, and K nuclei in the projection accurately, without the additional complications of the replaceable H or D atoms. This permits simplified calculation of cell structure factors, using the thus-derived coordinates. The positions of the two covalently-bound H atoms must of course be included in the calculations, when comparison of F_{obs} and F_{calc} are made for each (hk0) value.

If now we use as Fourier coefficients $\left(F^D_{obs} - F^H_{obs}\right)_{hk0}$, a map showing the locations of all replaceable H or D atoms should result. This is shown in Fig. 24; and of course we can again calculate F^D and F^H values from the coordinates so determined, evaluate the agreement and refine the coordinates.

Finally, we computed the two Fourier maps for all atoms, using first $\left(F^D_{obs}\right)_{hk0}$ and then $\left(F^H_{obs}\right)_{hk0}$ values as coefficients. The map corresponding to $\left(F^D_{obs}\right)_{hk0}$ is shown in Fig. 25. Here the negative contours for the two covalently-bonded H atoms are also included. Details of these calculations are shown in Figs. 26 and 27.

It is necessary to apply a field along the FE axis, in order to avoid the superposition of projections of the structure polarized along +a and -a. This field must of course be maintained throughout the collection of the scattered neutron intensities.

We have no reason to doubt that the symmetry of the HT phase is that listed in Table I: $P2_12_12$, with four formula units $NaK(C_4H_2O_6) \cdot 4H_2O$ per cell. The formula units must have identical conformations, related by the 2_1-axes parallel to a, the 2_1-axes parallel to b, and the 2-axes parallel to c. The symmetry is shown in projection along c in Fig. 28(a), in which the orientations of

---

FIG. 23

Rochelle salt: c-axis projection using coefficients $(F^D_{obs} - F^H_{obs} \times |f_D/f_H|)$ neutron data, polarized monoclinic cell. Vertical axis marked $b_{(monoclinic)}$ instead of $a_{(orthorhombic)}$; horizontal axis $a_{(monoclinic)}$ instead of $b_{(orthorhombic)}$. (b) Rochelle salt: X-RAC map from which 23(a) was obtained (Fourier map utilizing $D^D_{obs} + |f^D/f^H| f^H_{obs}$).

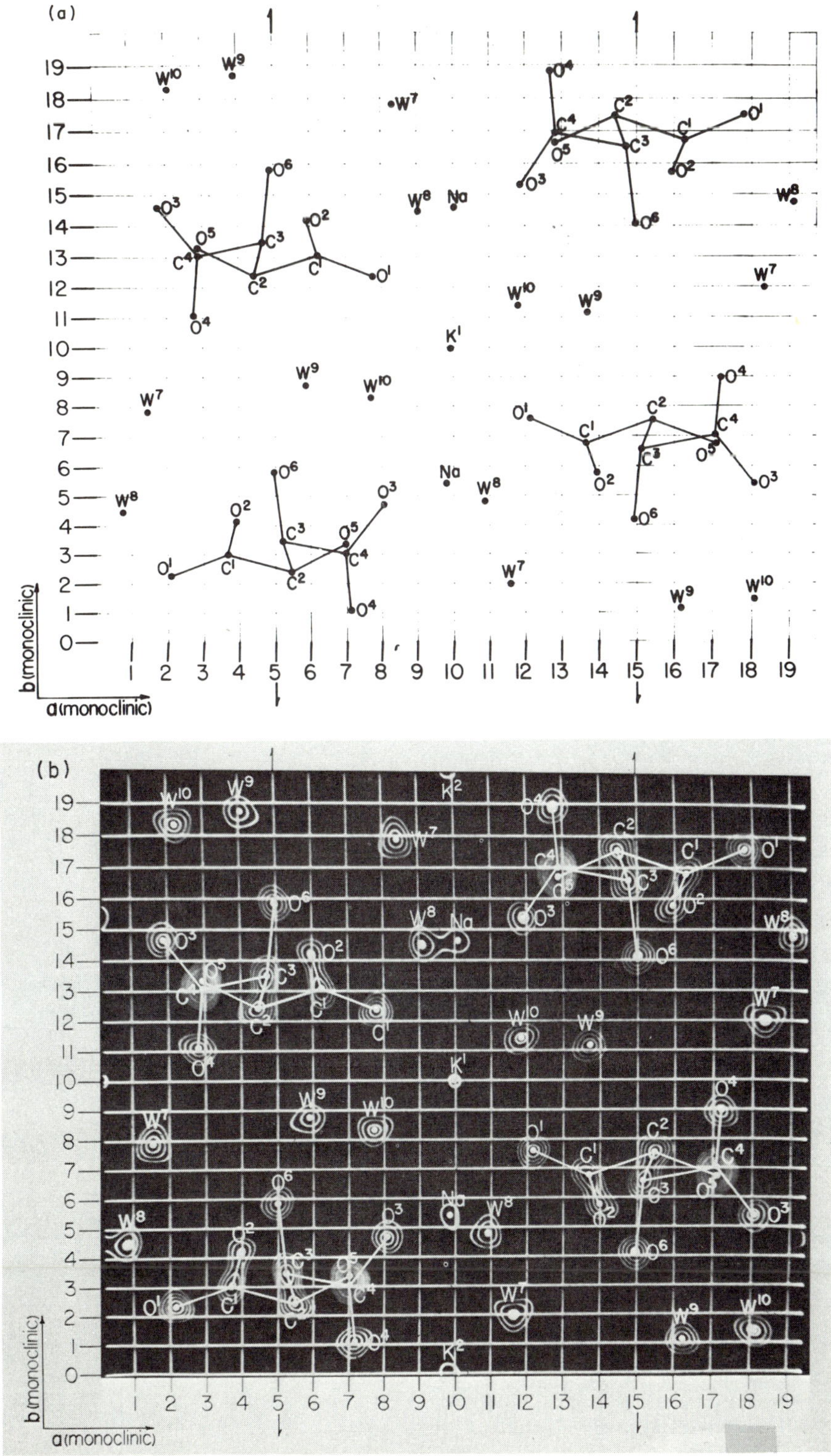
(a)
(b)
b (monoclinic)
a (monoclinic)
Na
K¹
K²
W⁷
W⁸
W⁹
W¹⁰
C¹
C²
C³
C⁴
O¹
O²
O³
O⁴
O⁵
O⁶

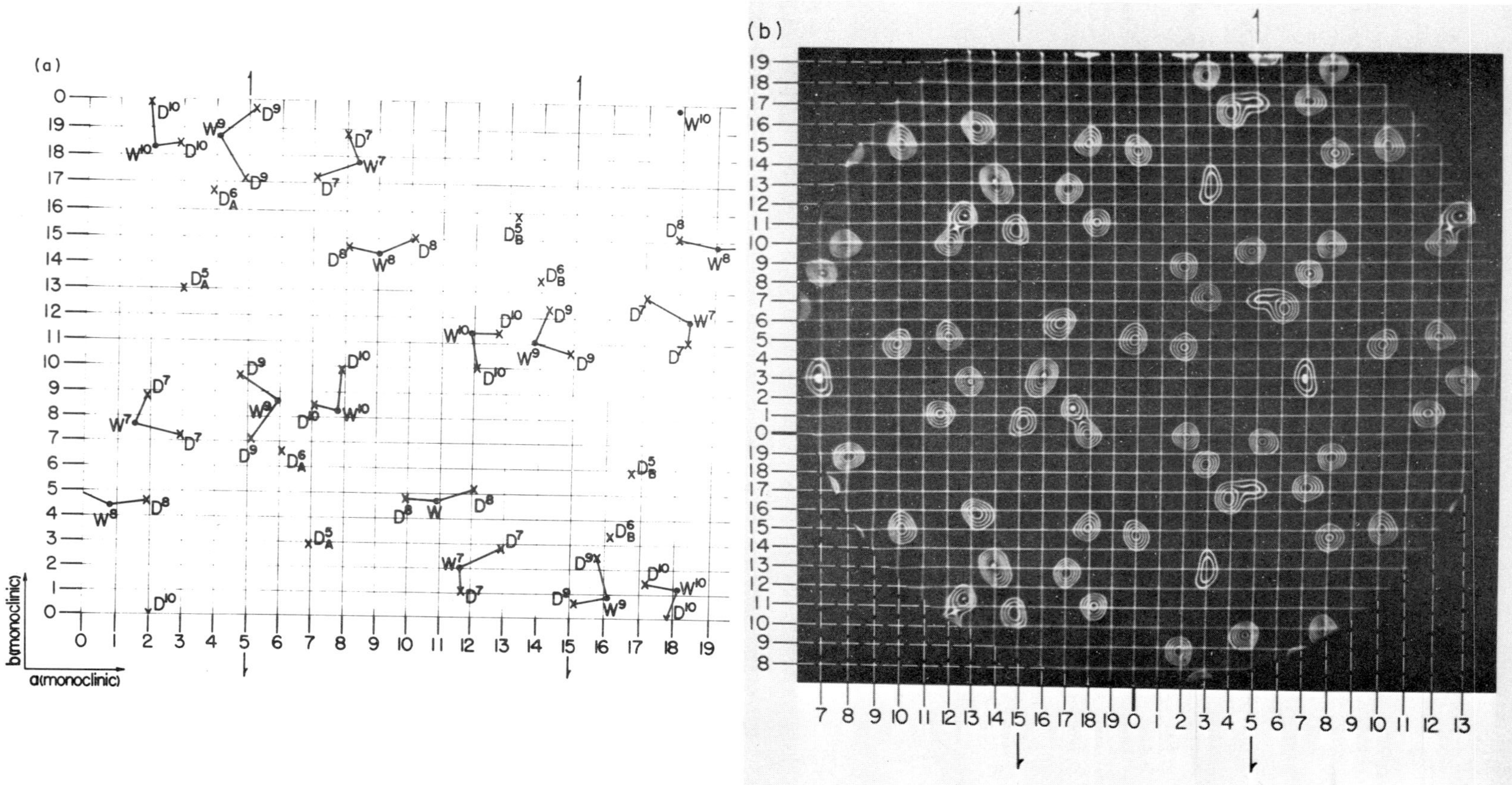
(a)
(b)
b(monoclinic)
a(monoclinic)

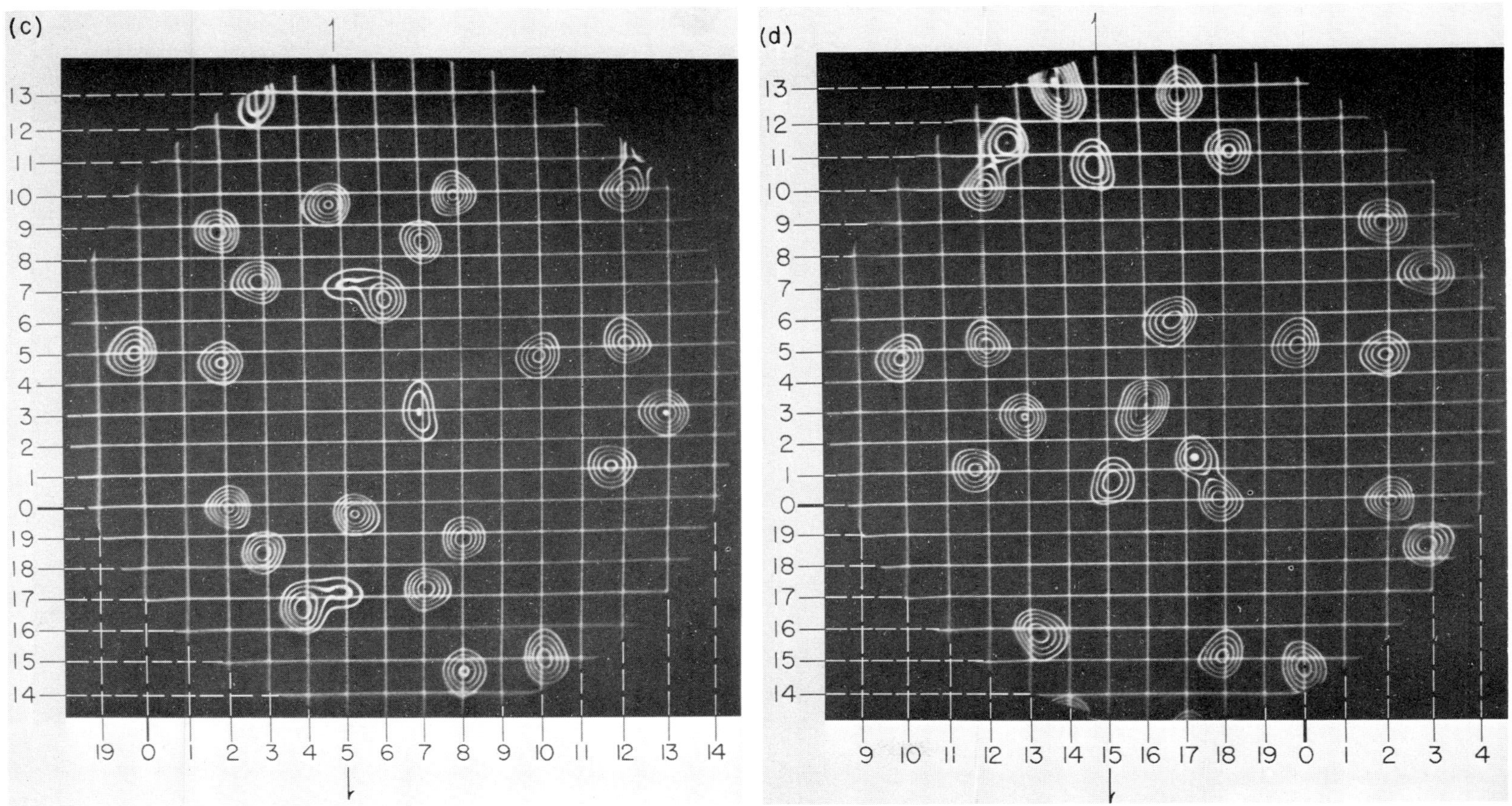

FIG. 24

(a) Rochelle salt: c-axis projection using coefficients $(F^D_{obs} - F^H_{obs})hk0$, other conditions as in Fig. 23 (deuterium only, $f^D - f^H$ synthesis); (b) Rochelle salt: X-RAC map from which 24(a) was obtained. Note that only contours for deuteriums appear; positions of oxygens shown in 24(a) determined from Fig. 23; (c) **detail** of (b): B-type region, Rochelle salt.

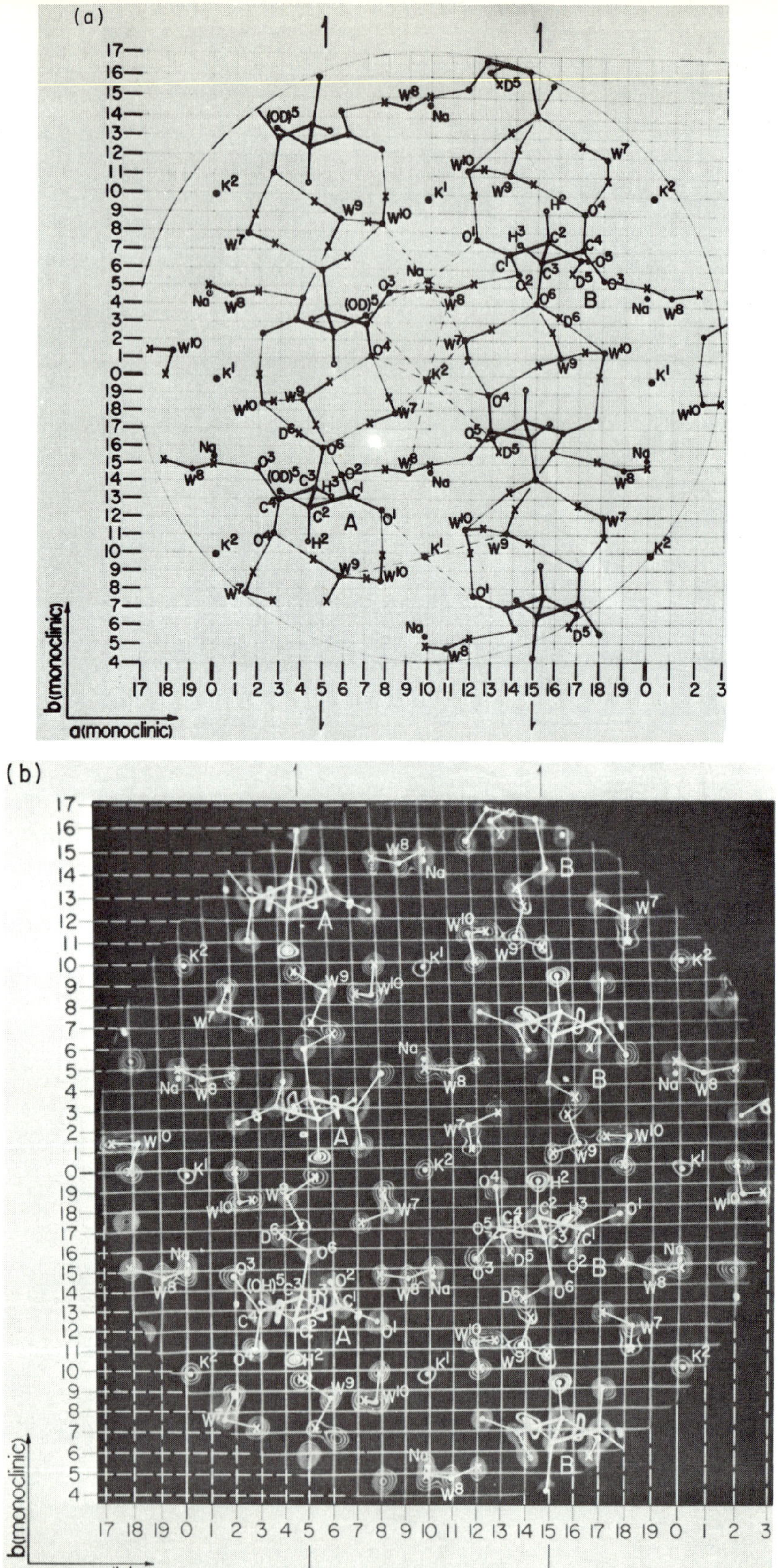
(a)
b(monoclinic)
a(monoclinic)
(b)
b(monoclinic)
a(monoclinic)

the formula units are represented by the asymmetric letter P. A plus sign beside the foot of the letter indicates that the foot is bent out of the plane toward the reader; a minus sign indicates the foot bent in the opposite direction. Thus the optical activity of the tartrate ion is simulated. A unit cell of the structure is outlined in the lower left corner of Fig. 28(a), in which the standard symbol for the twofold axes parallel to c appear; the arrows in the plane of the figure indicate 2_1-axes. The orientations of the letters indicate the orientations of the structurally equivalent quadrants of the cell.

There can be no pyroelectricity, and hence no FE, in a crystal with symmetry $P2_12_12$, since all axes are nonpolar. The FE phase symmetry corresponds to that symbolized in Fig. 28(b), in which only the 2_1-axes parallel to a are retained. The cell dimensions remain essentially the same, although the c-axes need no longer be perpendicular to a and b. The symmetry no longer requires that the four quadrants of the unit cell be structurally equivalent; they are now related in pairs only by the 2_1-axes parallel to a. The pairing is now indicated by the orientations of the letters P and R. The structure of a quadrant P in the FE phase differs slightly from that of the same quadrant in the HT phase.

FIG. 25
(a) Rochelle salt: (deuterated) c-axis projection using coefficients $(F^D_{obs})_{hk0}$: other conditions as in Fig. 23; (b) X-RAC map from which 25(a) was prepared: Rochelle salt: deuterated, negative contours double-exposed.

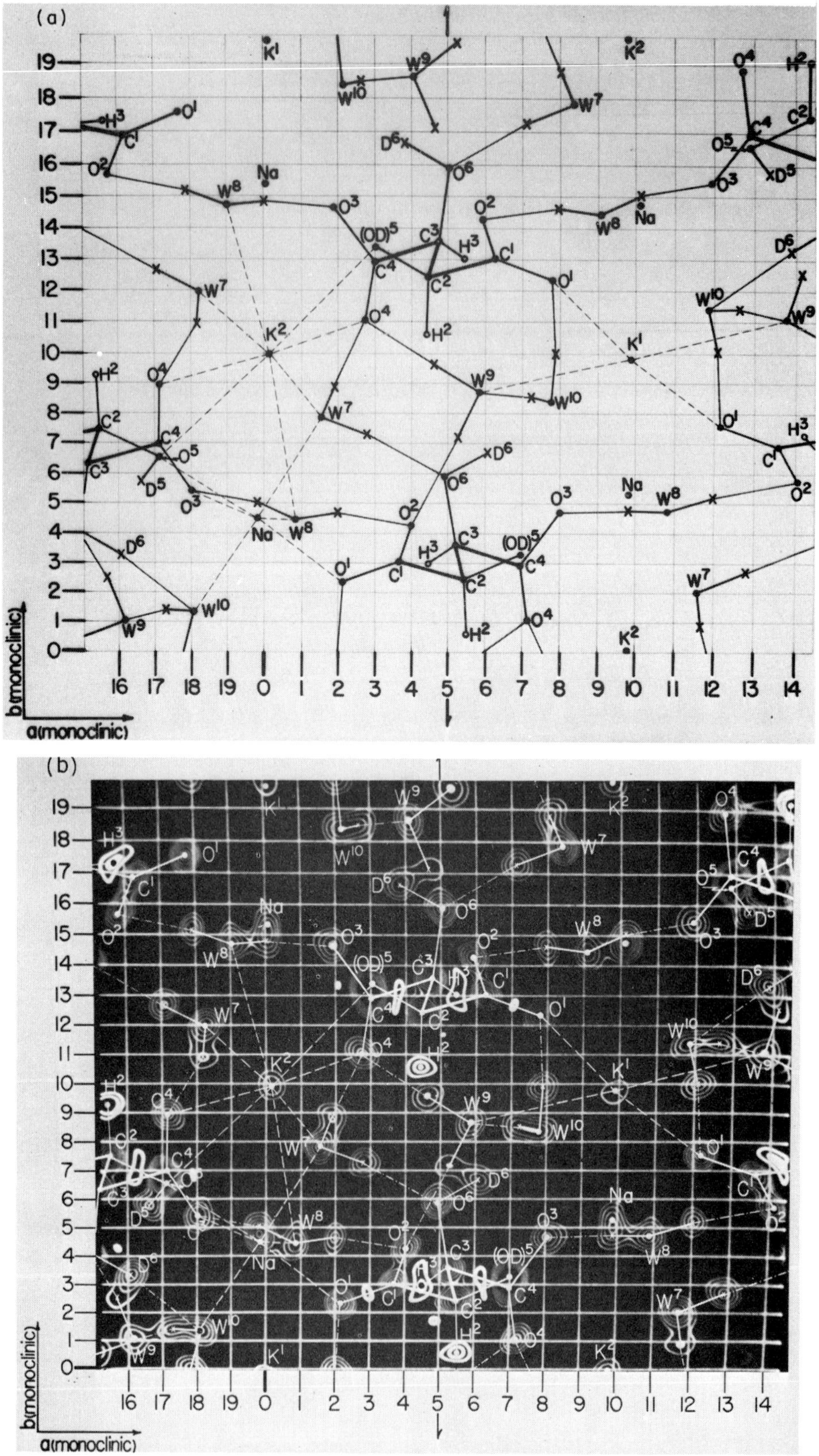
(a)
(b)
b(monoclinic)
a(monoclinic)

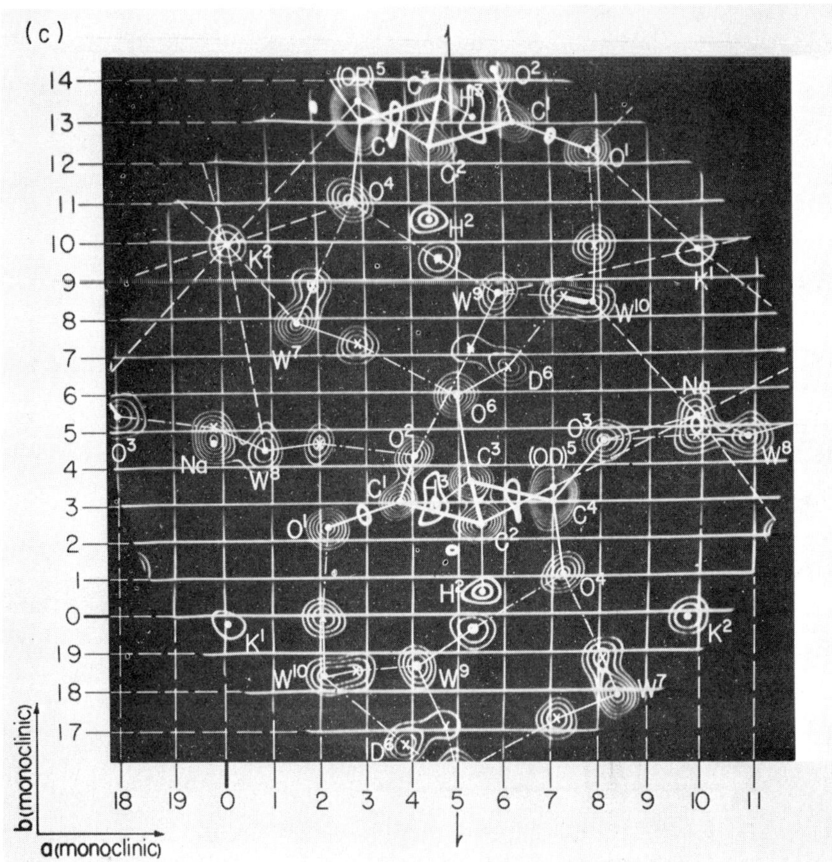

FIG. 26
(a) Rochelle salt: c-axis projection from Fig. 25; A-type molecules in central region; (b) Rochelle salt: X-RAC map corresponding to Fig. 26(a); negative contours showing H_1 and H_2 positions are double-exposed to facilitate their identification; (c) Rochelle salt: detail of 26(b).

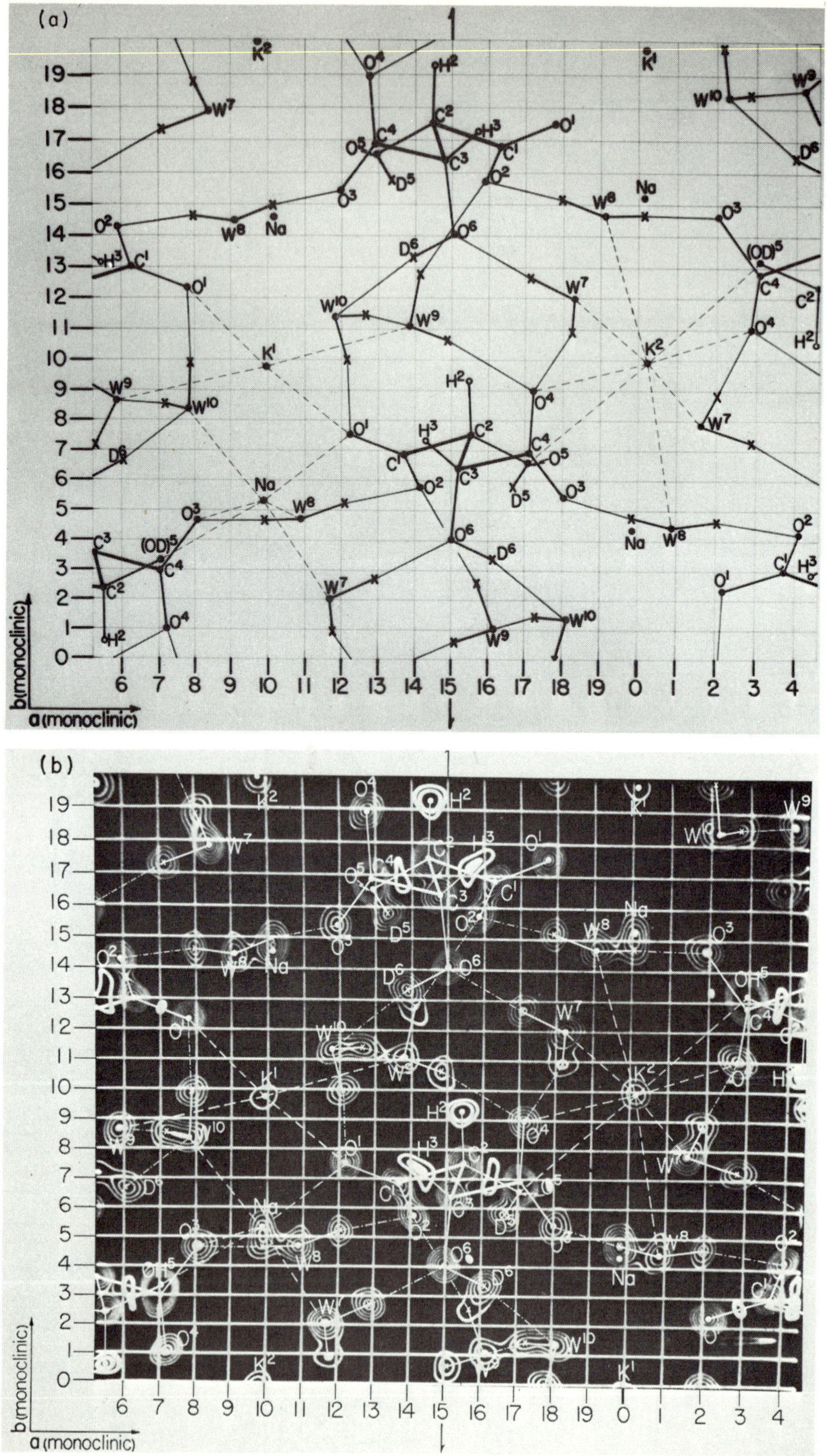
(a)
b (monoclinic)
a (monoclinic)
(b)
b (monoclinic)
a (monoclinic)

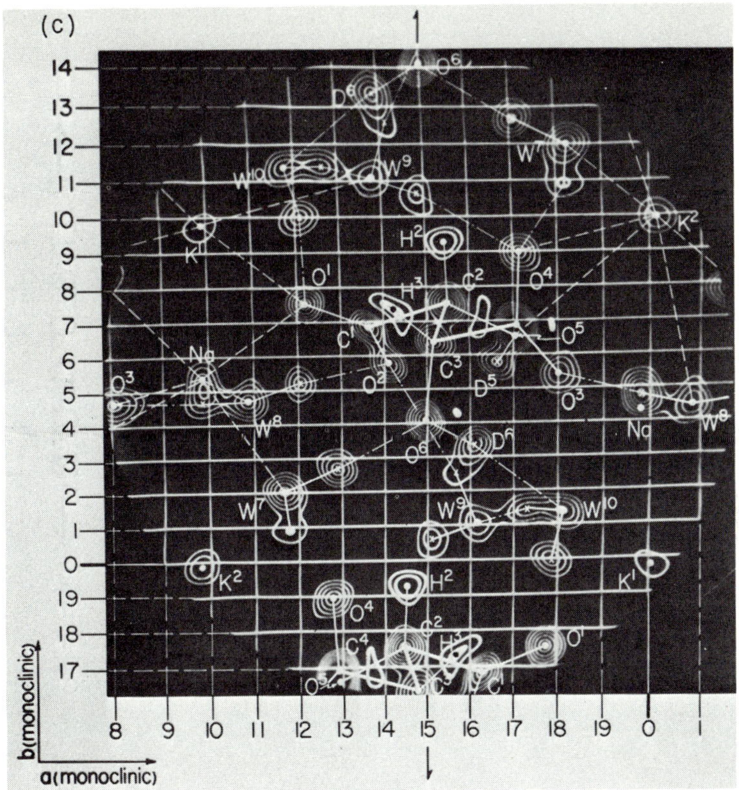

FIG. 27
(a) Rochelle salt: same as Fig. 26(a), but B-type molecules in central region; (b) X-RAC map from which Fig. 27(a) was obtained: Rochelle salt; (c) Rochelle salt: detail of 27(b).

Our immediate concern is with the structures of the P and R quadrants in the FE phase, and the mechanism of FE in these. The remaining parts of Fig. 28 have to do with possible symmetries of the LT phase. Table I assigns the same symmetry to the HT and LT phases. For reasons which will be discussed later, this is rather unlikely, and arguments will be offered which suggest that the symmetry of Fig. 28(c) is the most probable.

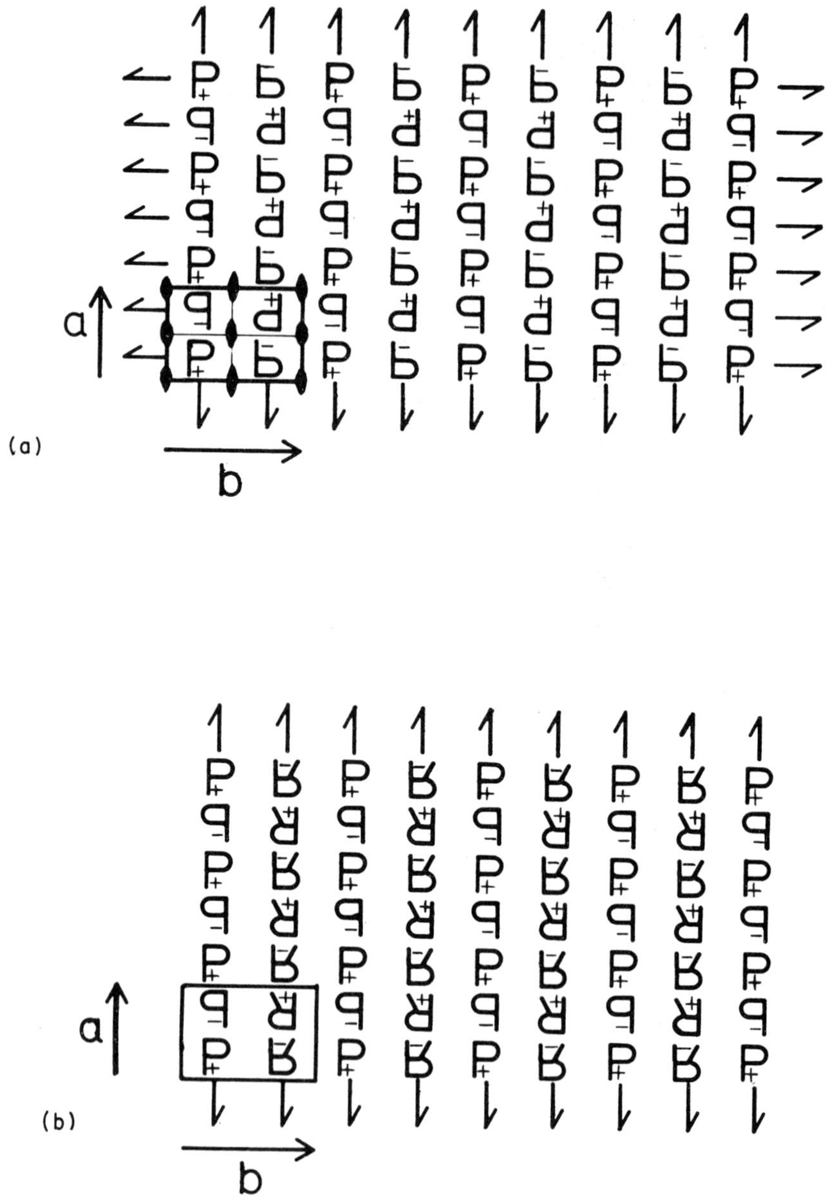
(a)
a
b
(b)
a
b

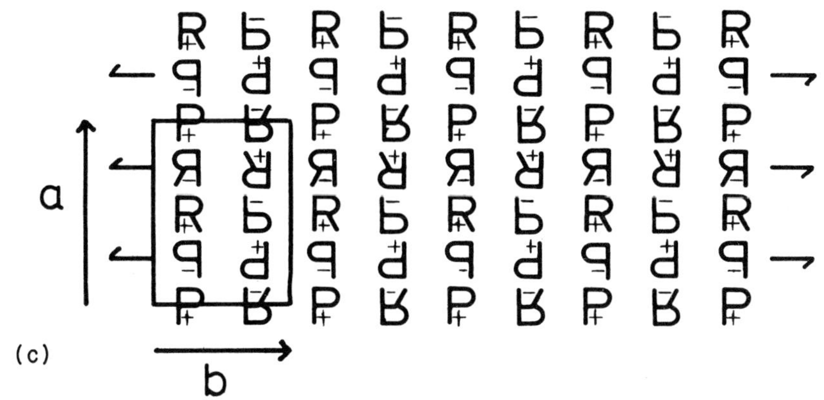

FIG. 28
Relations of asymmetric structure units in Rochelle salt. (a) Representation of HT phase of Rochelle salt, viewed along c-axis (two-fold axes parallel to c are shown only in cell outlined at lower left. P represents one molecular unit). (b) Rochelle salt: representation of symmetry of FE phase, viewed along c-axis (cell outlined at lower left. P represents A-molecular unit; R = B-molecular unit). (c) Proposed symmetry of LT phase of Rochelle salt (a axis doubled due to superstructure; 2_1-axis parallel to b. Cell outlined at lower left). (d) Another possible superstructure for LT phase of Rochelle salt (cell doubled along b; 2_1-axes parallel to a. Cell outlined at lower left).

Returning to the FE phase: The upright and reversed letters P correspond to the A_1 and A_2 quadrants, respectively, in Figs. 23—27; the R rotated about c by 180° corresponds to the B_1 quandrant, and the rotated and reversed R corresponds to B_2. In the figures coordinate lines appear at integral multiples of 0.05 of the cell edges, and these are numbered from 0 to 19. We will use these numbers at the outset, and later transform these to 1/10,000th's of cell edges. A_1 and A_2 atoms are related by the 2_1 axes at y = 5, such that

$$\left(x_{A2}, y_{A2}\right) = \left(x_{A1} + 10, 10 - y_{A1}\right) \tag{1}$$

This, and the corresponding relation for B atoms,

$$\left(x_{B1}, y_{B1}\right) = \left(x_{B2} + 10, 30 - y_{B2}\right) \tag{2}$$

hold for both the FE and the HT phase. In the HT phase we have in addition the 2-axes parallel to c at (x,y) = (0,0), (1/2,0), (0,1/2), and (1/2,1/2). In the (001) projection this relates quadrants A_1 and B_1, and also A_2 and B_2, such that

$$\left(x_{B1}, y_{B1}\right) = \left(x_{A1}, -y_{A1}\right) \tag{3}$$

$$\left(x_{B2}, y_{B2}\right) = \left(x_{A2}, -y_{A2}\right) \tag{4}$$

and by combining (1) and (2) with (3) and (4), or directly from the 2_1 axes parallel to be which result from these combinations, we have

$$\left(x_{B2}, y_{B2}\right) = \left(10 - x_{A1}, y_{A1} + 10\right) \tag{5}$$

$$\left(x_{B1}, y_{B1}\right) = \left(30 - x_{A2}, y_{A2} + 10\right) \tag{6}$$

Thus in the HT phase it suffices to determine the (x_{A1}, y_{A1}) coordinates for atom centers in the A_1 quadrant in order to establish the structure of the entire cell. In the FE phase we must determine the coordinates (x_{A1}, y_{A1}) and (x_{B1}, y_{B1}), since these are independent; but because of the pseudosymmetry (x_{B1}, y_{B1}) do not differ by large amounts from $(-x_{A1}, -y_{A1})$. The difference is responsible for the FE polarization. When the FE axis is reversed in direction by the application of an electric field, the A_1 and B_1 regions are interchanged, and so are the A_2 and B_2 regions. In Fig. 28(b) this amounts to interchanging letters P for R and conversely. This results in reversing the direction of the a axis.

It seems worthwhile to emphasize this presence of two types of constituents in the monoclinic cell because it may explain some effects which have been noted by investigators using magnetic resonance techniques for studies of the transitions[145]-[147][80]. (We will consider these resonance studies later.) When the number of nonequivalent atom sites is doubled, as the structure transforms into the monoclinic FE phase, the number of nonequivalent EFG's which occur and which may be resolvable in NQR (or NDR) spectra should be double the number in the HT phase.

The neutron diffraction analysis of the FE phase clearly shows the major contributions of atomic displacements to the FE polarization. Let us consider first the displacements of non-H and non-D atoms. These are shown in Tables II and III.

The atomic coordinates derivable from the neutron results are taken directly from the X-RAC maps[148] illustrated in Figs. 23-27. Readings are averaged for the two sets of symmetry related A-atoms (A1 and A2), and similarly for the two sets of atoms B_1 and B_2 in the B region. We write $\bar{x}_A$ for the average value of x_{A1} and $x_{A2} - 10$, and $\bar{x}_B$ for the average of x_{B1} and $x_{B2} + 10$. Similarly, $\bar{y}_A$ is the average of y_{A1} and $10 - y_{A2}$, and $\bar{y}_B$ is the average of y_{B1} and $30 - y_{B2}$.

A direct measure of the deviation of the monoclinic structure from the orthorhombic symmetry of the HT phase lies in deviation of coordinates ($\bar{x}_B$, $\bar{y}_B$) from a centrosymmetric equivalence to ($-\bar{x}_A$, $-\bar{y}_A$). We express this immediately in terms of the magnitudes of the sums $\bar{x}_A + \bar{x}_B$ and $\bar{y}_A + \bar{y}_B$. Values of these sums are shown in Tables III and IV.

TABLE II

ROCHELLE SALT, FE PHASE, NEUTRON ANALYSIS;
COORDINATES OF C, O, Na AND K ATOMS*,**,†

	x_{A1}	x_{A2}	y_{A1}	y_{A2}	$\bar{x}_A$	$\bar{y}_A$	x_{B1}	x_{B2}	y_{B1}	y_{B2}	$-\bar{x}_B$	$-\bar{y}_B$
C_1	3.05	13.05	3.70	6.20	3.05 (1525)	3.75 (1875)	16.85	6.85	16.25	13.65	3.15 (1575)	3.70 (1850)
C_2	2.40	12.40	5.50	4.50	2.40 (1200)	5.50 (2750)	17.55	7.55	14.50	15.50	2.45 (1225)	5.50 (2750)
C_3	3.60	13.60	5.20	4.80	3.60 (1800)	5.20 (2600)	16.40	6.40	14.75	15.20	3.60 (1800)	5.23 (2615)
C_4**	3.00	13.05	7.05	2.95	3.00 (1500)	7.00 (3500)	17.05	7.05	13.90	17.10	2.95 (1475)	7.10 (3550)
O_1	2.30	12.30	2.15	7.80	2.30 (1150)	2.18 (1090)	17.55	7.55	17.75	12.25	2.45 (1225)	2.25 (1125)
O_2	4.20	14.20	4.05	5.90	4.20 (2100)	4.08 (2040)	15.75	5.75	15.85	14.15	4.25 (2125)	4.15 (2075)
O_3	4.70	14.70	8.05	1.90	4.70 (2350)	8.08 (4040)	15.40	5.40	11.90	18.05	4.60 (2300)	8.08 (4040)
O_4	1.05	11.05	7.25	2.75	1.05 (0525)	7.25 (3625)	18.95	8.95	12.75	17.15	1.05 (0525)	7.20 (3600)
O_5**	3.30	13.40	7.00	3.00	3.35 (1675)	7.00 (3500)	16.60	6.70	12.85	17.15	3.35 (1685)	7.15 (3575)

	x_{A1}	x_{A2}	y_{A1}	y_{A2}	$\bar{x}_A$	$\bar{y}_A$	x_{B1}	x_{B2}	y_{B1}	y_{B2}	$-\bar{x}_B$	$-\bar{y}_B$
O_6	5.90	15.90	4.95	5.05	5.90 (2950)	4.95 (2475)	14.05	4.05	15.00	14.95	5.95 (2975)	4.98 (2490)
W_7	7.85	17.85	1.60	8.35	7.85 (3925)	1.63 (0815)	12.05	2.00	18.30	11.60	7.98 (3990)	1.65 (0825)
W_8	4.40	14.40	0.80	9.10	4.40 (2200)	0.85 (0425)	14.70	4.70	19.15	10.85	5.30 (2650)	0.85 (0425)
W_9	8.70	18.70	5.80	4.15	8.70 (4350)	5.88 (2940)	11.05	1.05	13.80	16.20	8.95 (4475)	6.20 (3100)
W_{10}	8.35	18.35	7.75	2.20	8.35 (4175)	7.78 (3890)	1.40	11.40	11.75	18.15	8.60 (4300)	8.20 (4100)
Na		14.70		10.10	4.70 (2350)	-0.10 (-0050)		5.40		9.80	4.60 (2300)	-0.20 (-0100)
K_1	-0.20	9.80	0.10	9.85	-0.20 (-0100)	+0.13 (0065)						
K_2	-0.15	9.95	9.80	0.20	-0.10 (-0050)	9.80 (4900)						

*Read from Figs. 24, 25, 26, and 27.

**Accuracy low due to overlapping of C_4 and O_5 in the projection.

†Figures in parentheses are in 1/10,000 of cell edge; other figures are in 1/20 of cell edge.

TABLE III

ROCHELLE SALT, FE PHASE, NEUTRON ANALYSIS: DISPLACEMENT OF B ATOMS WITH RESPECT TO A ATOMS*

	C_1	C_2	C_3	C_4**	O_1	O_2	O_3	O_4	O_5	O_6	W_7	W_8	W_9	W_{10}	†Na	K_1	K_2
$\bar{x}_A+\bar{x}_B$	-50	-25	0	25	-75	-25	50	0	0	-25	-65	-450	-125	-125	50	-100	-50
(Å×10^{-2})	(-6)	—	—	—	(-9)	—	(6)	—	—	—	(-8)	(-53)	(-15)	(-15)	(6)	(-12)	(-6)
$\bar{y}_A+\bar{y}_B$	25	0	-15	-50	-35	-35	0	25	-75	-15	-10	0	-160	-210	-50	65	-100
(Å×10^{-2})	—	—	—	(-7)	—	—	—	—	(-10)	—	—	—	(-23)	(-30)	(-7)	(9)	(-14)

*Numbers not in parentheses are in 1/10,000th's of cell edge; numbers in parentheses are significant displacements in 10^{-2} Å.

**Values for C_4 and O_5 are not reliable because of overlapping of these atoms in (001) projection.

†Na positions taken from Fig. 23, to avoid overlapping of D_8 with Na in Figs. 25-27.

The X-RAC coordinate lines, at 1/20th of the cell edges, are generated accurately and simultaneously with the electronic frequencies which ultimately produce the contour lines for the distribution of scattering material. Thus any distortions of the maps which occur due to nonlinear displacement of the electron beam on the face of the cathode ray tube are not destructive of the accuracy of atomic coordinate determination. Coordinate accuracy can be read reliably to about 1/20th of the distance between coordinate lines, and hence to about 1/400th of a cell edge: i.e., to about 0.0025 of a cell edge. In the case of Rochelle salt this corresponds to about 0.03Å. This error is about three times that achievable from a least-squares refinement of the coordinates, involving observed and calculated structure factors. Data for such refinement are not immediately available; but they will be disinterred and published elsewhere[149].

The decoupling of the A and B regions is immediately clear from Figs. 23-27, in the displacement of K_1^+ and K_2^+ ions from their special positions, in shifts in positions of W_8, W_9, and W_{10}, and in lesser shifts of the Na^+ ions and atoms C_1, O_1, and O_3. Values of the coordinates of the non-H and non-D atoms are shown, in 1/20th's and decimal parts of the cell edges, in Table II; and $\bar{x}_A + \bar{x}_B$ values and $\bar{y}_A + \bar{y}_B$ values appear in Table III.

Values of $(\bar{x}_A - \bar{x}_B)/2$, and $(y_A + y_B)/2$, are averages of $|\bar{x}_A|$ and $|\bar{x}_B|$, and $|\bar{y}_A|$, and $|\bar{y}_B|$, respectively. These should represent a good approximation to the values for x_A and y_A found in the three-dimensional X-ray analysis of the HT phase, which is reported in the next section.

TABLE IV

ROCHELLE SALT, FE PHASE, NEUTRON ANALYSIS; COORDINATES OF D NUCLEI IN DEUTERATED CRYSTAL*

															In Å×10^{-2}	
	x_{A1}	x_{A2}	y_{A1}	y_{A2}	$\bar{x}_A$	$\bar{y}_A$	x_{B1}	x_{B2}	y_{B1}	y_{B2}	$-\bar{x}_B$	$-\bar{y}_B$	$\bar{x}_A+\bar{x}_B$	$\bar{y}_A+\bar{y}_B$	$\bar{x}_A+\bar{x}_B$	$\bar{y}_A+\bar{y}_B$
D_5	2.95	12.95	6.90	3.10	2.95 (1475)	6.90 (3450)	15.85	5.85	13.28	16.72	4.15 (2075)	6.72 (3360)	-1.20 (-0600)	0.18 (0090)	-71	13
D_6*	6.65	16.65	6.10	3.90	6.65 (3325)	6.10 (3050)	13.50	3.50	13.85	16.15	6.50 (3250)	6.15 (3075)	0.15 (0075)	-0.05 (-0025)	9	—
D_7	8.80	18.80	1.95	8.05	8.80 (4400)	1.95 (0975)	11.10	1.10	18.30	11.70	8.90 (4450)	1.70 (0850)	-0.10 (-0050)	0.25 (0125)	-6	18
D_7'	7.21	17.21	2.88	7.12	7.21 (3605)	2.88 (1440)	12.78	2.72	17.05	12.95	7.28 (3640)	2.95 (1475)	-0.07 (-0035)	-0.07 (-0035)	—	—
D_8	5.00	15.00	-0.19	10.19	5.00 (2500)	-0.19 (-0095)	14.74	4.74	-0.04	9.96	5.26 (2630)	+0.04 (0020)	-0.26 (-0130)	-0.23 (-0115)	-15	-16
D_8'	4.64	14.64	1.95	8.05	4.64 (2320)	1.95 (0975)	15.11	5.11	17.94	12.06	4.89 (2445)	2.06 (1030)	-0.25 (-0125)	-0.11 (-0055)	-15	-8
D_9	9.68	19.68	4.73	5.27	9.68 (4840)	4.73 (2365)	10.63	0.63	14.85	15.15	9.37 (4685)	5.15 (2575)	0.31 (0155)	-0.42 (-0210)	18	-30
D_9'†	7.15	17.15	5.15	4.85	7.15 (3575)	5.15 (2575)	12.50	2.50	14.25	15.75	7.50 (3750)	5.75 (2875)	-0.35 (-0175)	-0.60 (-0300)	-21	-43

	x_{A1}	x_{A2}	y_{A1}	y_{A2}	$\bar{x}_A$	$\bar{y}_A$	x_{B1}	x_{B2}	y_{B1}	y_{B2}	$-\bar{x}_B$	$-\bar{y}_B$	$\bar{x}_A+\bar{x}_B$	$\bar{y}_A+\bar{y}_B$	$\bar{x}_A+\bar{x}_B$ (In Å×10^{-2})	$\bar{y}_A+\bar{y}_B$ (In Å×10^{-2})
D_{10}	9.89	19.89	7.88	2.12	9.89 (4945)	7.88 (3940)	10.00	0.00	12.15	17.85	10.00 (5000)	7.85 (3925)	-0.11 (-0055)	0.03 (0015)	-7	—
D_{10}'	8.45	18.45	7.09	2.91	8.45 (4225)	7.09 (3545)	11.38	1.38	12.73	17.27	8.62 (4310)	7.27 (3635)	-17 (-0085)	-0.18 (-0090)	-10	-13

*As read from Figs. 23, 25, 26, and 27.

[b]Numbers in parentheses are in 1/10,000th's of cell edge; other numbers are in 1/20th's of cell edge, except for last two columns.

**[c]x_B and y_B values of D_6 are less accurate due to proximity of D_9' in projection.

†[d]x_B and y_B values of D_9' are less accurate due to proximity of D_6 in projection.

The positions and shifts of C_4 and O_5 are not accurately determinable from the (001) neutron projections because of the overlapping of these two atoms. The values selected were taken from earlier calculations. The positions of the Na ions are obscured in Figs. 25–27 by overlapping of D_8; the peak due to D_8 is eliminated in Fig. 23, however, and in this the Na positions and shifts are readily seen.

The largest shift is that of W along x. Other large shifts are apparent in the $\bar{y}_A + \bar{y}_B$ values of W_9 and W_{10}. But before analyzing the results further, we turn our attention to the shifts in deuteron positions. These are read from the X-RAC maps of Figs. 24 and 25–27. Table IV lists the D-atom coordinates as obtained from Fig. 24 checked against Figs. 25–27. The largest shifts here are those of D_5, D_9, and D_9'.

The numbering of the D nuclei is referred to the water oxygen or hydroxyl to which they are attached. In each water molecule, the D with the highest x coordinate is unprimed; the second D is then primed.

The major D shift is that of D_5, which is the hydroxyl D on O_5. For the A molecules, D_5 lies directly below or above C_4 in the (001) projection. In Figs. 26(a) and (b) the contribution of the hydroxyl to the unresolved C_4-O_5-D_5 peak is designated as $(OD)_5$. In Figs. 27(a) and (b), O_5(b) still overlaps C_4(B); but D_5 is very clearly resolved at coordinates (x,y) = (15.85, 13.28) and (5.85, 16.72). The shift from the A to B position for D_5 is 0.71 Å along x and 0.13 Å along y.

The three-dimensional X-ray analysis (which will be reported in the next section) shows that O_5 is usually close to O_2 of the same molecule, due to the bond conjugation which causes each half of the tartrate molecule to be essentially coplanar; O_6 is correspondingly close to O_3, for the same reason. No H nuclei lie between or nearly between O_2 and O_5 or O_3 and O_6; of course the only H atoms which might be available for such H-bonding are H_5 (or D_5) on O_5 and H_6 (or D_6) on O_6. The D_6 is involved in an H-bond to W_{10}; what about D_5: is it involved in an H-bond?

In the A configuration, D_5 is clearly not H-bonding, but in the (001) projection of the B configuration, D_5 seems to lie on a line between O_5 and O_6. One must explain why there are only two distinct positions for D_5, depending upon the A or B configuration of the tartrate ion. It appears that $(O_5\text{-}D_5 \ldots O_6)_B$ represents an H-bond, and $(O_5 \ldots O_6)_A$ is only a Van der Waals contact. The X-ray study (next section) indicates 2.91 Å for the $O_5 \ldots O_6$ separation. This corresponds to a weak H-bond; but the X-ray analysis does not distinguish between A and B configurations. It is likely that 2.91 Å is an average of two separations: one somewhat longer, for A ions; and a second somewhat shorter for the B ions. It is impossible to find any support for this latter conjecture from the neutron projection, because of the overlapping of C_4 and O_5. A three-dimensional neutron analysis is essential.

The H-bonding between O_5 and O_6 affects the conformation of the tartrate ion, and particularly the angle α between the planes of the two halves of the ion (see Fig. 19). The conformation in solution should vary with pH, concentration of nonchelated metal ions, etc. It becomes understandable, why the tartrate ion conformation is so closely similar for Rochelle salt and for $LiNH_4$tartrate · H_2O[142].

It is probable that the position of D_5 is influenced by neighboring K^+ and Na^+ positions. This will be considered further in the next section.

Other shifts in D nuclei are listed, in Å, in the last two columns of Table IV. Both D_8 and D_8' shift, and by similar amounts; but the motion of W_8 itself is very much larger along the x-direction. As Fig. 24(a,b) indicate, W_8 is at the vertex of a V in the projection, with the vertex down for the FE polarization condition shown in Figs. 23-27. When the polarization is reversed, the V's are inverted.

Just as W_9 is significantly shifted with polarization reversal (Table III), so are its two D-atoms. The molecule as a whole shifts toward -x and -y, and also rotates. The W_{10} itself moves noticeably, but its two D-atoms shift only slightly. The effects are summarized in Table V.

Molecule W_9 bears watching. The peaks corresponding to D_9 and D_9' positions are anomalously weak, and the peak for W_9 itself seems low. W_{10} itself moves, particularly along y; but D_{10} motion is unobservable and D_{10}' shifts only slightly. W_7 moves very slightly but D_7' does not and D_7 shifts most of the three.

It is clear that all four water molecules rotate, with water 7 showing the most stability and W_8 and W_9 the greatest movements. Water 9 betrays a sort of shadowy character, which we will discuss later. It is clearly the water which is most readily lost in low-humidity conditions, and is most readily disturbed by radiation effects. Water 10 rotates but leaves its D atoms relatively undisturbed, particularly along the FE direction.

TABLE V

ROCHELLE SALT, FE PHASE, NEUTRON ANALYSIS:
SUMMARY OF ALL SIGNIFICANT DIFFERENCES
BETWEEN A AND B POSITIONS[a]

A. Shifts below level of discernability in Fourier projections:

atoms C_2, C_3, C_5, O_2, O_4, O_5*, O_6, and D_7'.

B. Shifts small but discernable:

atoms C_1, C_4*, O_1, O_3, W_7, Na, D_6**, and D_{10}

C. Moderate shifts:

Atom:	W_9	W_{10}	K_1	K_2	D_7	D_8	D_8'	D_9	D_{10}'
$\bar{x}_A + \bar{x}_B$	-15	-15	-12	-6	-6	-15	-15	18	-10
$\bar{y}_A + \bar{y}_B$	-23	-30	9	-14	18	-16	8	30	-13

D. Large shifts:

Atom:	W_8	D_5	D_9'**
$\bar{x}_A + \bar{x}_B$	-53	-71	-21
$\bar{y}_A + \bar{y}_B$	—	13	-43

[a]In Å x 10^{-2}.

*Positions obscured somewhat by near overlapping of C_4 and O_5 in projection.

**Positions less accurate due to proximity of atoms D_6 and D_9'.

Thus it is obvious that the noble effort represented by the Mason theory[121] for the $O_1 \ldots H - W_{10}$ bond is demolished, and that the original proposal by Kourtschatov[120][132] is a large part of the truth. The invitation to the Lobster Quadrille[139] is answered by all the D's except for D_6, D_7', and D_{10} (which remain wallflowers). The search of the NQR investigators[145]-[147], which focused from one D to another, it quite understandable and can now become somewhat more rational. We have promised to consider the resonance results later.

We are now in a position to calculate the spontaneous polarization which is produced by these shifts. We compute $\Sigma Q_i \cdot \Delta x_i$; Q_i is the charge carried by the *ith* atom and Δx_i is $(\bar{x}_A + \bar{x}_B)_i$, which is the shift of the *ith* atom along -x in the B quadrants of the FE cell, with respect to the corresponding atoms in the A quadrants; or it may be viewed as the shift of A-quadrant atoms along +x, with respect to the positions of the B-region atoms.

The Na, K_1, and K_2 ions carry a charge ε, where $\varepsilon = 4.80 \times 10^{-10}$ esu. The O_1 and O_3 are carboxylic oxygens; the carboxyl group carries a charge $-\varepsilon$, and we may assume that O_1 and O_3 carry -0.5ε each (the X-ray analysis, next section, shows that C-O distances are about the same for C_1-O_1, C_1-O_2, and C_4-O_3).

In order to arrive at a reasonable estimate of the charge carried by the O and H atoms of a water molecule, we recall that the dipole moment of a water molecule in the gaseous state is 1.84 esu · Å. The two O-H bonds make an angle of about 105° at the oxygen, and the O-H separation is about 1.0 Å. Assigning a charge $-2p\varepsilon$ to the water O, and $+p\varepsilon$ to each H atom, it is seen that $p \sim -0.315$. Thus the charge on a water O must be $\sim -0.63\varepsilon$, and on each water H (or D) $\sim 0.315\varepsilon$. We will assume, without more justification, that the H atoms of a hydroxyl group (D_5 and D_6, in the present case) also carry a charge of $+0.315\varepsilon$.

These values of Q_i are introduced in Table VI to obtain values of $Q_i \cdot \Delta x_i$, and we find

$$\Sigma Q_i \cdot \Delta x_i = 0.097\varepsilon \cdot \text{Å}$$

In calculating P_s, we must take account of the fact that the above summation represents the contribution of one molecular unit out of the two in each A and B region. The P_s is then obtained by dividing $\Sigma Q_i \cdot \Delta x_i$ by half the volume of a unit cell (in cm^3), converting ε to coulombs and Å to 10^{-8}cm ($\varepsilon = 1.5921 \times 10^{-19}$ coulombs). From Table I, $V/2 = 526.5 \times 10^{-24}$ cm^3. Thus we obtain P_s(calc) = 0.308×10^{-6} coulombs/cm^2 = 0.308 $\mu C/cm^2$.

The observed value of P_s for deuterated Rochelle salt is shown in Fig. 17. At 18°C (which is the temperature at which our neutron observations were made at the Brookhaven reactor) P_s (observed) ≃ 0.30 $\mu C/cm^2$. The agreement between observed and calculated values is embarrassingly good. No effects of electronic polarization have been included, since we do not know how to do so; the values of atomic shifts are reliable to only about 0.03 Å, and only those as great as or exceeding 0.06 Å have been included in the calculation; and our estimates of the charges on O and D atoms are also approximate.

In calculating $\Sigma Q_i \cdot \Delta x_i$ from the values of Table VI, we see that the summation is comprised of a large positive and a large negative contribution: 76.4 - 66.7 = 9.7, in $\varepsilon \times$ Å $\times 10^{-2}$. Thus P_s is the consequence of a small inbalance between large positive and negative components. It is not critical to the present discussion, that the atomic shifts could be improved in accuracy or the charge assignments made with greater precision. This small imbalance between large positive and negative contributions is to be expected because the a direction is polar and the structure deviates only

TABLE VI

CALCULATION OF $\Sigma Q_i \cdot \Delta x_i$ FOR FE PHASE OF DEUTERATED ROCHELLE SALT*

Atom (i)	O_1	O_3	W_7	W_8	W_9	W_{10}	Na	K_1	K_2	D_5	D_6	D_7	D_8	D_8'	D_9	D_9'	D_{10}	D_{10}'
ΔQ_i**	-50	-50	-63	-63	-63	-63	100	100	100	31.5	31.5	31.5	31.5	31.5	31.5	31.5	31.5	31.5
Δx_i†	-9	6	-8	-53	-15	-15	6	-12	-6	-71	9	-6	-15	-15	18	-21	-7	-10
$Q_i \cdot \Delta x_i$††	4.5	-3.0	5.0	33.4	9.5	9.5	6.0	-12.0	-6.0	-22.4	2.8	-1.9	-4.7	-4.7	5.7	-6.6	-2.2	-3.2

*$\Sigma Q_i \cdot \Delta x_i = 9.7\varepsilon \cdot \text{Å} \times 10^{-2} = 0.097\varepsilon \cdot \text{Å}$

**Magnitude of Q_i is in number of electrons x 100; see text for charge assignments.

†In $\text{Å} \times 10^{-2}$.

††$\varepsilon \cdot \text{Å} \times 10^{-2}$; ε = charge on electron.

slightly from nonpolar symmetry. The major contributions to the summation come from W_8 and D_5, with intermediate contributions from W_9, W_{10}, and K_1; the sum of $Q_i \cdot \Delta x_i$ terms for these five atoms is $+18.0\varepsilon \cdot \text{Å} \cdot 10^{-2}$; this is reduced by the sum of the smaller contributions, but it is unlikely that the latter could reverse the sign of the total P_s. It will be interesting to see how a future, accurate analysis will alter these results; we can predict that the corrections will be minor.

We should think a moment about the meaning of the Δy_i values. These do indicate the shift in values of $\bar{y}_B$ with respect to $\bar{y}_A$, corresponding to the nonequivalence of A and B regions. But there is no polarization along the b axis, because $y_{A2} = 10 - y_{A1}$, and $y_{B2} = 30 - y_{B1}$, because of the 2_1-axes parallel to a and lying at $y = 5$ and $y = 15$. (The y values here are in 1/20th's of the b periodicity.) Because of the 2_1-axes, $y_{A1} + y_{A2} = 10$, and $y_{B1} + y_{B2} = 30$, and hence $\Sigma Q_i \cdot (y_{A1} + y_{A2} + y_{B1} + y_{B2}) = 0$. Any direction perpendicular to a 2-axis or 2_1-axis is never polar.

A three-dimensional neutron analysis is capable of providing the absolute configuration of the polarized structure. Ordinarily one determines an absolute configuration by means of the anomalous scattering of X-rays; such determinations were first accomplished for FE's by Unterleitner, Okaya, Vedam, and Pepinsky[150] on KD_2AsO_4 and (glycine)$_3 \cdot H_2SO_4$, and by Okaya, Pepinsky and Unterleitner for $BaTiO_3$[151]; a review of the technique has been published by Okaya and Pepinsky[152]. But the writer first pointed out, earlier, that the absolute configuration of a structure can be established without the use of anomalous dispersion if the absolute configuration of some component in the structure is known[153]; and this is of course the case for Rochelle salt since the absolute configuration of the d-tartrate ion is already known, thanks to the anomalous dispersion technique as applied to tartrate salts by Bijvoet in 1949[154]. The direction of the +x axis in Rochelle salt can be assigned from the hand of the tartrate ion in three dimensions; and the FE coercive field, when applied along this +x direction, then permits a full determination of the polarized structure configuration. This will also permit us to determine whether the angle by which the c axis deviates from 90° with respect to the +b axis is + or - 3 minutes of arc.

The observed spontaneous polarization P_s of Rochelle salt is smaller than that of almost all other FE's. Only natural boracite has a significantly smaller value[155]. The observed value of P_s for tetragonal $BaTiO_3$ is 26 $\mu C/cm^2$, which is over 100 times larger than that for Rochelle salt; and yet the largest ion displacement in $BaTiO_3$ is 0.09 Å[156] for one oxygen. The value of P_s for KH_2PO_4 is $\sim 4 \mu C/cm^2$; in this crystal the largest atom shift is less than any considered significant in our present analysis[157], and even so the two positively charged ions (P and K) move in opposite directions and hence their contributions to P_s partially cancel. The contributions in Rochelle salt are indeed delicately balanced. One begins to understand why its dielectric properties must be examined with small fields.

F. Three-Dimensional X-Ray Analyses

Attention can now be turned to the three-dimensional X-ray studies of all three phases, by Krstanovic, Okaya, and the present writer[137]. These were accomplished in 1958-1959, and were not published in detail because certain problems arose in their interpretation for which explanations were not found. The present account summarizes the results obtained, suggests some interpretations, and also suggests the source of the earlier difficulties.

Crystals from aqueous solution were ground into cylinders of about 0.2 mm diameter for these studies. CuKα radiation was used except as indicated below for some aspects of the FE phase analysis. Data for the two PE phases (HT and LT) were collected at +32°C and -50°C, on nondeuterated crystals. Data for the FE phase analysis were taken from a deuterated sample, because the upper Curie point for nondeuterated material is so close to room temperature. For the latter analysis an electric field of 500 V/cm was applied along the FE axis to obtain a single-domain specimen. The electrodes limited observation of some of the higher layers with CuKα radiation; and in order that the same reciprocal space be covered for all three phases, the CuKα data was supplemented by MoKα reflections for the FE phase. Corrections for the Lorenz-polarization factors and absorption were accomplished by IBM 704 computations. It was later found that no extinction correction was necessary.

The problem of the apparent symmetry of the FE phase arose at the outset. Even greatly overexposed X-ray photographs, using crystal-monochromatized radiation, failed to reveal odd (0K0) reflections from the monoclinic structure. The FE structure was therefore treated as orthorhombic and refined in space group $P2_12_12$, although the polarizing electric field was maintained nevertheless.

It was presumed that at least some electron density peaks in the Fourier maps would then appear nonspherical, and refinement could proceed in $P2_1$ in an attempt to eliminate these distortions. But no such peak distortion appeared: nor was any observable in F_o-F_c difference maps. The only indication of anomalous conditions appears in the Debye-Waller temperature factors for W_9, W_{10}, and K_1 in the FE phase; these are unusually high: 3.85×10^{-16} cm^2 for W_9, 3.617 for W_{10}, and 3.487 for K_1; corresponding factors for other atoms are about half as large.

Atomic coordinates (in decimal parts of cell edges), their standard deviations calculated accordingly to Cruickshank's method[158], and individual (but isotropic) temperature factors, are presented for the three phases in Table VII, VIII, and IX. The computing capacity of the IBM 704 machine available to us was not adequate for computation of anisotropic temperature factors, at the time of these analyses. The R-factors achieved for the HT, FE, and LT phases were 0.152, 0.142, and 0.100 respectively.

The results for the LT phase are more satisfactory, in terms of lower standard deviations, temperature factors and R-factor. The temperature factors for all four water oxygens and K_1 are still remarkably high, as are those for O_3 and O_4. In the HT phase, the temperature factors for W_9 and W_{10} are again anomalously high, and that for K_1 is quite unacceptably so.

Let us examine these results, taking them at face value at first.

The analysis of Beevers and Hughes[118] was confirmed as basically correct; but significant differences appear in details of the structures.

TABLE VII. ROCHELLE SALT: HIGH TEMPERATURE PHASE*

Atom	x_2	σ(x)	y	σ(y)	z	σ(z)	B
C_1	0.1550	0.00078	0.1888	0.00080	0.2879	0.00075	1.342
C_2	0.1248	0.00085	0.2727	0.00085	0.4234	0.00081	1.655
C_3	0.1777	0.00084	0.2646	0.00086	0.6456	0.00081	1.629
C_4	0.1563	0.00092	0.3499	0.00093	0.7827	0.00089	2.008
O_1	0.1190	0.00058	0.1084	0.00059	0.3559	0.00056	1.495
O_2	0.2096	0.00059	0.2036	0.00060	0.1212	0.00057	1.573
O_3	0.2376	0.00068	0.4052	0.00068	0.8136	0.00066	2.124
O_4	0.0525	0.00068	0.3611	0.00068	0.8471	0.00067	2.150
O_5	0.1625	0.00063	0.3569	0.00063	0.3173	0.00060	1.784
O_6	0.2963	0.00058	0.2487	0.00059	0.6268	0.00056	1.483
O_7	0.3940	0.00067	0.0828	0.00070	0.4855	0.00066	2.112
O_8	0.2456	0.00075	0.0412	0.00071	0.8882	0.00076	2.722
O_9	0.4373	0.00087	0.3012	0.00086	0.0372	0.00086	3.244
O_{10}	0.4224	0.00094	0.3958	0.00093	0.4294	0.00093	3.604
Na	0.2318	0.00032	0.9940	0.00032	0.5218	0.00030	1.295
K_2	0.0000	—	0.5000	—	0.1584	0.00039	1.948
K_1	0.0000	—	0.0000	—	0.0485	0.00059	3.644

*Atomic coordinates (in fractions of cell edges); their standard deviations in Å unit and individual temperature factors (in 10^{-16} cm^2).

TABLE VIII. ROCHELLE SALT: FERROELECTRIC PHASE*

Atom	x	σ(x)	y	σ(y)	z	σ(z)	B
C_1	0.1530	0.00083	0.1875	0.00064	0.2863	0.00057	1.274
C_2	0.1251	0.00080	0.2728	0.00061	0.4220	0.00054	1.062
C_3	0.1769	0.00092	0.2647	0.00072	0.6426	0.00064	1.716
C_4	0.1569	0.00093	0.3483	0.00072	0.7828	0.00064	1.720
O_1	0.1186	0.00062	0.1092	0.00049	0.3526	0.00044	1.472
O_2	0.2072	0.00065	0.2034	0.00052	0.1198	0.00047	1.677
O_3	0.2356	0.00071	0.4064	0.00057	0.8135	0.00052	2.083
O_4	0.0517	0.00073	0.3637	0.00060	0.8485	0.00054	2.259
O_5	0.1635	0.00062	0.3566	0.00050	0.3223	0.00044	1.494
O_6	0.2962	0.00066	0.2476	0.00053	0.6272	0.00047	1.756
O_7	0.3952	0.00072	0.0827	0.00058	0.4863	0.00052	2.158
O_8	0.2450	0.00072	0.0395	0.00059	0.8869	0.00052	2.171
O_9	0.4389	0.00086	0.3020	0.00086	0.0358	0.00078	3.850
O_{10}	0.4244	0.00097	0.3966	0.00082	0.4234	0.00073	3.617
Na	0.2317	0.00035	0.9927	0.00028	0.5207	0.00025	1.481
K_2	0.0000	—	0.5000	—	0.1581	0.00045	1.961
K_1	0.0000	—	0.0000	—	0.0469	0.00031	3.487

*Atomic coordinates (in fractions of cell edges); their standard deviations in Å unit and individual temperature factors (in 10^{-16} cm^2).

TABLE IX. ROCHELLE SALT: LOW TEMPERATURE PHASE*

Atom	x	σ(x)	y	σ(y)	z	σ(z)	B
C_1	0.1551	0.00056	0.1894	0.00048	0.2823	0.00062	1.587
C_2	0.1265	0.00059	0.2736	0.00050	0.4230	0.00065	1.798
C_3	0.1768	0.00060	0.2645	0.00051	0.6491	0.00066	1.850
C_4	0.1494	0.00062	0.3502	0.00053	0.7828	0.00069	1.942
O_1	0.1198	0.00045	0.1072	0.00038	0.3476	0.00050	1.580
O_2	0.2125	0.00041	0.2036	0.00035	0.1171	0.00045	1.676
O_3	0.2311	0.00045	0.4068	0.00040	0.8187	0.00051	2.241
O_4	0.0485	0.00045	0.3595	0.00039	0.8503	0.00051	2.189
O_5	0.1667	0.00042	0.3577	0.00036	0.3232	0.00031	1.778
O_6	0.2967	0.00041	0.2490	0.00036	0.6346	0.00046	1.750
O_7	0.3965	0.00045	0.0838	0.00039	0.4838	0.00050	2.130
O_8	0.2386	0.00049	0.0418	0.00043	0.8805	0.00055	2.586
O_9	0.4403	0.00050	0.3073	0.00044	0.0293	0.00057	2.676
O_{10}	0.4223	0.00049	0.3923	0.00043	0.4287	0.00056	2.556
Na	0.2325	0.00025	0.9930	0.00022	0.5175	0.00028	1.626
K_2	0.0000	—	0.5000	—	0.1598	0.00036	1.725
K_1	0.0000	—	0.0000	—	0.0424	0.00029	2.658

*Atomic coordinates (in fractions of cell edges); their standard deviation in Å unit and individual temperature factors (in 10^{-16} cm^2).

1. Coordination Around the Metal Ions

The two K ions are nonequivalent, and in $P2_12_12$ they are in special positions [at $(0,0,z_1)$ and $(\frac{1}{2},\frac{1}{2},\frac{1}{2} - z_1)$ for K_1, and at $(\frac{1}{2},0,z_2)$ and $(0,\frac{1}{2},\frac{1}{2}, - z_2)$ for K_2]. The Na ions are in general positions. The coordination of K_1, K_2 and Na are shown in Fig. 29. Table X lists K-O separations for all three phases, and Table XI gives Na-O separations for these phases. It will be noted that W_{10} does not coordinate to K_1, although it does to Na; W_9 coordinates

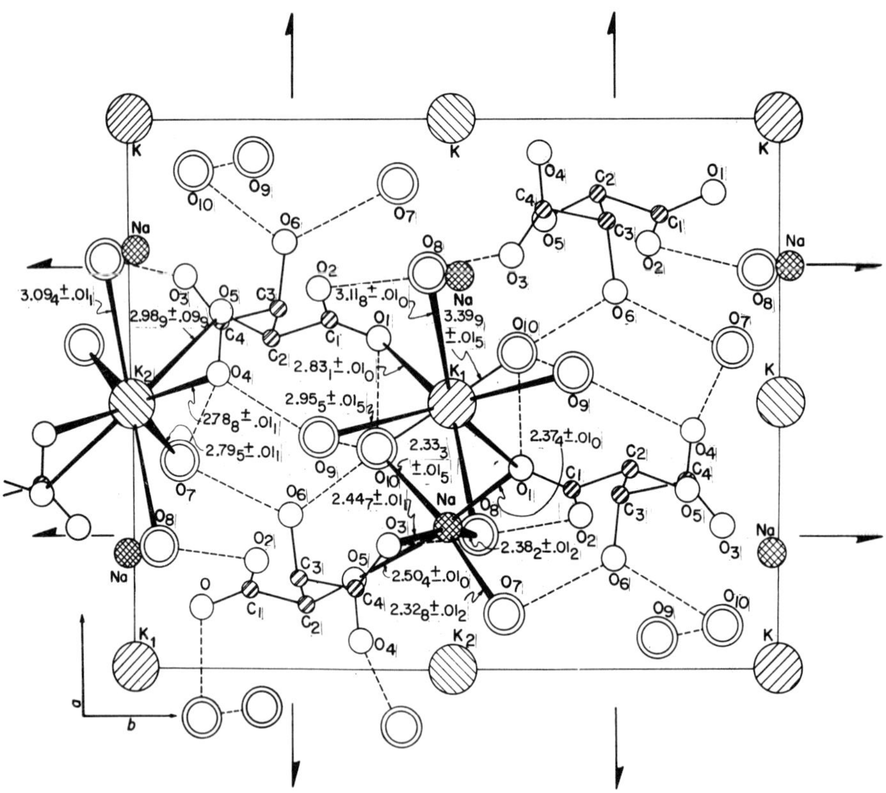

FIG. 29
Rochelle salt, FE phase: coordination around the metal ions, from X-ray analysis; distances for all three phases shown in Tables X and XI.

TABLE X

ROCHELLE SALT: K–O DISTANCES*

		High temperature	Ferroelectric	Low temperature
(1)	Around K_1			
To	O_1	$2.83_3 \pm 0.01_0$	$2.83_1 \pm 0.01_0$	$2.81_0 \pm 0.00_9$
	O_8	$3.13_6 \pm 0.01_3$	$3.11_8 \pm 0.01_1$	$3.05_1 \pm 0.01_0$
	O_9	$2.97_4 \pm 0.01_1$	$2.95_5 \pm 0.01_5$	$2.86_1 \pm 0.01_0$
	O_{10}**	$3.44_5 \pm 0.01_6$	$3.39_9 \pm 0.01_5$	$3.42_0 \pm 0.00_9$
(2)	Around K			
To	O_4	$2.83_5 \pm 0.01_4$	$2.78_8 \pm 0.01_1$	$2.82_6 \pm 0.00_8$
	O_5	$2.97_4 \pm 0.01_3$	$2.98_9 \pm 0.00_9$	$2.99_8 \pm 0.00_7$
	O_7	$2.80_5 \pm 0.01_4$	$2.79_7 \pm 0.01_1$	$2.79_1 \pm 0.00_8$
	O_8	$3.08_8 \pm 0.01_3$	$3.09_4 \pm 0.01_1$	$3.15_5 \pm 0.00_9$

*In Å.

**O_{10} does not coordinate to K_1.

TABLE XI

ROCHELLE SALT: Na-O DISTANCES*

	High temperature	Ferroelectric	Low temperature
Na—O_1	$2.34_7 \pm 0.01_1$	$2.37_4 \pm 0.01_0$	$2.34_7 \pm 0.00_9$
Na—O_3	$2.46_4 \pm 0.01_3$	$2.44_7 \pm 0.01_1$	$2.45_4 \pm 0.00_9$
Na—O_5	$2.52_6 \pm 0.01_3$	$2.50_9 \pm 0.01_0$	$2.46_7 \pm 0.00_8$
Na—O_7	$2.31_3 \pm 0.01_3$	$2.32_8 \pm 0.01_2$	$2.33_7 \pm 0.00_9$
Na—O_8	$2.37_9 \pm 0.01_4$	$2.38_2 \pm 0.01_2$	$2.35_4 \pm 0.01_0$
Na—O_{10}	$2.32_3 \pm 0.01_7$	$2.33_3 \pm 0.01_5$	$2.34_6 \pm 0.01_0$

*In Å.

only to K_1; W_8 coordinates to K_1, K_2, and Na; O_2 and O_6 do not coordinate to any of the metal ions. (The bonding of the O atoms will be considered separately, below.) The results for K_1 differ significantly from those of Beevers and Hughes, who found only four O's coordinated to K_1. The K_1-O_8 bond is 0.2 Å shorter than that found by the early investigators; thus K_1 has six O's in its coordination sphere, which form a distorted octahedron. The K-O separations, between 2.7 and 3.2 Å, are similar to those found in various other K salts of organic acids.

K_1-O_8 and K_1-O_9 seem to shorten significantly in the LT phase; K_2-O_8 shows a significant increase in the LT phase.

Coordination around Na is octahedral, as commonly observed; distances are given in Table XI. Na-O_1 seems to increase somewhat in the FE phase, and Na-O_8 to decrease in the LT phase. The bond distances are all similar to those found in other organic Na salts.

2. Tartrate Ions

The two halves of the tartrate ion, each half essentially coplanar and with the angle α between the planes, was shown in Fig. 19. The linear and angular dimensions are shown in Figs. 30 and 31, for all three phases. The planarity of each half-molecule group was first evaluated by selecting a plane, by least squares, closest to the α-C and carboxyl group (but excluding the hydroxyl O). Table XII shows the displacements of the atoms from these planes; plus values indicate that the atom lies on the side of the plane away from the origin.

Table XII shows that the planarity of Group I (C_1, C_2, O_1, O_2) is maintained in all three phases, with O_5 only about 0.07 Å from

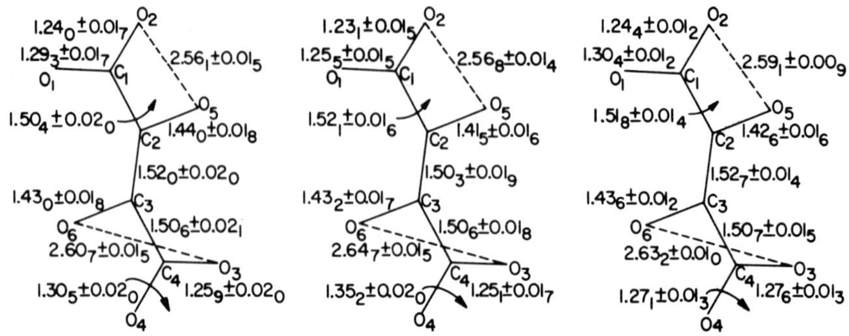

FIG. 30
Rochelle salt: bond distances in tartrate ion, from X-ray analyses.

the plane; in Group II, C_4 is out of the plane of C_3, O_3, and O_4 by 0.03 Å in the HT phase, and by 0.07 Å in the FE phase, but is coplanar in the LT phase. O_6 is 0.33 to 0.37 Å out of the plane of Group II. Why?

Table XIII lists intramolecular oxygen separations, some of which were not computed at the time when Fig. 30 was prepared. Four of these are important: O_5-O_6, which has been mentioned in connection with the neutron analysis of the FE phase, and which seems of particular importance insofar as possible involvement of $(D_5)_B$ in H-bonding of $(O_5)_B$ and $(O_6)_B$; and O_1-O_6, O_2-O_6, and O_3-O_5, all of which correspond to Van der Waals contacts. We record these intramolecular distances and conformations, but will withhold further discussions of oxygen interactions until we have considered H-bonding between water molecules, carboxyl and hydroxyl oxygens.

The angle α between Group I and Group II of the tartrate ion is ≃47°, which is much smaller than the angle of 67° as found in tartaric acid by Stern and Beevers[159]. The C-C separations are normal. In the original (unpublished) discussion of these X-ray

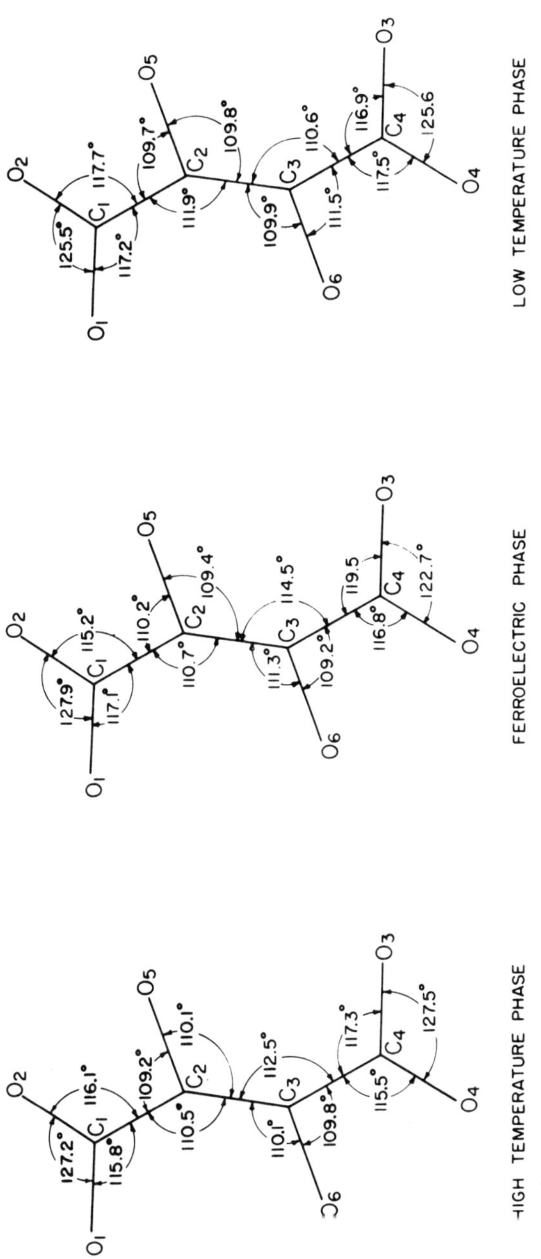

FIG. 31
Rochelle salt: bond angles in tartrate ion, from X-ray analyses.

TABLE XII

ROCHELLE SALT: THE TARTRATE ION*

	Phase		
	High temperature	Ferroelectric	Low temperature
Displacements from Plane I:			
O_1	-0.012	+0.001	+0.003
O_2	-0.001	+0.001	+0.003
C_1	+0.005	-0.003	-0.009
C_2	+0.007	+0.001	+0.003
O_5	-0.070	-0.068	-0.075
Displacements from Plane II:			
O_3	-0.006	-0.016	+0.001
O_4	-0.006	-0.016	+0.001
C_3	-0.006	-0.023	+0.000
C_4	+0.020	+0.053	-0.002
O_6	+0.329	+0.350	+0.367
Angle between Plane I and Plane II:			
	47.9°	47.7°	45.9°

*Displacements (in Å) of the atoms from the two basic planes in the ion.

TABLE XIII

INTRAMOLECULAR OXYGEN SEPARATIONS IN TARTRATE ION OF ROCHELLE SALT*

Phase	O_2–O_5	O_3–O_6	O_5–O_6	O_3–O_5	O_1–O_6	O_2–O_6	O_4–O_5	O_1–O_5	O_4–O_6
Ferroelectric	2.57	2.63	2.91	3.25	3.35	3.39	3.53	3.57	3.61
Low temperature	2.59	2.63	2.91	3.24	—	—	—	—	—
High temperature	2.56	2.61	2.93	3.29	—	—	—	—	—

*In Å.

analyses, the investigators[137] placed much emphasis on intramolecular resonances and consequent intramolecular bond variations, which it was felt played a significant part in the polarization mechanism in the FE phase. The present writer now feels that these changes are of secondary importance, and are the consequences of H-bond and metal ion shifts rather than the cause of these.

3. The Hydrogen-bond System

Assignments of H-bonds depends not only on inter-oxygen distances as determinable from Tables VII, VIII, and IX, but also on D-locations as determined from the neutron analysis reported in the previous section. The H-bonds are listed in Table XIV and illustrated in Fig. 32. [The dashed line between W_8 and O_1 is erroneous; no D atom is found between these (none is available), and the separation between these atoms is far in excess of H-bond distance.] Three rather short oxygen contacts were overlooked in the early report[137], and do not appear in the Fig. 32:

(a) $(O_5)_B$ and $(O_6)_B$, which from the neutron analysis appear to be H-bonded

(b) O_2-W_9

(c) O_6-W_9.

The neutron analysis reveals D_9' in the general region between W_9, O_2, and O_6; D_9' may contribute to a "bifurcated" bond.

4. Changes at the Transition Points

Table XV summarizes information on atomic shifts at the transition points, according to the X-ray analyses of the three phases. This indicates a remarkable similarity between the HT and FE phases: from which we may conclude that radiation damage tends

TABLE XIV. ROCHELLE SALT: HYDROGEN-BOND DISTANCES*

	High temperature	Ferroelectric	Low temperature
$O_2 - W_8$	$2.76_2 \pm 0.01_7$	$2.77_9 \pm 0.01_4$	$2.74_3 \pm 0.01_3$
$O_6 - W_7$	$2.77_4 \pm 0.01_5$	$2.76_3 \pm 0.01_4$	$2.78_7 \pm 0.01_0$
$O_6 - W_{10}$	$2.85_1 \pm 0.01_9$	$2.89_2 \pm 0.01_7$	$2.82_4 \pm 0.01_1$
$O_3 - W_8$	$2.68_7 \pm 0.01_8$	$2.65_7 \pm 0.01_6$	$2.69_8 \pm 0.01_2$
$O_4 - W_9$	$2.78_0 \pm 0.01_9$	$2.80_2 \pm 0.01_8$	$2.79_3 \pm 0.01_2$
$O_4 - W_7$	$2.90_7 \pm 0.01_6$	$2.89_3 \pm 0.01_5$	$2.85_7 \pm 0.01_1$
$O_1 - W_{10}$	$2.68_8 \pm 0.01_9$	$2.69_5 \pm 0.01_7$	$2.71_4 \pm 0.01_2$
$W_9 - W_{10}$	$2.78_9 \pm 0.02_2$	$2.75_3 \pm 0.02_0$	$2.76_1 \pm 0.01_2$
$O_5 - O_6$**	2.94	2.91	2.91
$O_2 - W_9$**	3.09	3.13	3.11
$O_6 - W_9$**	3.14	3.15	3.09

*In Å.

**From the neutron analysis of the FE phase, O_{5B} and O_{6B} are H-bonded by D_{5B}; and D_9' may be involved in bonding $O_2 - W_9$ or $O_6 - W_9$, or in a bifurcated bond involving all three atoms.

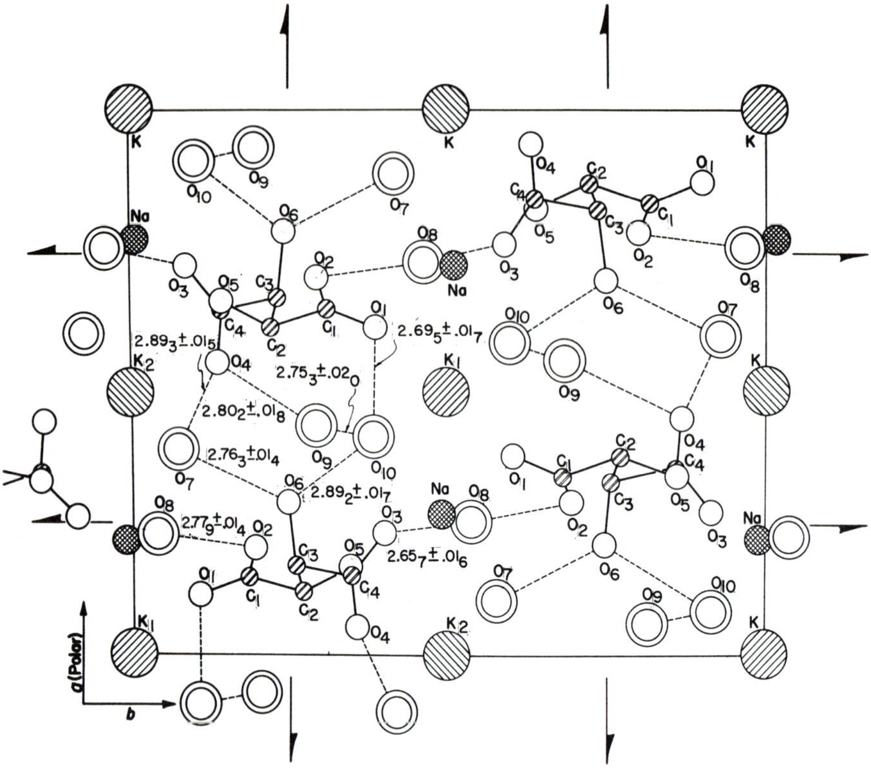

FIG. 32
Rochelle salt: hydrogen bond system, shown in c-axis projection; distances for three phases shown in Table XIV.

to emasculate differences between these. The changes which occur at the lower Curie point tend to persist, however: perhaps because atoms are less readily displaced by the radiation at the lower temperature.

It is unfortunate that the accuracy with which we have been able to read coordinates for the FE phase from the neutron analysis is so much lower than the analytical values achieved in the X-ray studies. The neutron values do indicate that departure in the FE phase from $P2_12_12$ positions is small for all atoms of the tartrate

TABLE XV

ROCHELLE SALT: SIGNIFICANT DIFFERENCES BETWEEN ATOMIC POSITIONS AS INDICATED BY X-RAY ANALYSES OF FE, HT, AND LT PHASES*

	C_1	C_2	C_3	C_4	O_1	O_2	O_3	O_4	O_5	O_6	W_7	W_8	W_9	W_{10}	Na	K_1	K_2
$x_{HT}-x_{FE}$	2	—	—	—	—	3	2	—	—	—	—	—	-2	-2	—	—	—
$y_{HT}-y_{FE}$	—	—	—	2	—	—	—	4	—	—	—	2	—	—	2	—	—
$z_{HT}-z_{FE}$	—	—	2	—	2	—	—	—	3	—	—	—	—	4	—	—	1
$\|r_{HT}-r_{FE}\|$	2	—	2	2	2	3	2	4	3	—	—	2	2	4	2	—	1
$x_{LT}-x_{FE}$	2	—	—	9	—	6	-5	-4	4	—	—	-8	—	-2	—	—	—
$y_{LT}-y_{FE}$	1	—	—	2	-2	—	—	-5	—	—	—	3	8	-6	—	—	—
$z_{LT}-z_{FE}$	-2	—	4	—	-3	-1	3	1	—	5	-2	-4	-4	3	-2	1	3
$\|r_{LT}-r_{FE}\|$	3	—	4	9	4	6	6	6	4	5	2	9	9	7	2	1	3

*In Å × 100.

ions except for C_1, O_1 and O_3 (values for C_4 and O_5 are not reliable because of overlapping); and the deviations for C_1, O_1, and O_3 are very much less than the shifts for W_8, W_9, and W_{10} (see Table III). We may conclude, from the slightness of shifts which the X-ray analyses show between the HT and FE phases, and from the very high values of the temperature factors for W_9, W_{10}, and K_1 in both the HT and FE phases, that these are the three atoms most affected by radiation.

Table XVI summarizes the bonding and coordination of the water oxygens. According to the X-ray and neutron results W_7 is in relatively stable situation: H-bonded by its own H's to O_4 and O_6; coordinating to both Na and K_2; and tightly pressed by W_8 but not H-bonded to W_8 (no H is available). Any shifts in K, Na, or W_8 must affect W_7 somewhat; but the strong bonds are those to O_4 and O_6.

W_8 is less strongly coordinated, to K_1, K_2, and Na; it hangs by H-bonding between O_2 and O_3. The X-ray results give no indication of the shifts in W_8 positions as revealed so clearly in the neutron study of the FE phase. Since from the neutron results the $(x_A + x_B)$ value for W_8 is so large (Table III), we must again conclude that X-rays have altered the structure. Note (Tables VII-IX) that there are no marked changes in the temperature factor for W_7; but B for W_8 is lowest in the FE phase.)

The W_9 has a very high B value in the FE-temperature state, as does W_{10}. The W_9 receives one H-bond from D'_{10} of water 10, and is H-bonded to O_4 through its own D_9. The second D on W_9, D'_9, shows up weakly in the neutron Fourier maps. The W_9 - D'_9 bond points in the general direction of O_2 and O_6; the W_9 - O_2 and W_9 - O_6 separations

TABLE XVI

ROCHELLE SALT: BONDING AND COORDINATION OF WATER OXYGENS

Oxygen	Contact to	H-bonded through	Distances in Phases (in Å)		
			HT	FE	LT
W_7	O_6	D_7'	2.77	2.76	2.79
	O_4	D_7	2.91	2.89	2.86
	W_8	—	3.12	3.12	3.14
	Na	—	2.31	2.33	2.34
	K_2	—	2.81	2.80	2.79
W_8	O_3	D_8	2.69	2.66	2.70
	O_2	D_8'	2.76	2.78	2.74
	W_7	—	3.12	3.12	3.14
	O_1*	—	3.41	3.39	3.35
	Na	—	2.38	2.38	2.35
	K_2	—	3.09	3.09	3.16
	K_1	—	3.14	3.12	3.05
W_9	O_4	D_9	2.78	2.80	2.79
	O_2	D_9' (possibly)	3.09	3.13	3.12
	W_{10}	D_{10}'	2.79	2.75	2.76
	O_6**	D_9' (possibly)	3.14	3.15	3.09
	O_1(A2)	—	3.50	3.46	3.38
	K_1	—	2.97	2.96	2.86
W_{10}	O_1	D_{10}	2.69	2.70	2.71
	W_9	D_{10}'	2.79	2.75	2.76
	O_6	D_6	2.85	2.89	2.82
	O_5	—	3.21	3.21	3.13
	O_3	—	3.25	3.30	3.31
	Na	—	2.32	2.33	2.35

*O_1 - W_8 bond: to O_1 in next cell along c direction.

**O_6 - W_9 bond: to O_6 in next cell along c direction.

fluctuate from 3.09 to 3.15 Å, and suggest a weak H-bond. The W_9 is coordinated to K_1.

The W_{10} is H-bonded to O_1 and W_9 by its own D atoms, and receives a weak H-bond from O_6 through D_6. It makes Van der Waals contacts to O_3 and O_5, and is coordinated to Na. Its B-value is slightly lower than that of W_9 in the FE temperature range, and somewhat higher in the HT range.

5. Conclusions from the Structural Studies

We have encountered four different structures, in the preceding discussion.

(a) The nondamaged FE structure, symmetry $P2_1$, examined in (001) projection by neutron diffraction, which has no 2_1-axes parallel to b or 2-axes parallel to c within the FE temperature range.

(b) The X-ray damaged structure of the FE-temperature phase, which remains monoclinic in that temperature range[138] but shows no evidence of loss of 2_1-axes parallel to b or of 2-axes parallel to c. It is unfortunate that no attempt was made in the X-ray refinement to free the A and B regions by abandoning the coordinate coupling required by $P2_12_12$ symmetry, but there is no certainty that assumption of the lower symmetry would be incontrovertible. Such a treatment would no doubt have reduced the R factor for this phase: that is the general result of abandoning restrictive symmetry elements; but interpretation of the results would still be open to question unless it markedly reduces the unusually high B-factors for W_9, W_{10}, and K_1.

(c) The X-ray HT structure, which differs only slightly from the FE-temperature X-ray structure (Table XV).

(d) The X-ray LT structure, which differs significantly from the other two X-ray structures (Table XV).

The NQR studies of Bjorkstam and Wilmorth[145][146] are concomitant with the change from $P2_12_12$ to $P2_1$ symmetry at the upper Curie point: i.e., the doubling of EFG tensors corresponds to the doubling of general and special atomic positions in the FE phase. The present writer has not been able to understand the arguments of these investigators, which have been repeatedly echoed by Blinc[80][147][160], that the number of molecules/cell must be doubled in the FE phase (from 4 to 8). This conclusion appears to depend upon a dangerous assumption: that one can differentiate between resonances involving D_5 and D_6 (from the tartrate hydroxyls) and all D atoms of water molecules. The most important influence on the EFG tensor about each D atom must be the extent of the deuteron's involvement in the H-bond. The D_{5A} is not H-bonding at all, and quite possibly D_9' is also not so involved (if bonds W_9- $W_9'\cdots O_6$ or W_9-$D_9'\cdots O_2$ exist at all, they are certainly very weak). The bonds O_6-$D_6\cdots W_{10}$ and W_7-$D_7\cdots O_4$ are clearly weaker than other H-bonds; the O_6-W_{10} and W_7-O_4 separations are about 2.9 Å, while the other H-bonded oxygens have separations between ~2.6 and 2.8 Å. It will be noted that we have not derived z-coordinates for the D positions, nor have we computed Donahue angles since without z-values calculations of these are impossible.

One cannot blame the resonance investigators for not withholding the large and often contradictory outpouring of interpretations of resonance observations on Rochelle salt until three-dimensional neutron analyses were available—of at least the FE phase and preferably all three phases; but there does seem to be too little awareness of structural facts, possiblities and pitfalls. Each paper adds a little confusion along with whatever light it might also shed. As late as last year we find Blinc, Peterson, and O'Reilly[161] concluding from spin-relaxation studies that their results "clearly show that the H_2O molecules, though exhibiting a curious and not well understood two-sublattice behavior, are not the main FE dipoles in the structure." A bad guess!

There is a lesson in this investigation which every student of ferroelectric structures should be eager to recite. We have taken such pains to elucidate the atomic interactions in Rochelle salt because there is no one atomic shift which is the seat of the FE behavior; the entire structure, with its profoundly complex system of interlocking bonds, is involved. The switching of attention, by dielectric theorists, resonance practioners and others, from one atom to another, has long missed this point. Based upon the early, partial reports of our neutron analysis, calculations of the spontaneous polarization have been made which are based entirely upon the motions we assigned to W_8 and D_5. But P_s is to be computed from $\Sigma Q_i \cdot \Delta x_i$, which is to be summed over all atoms; and in Rochelle salt as well as in most other FE's this summation is the difference between two large components: one positive and one negative (as shown in Table VI for Rochelle salt). The $\Sigma Q_i \cdot \Delta x_i$ is not merely $0.097\varepsilon \cdot \text{Å}$; it is $0.764\varepsilon \cdot \text{Å} - 0.667\varepsilon \cdot \text{Å}$; and even though the $Q \cdot \Delta x$ contributions for W_8 and D_5 are the largest, it is nonsense to ignore the total contributions of at least 16 other atoms in each molecular unit because if these atoms did not shift neither would W_8 or D_5.

Such a profoundly complex system of interacting bonds as we have encountered in Rochelle salt is present in many other FE's. Tris-glycine sulfate is another good example. Our analysis of this structure[102] revealed another network of H-bonds, not less complex than that in Rochelle salt. Again, P_s is the resultant of many contributions, not any one of which can be overlooked: because each atomic shift implies very many other displacements. We cannot extend the present already overlong recital with a report on the tris-glycine sulfate mechanism, or of any one of many others; a treatise on this general topic is in preparation by the writer.

To miss the point of these complex interactions, as is done when nonstructural investigations focus sole attention on one or another atomic motion, is to miss the central motivation for investigation of FE transitions and polarization reversals. Every crystal represents a delicate balance of complex atomic interactions. When a crystal undergoes a temperature change, these interactions are altered. The changes may be revealed as new vibrational modes, or thermal or dilatometric or optical or dielectric or magnetic anomalies; and these anomalies may or may not be premonitory of a structural transition. A transition represents a reorganization of the balance of forces; its study—structurally—is important because through an understanding of the structural mechanism we gain deeper insight into the complex of interactions. The FE's represent one of several particularly interesting types of readjustments: because in FE's the delicate balance can be altered by application of an electric field, and the relations between charge displacements, dielectric constants, pyroelectric charge and spontaneous polarization can be exploited to deepen understanding of the structural mechanisms: all to the end that we may eventually understand why under given conditions of temperature and pressure the atoms and atomic groups have elected this particular structure, and not some other one, and why and how the structure readjusts when the conditions are altered. Out of such knowledge, eventually, comes the ability to predict what sorts of structures we should seek, or design, to obtain physical properties which we want, under a variety of conditions.

IV. DEHYDRATION AND RADIATION DAMAGE IN ROCHELLE SALT

From the earliest days of its physical measurement, the ease with which Rochelle salt dehydrated—with consequent alteration of properties under study—was a subject of concern. Irving Vigness, as a student of Valasek, first remarked on the sensitivity of the material to X-rays[162], as evidenced by dehydration, coloration, and alteration of dielectric properties. A large number of radiation damage studies have been accomplished since Vigness' original work. Among the most important of these for the present discussion are the papers by Okada[163][164][165]; Toyoda, Shimada, and Tanaka[166]; Krueger, Cook, Sartain, and Yockey[167]; and Boutin, Frazer, and Jona[138]. The literature is much more extensive, and reference should be made to Okada[165] or Boutin et al.[138] for proper listings.

A. The Dehydration Mechanism

Let us consider the dehydration problem first: whether this results from low humidity at normal temperature, elevated temperature, or radiation damage, or a combination of these.

To facilitate understanding of the dehydration process, the reader should refer to Figs. 25-27, 29, and 32, to Tables III, IV, V, and XVI, and to the discussion accompanying Table XVI concerning the bonding and coordination of the water molecules.

The above-listed figures show that the waters lie within channels parallel to the c-axis; and Table XVI suggests that waters 9 and 10 are most readily lost by escape along these channels. The temperature factors for W_9 and W_{10} are extraordinarily high, according to the X-ray analyses, and the bonding of these two oxygens to other atoms in the structure is weaker than that of W_7 and W_8. A high B-value can result from large temperature oscillations, or because of statistical static displacements from a mean position, or because of statistical absences of an atom from a position which it should occupy.

The Rochelle salt structure resembles a channel inclusion complex, with the channels parallel to the c-axis; and all of the waters are found in such channels. Because of the weak bonding of waters 9 and 10 to the atoms lining the channel, these molecules can move along the channels and escape from them into domain boundaries or other defect regions, or through the crystal surface. The process should be partially reversible: up to the point where a concentration of displaced waters is sufficient to solvate a cation or tartrate ion and channel-blocking results, or where water molecules escape into domain boundaries or from the surface and for one reason or another cannot find their way back to a channel opening.

If water 9 escapes, water 10 can shift to its position, and conversely; but the intermediate equilibrium condition is probably to lower occupation of the position of W_9. One recalls, from the neutron Fourier maps, the abnormally lower peaks for W_9, D_9, and D_9'. The X-ray analyses reveal a very high B-factor for K_1 as well as for W_9 and W_{10}. It will be recalled that W_{10} is not coordinated to K_1; the separation $K_1 - W_{10}$ is much too large. From Fig. 29 we see that even if W_9 and/or W_{10} are in their proper positions, K_1 has more freedom of thermal movement than K_2. If W_{10} is missing, the position of the adjacent W_9 is less well-defined, and K_1 has still more freedom; and if W_9 is absent K_1 can shift even more easily. Since the B-factor for Na is low (lowest in the HT phase, among all the atoms), one again suspects that if W_{10} is removed, W_9 shifts to take the W_{10} position and thus the coordination around Na tends to be maintained. Hence the low B-value for Na, and the large difference between those values for K_1 and K_2.

Cady's discussion of the preservation of Rochelle salt has been mentioned earlier, and is again pertinent. Reference should be made to pages 521-528 of Ref. [119]. Valasek in his second paper [1921, p. 479][63] noted that the weight of a crystal of the material may vary as much as 5% with differences in humidity of the surrounding air. The loss of one water molecule per molecular unit would decrease the weight by ~6.4%. Valasek's observation implies that almost one water per molecular unit can be lost, without destruction of the crystal.

Beginning with a crystal of non-deuterated Rochelle salt, can one by gentle dehydration followed by exposure to D_2O vapor, and with the process repeated many times, preferentially replace normal waters 9 and 10 with heavy waters? The obvious way to test this possibility is by a deuteron quadrupole resonance study. This is one of a good number of experiments suggested by the present discussion, which are now in progress or planned for the immediate future in our laboratories. The H_2O-to-D_2O exchange should be examined under a wide variety of conditions: in liquid as well as gaseous environments, in pure and substituted Rochelle salt; and of course the important question is what effect such replacement has on physical properties.

D. Radiation Damage Studies and Influence of Foreign Cations

Krueger et al.[167] have discussed the physics of the interaction of Co^{60} γ-rays on Rochelle salt. The primary interaction results in a recoil (Compton) electron with average energy of ~480 keV. These produce secondary electrons with ~70 eV energy, which energy is released in a region of volume of the order of that of one unit cell. The regions of energy release are independent of one another, and are thus randomly distributed in the crystal. The release is describable on the thermal spike model, with a temperature rise calculable as 160°C in 10^{-12} sec within a region of ~8 unit cells. At 10^6 R, approximately 3% of the cells are affected; this dosage is about that sufficient to destroy FE behavior[167]. It has been noted repeatedly that the effects of radiation are similar in many respects to the effects of substitution of NH_4 for K ions (Fig. 18). The dielectric peaks are reduced in magnitude and move toward one another, until they coalesce at the dosage indicated above. Krueger et al.[167] demonstrated that there are definite differences in dielectric behavior of irradiated and substituted crystals, however.

Whereas substitution of NH_4 narrows and at ~2.5% destroys the range of monoclinic symmetry (see Fig. 18), crystals which have been subjected to sufficient radiation to just destroy FE behavior still show optical evidence of monoclinic domains, when the irradiation is condicted at a temperature within the normal FE range; when irradiation is carried out at a temperature outside of the normal FE range, the orthorhombic symmetry is retained.

Insufficient attention has been given to the effects of crystallization of substituted Rochelle salt within or outside of the FE temperature range. Makita and Takagi[168] reexamined our earlier phase studies of the "Kourtschatov salts" (Melmed et al.[133]), obtained results in essential agreement, and extended the studies to a more careful examination of the phases at high NH_4 content. The hydrogenous crystals used in these studies were grown above 26°C, and hence always in the orthorhombic phase. One should return to this matter, in order to determine whether crystallization of the low-NH_4-content material within the FE temperature range tends to concentrate either NH_4 or K ions in the domain wall regions, and whether the magnitude of the difference between the upper Curie temperature and the temperature of crystallization is of much significance. This should be accomplished for small concentrations of Rb and Tl as well as for NH_4, above and below the upper transition temperature. Concentration of Rb or Tl in domain wall regions could be established by use of the scanning electron microscope equipped with an X-ray wavelength analyzer, as available in our Department of Metallurgical and Materials Engineering. When crystallization is carried out in the orthorhombic phase, the foreign cations do tend to stabilize that phase; and when the concentration of these ions is low enough to permit occurrence of FE in a narrowed temperature range, the foreign atoms probably serve as points of nucleation of domain walls. Optical examination of domain patterns at low concentrations of foreign atoms is not reported.

We did obtain some X-ray evidence, in our studies of NH_4-substituted Rochelle salt, that NH_4 preferentially replaced K_1 rather than K_2. This is in keeping with the looser coordination of K_1 to its surrounding oxygens (6 at coordinating distances, 2 beyond), whereas K_2 is 8-coordinating and has a much lower B-factor than K_1. The problem of preferential replacement could be readily solved by neutron diffraction; not only is the scattering factor for N almost three times that for K, but one would also observe the H or D nuclei on the ammonium ion. The effect of deuteration upon the "Kourtschatov salts" phase diagram has also not been properly explored.

Introduction of Rb or $T\ell^{I}$ ions in small concentrations has n effect similar to that of small concentrations of NH_4 ions; but no new FE phase appears when the concentrations of Rb or $T\ell^{I}$ is much greater.

Buckley first demonstrated that mixed aqueous solutions of NaK-and $NaNH_4$-tartrate · $4H_2O$ crystallized to give mixed tartrates with essentially the same composition as that of the solutions[169][170][171]. Buckley established the property approximately for $Na(NH_4)_xK_{(1-x)}$ · tartrate · $4H_2O$, with $0 < x < 0.5$; Kourtschatov and Eremejev[132] confirmed it, also approximately, for x as large as 0.7. In conjunction with our own studies of this system (Melmed et al.[133]) we analyzed both solutions and crystals prepared from them more accurately, and found that there is a tendency toward a higher NH_4 concentration in the crystals than in solution, over the range $0 < x < 1$; at 40% NH_4 - 60% K in solution, x in the crystals was ~53%. Makita and Takagi[168] confirmed these results. This behavior is rather unusual in mixed crystal systems. The fact that somewhat higher NH_4 concentrations appear in the crystal than are present in the mixed solutions indicates that the NH_4-containing crystals have somewhat lower solubility than pure Rochelle salt. This is reasonable in terms of the tighter bonding one might expect for NH_4 rather than K in the K_1 position.

Iurin[172] mentioned the effect of small concentrations of Mo, Cu, or B on the dielectric behavior of Rochelle salt. These ions must be chelated by the tartrate ions. For instance, Mo might appear in place of K_2, chelated by O_4 and O_5 from each of two tartrate ions; Cu or B might appear in place of Na, chelated by one tartrate through O_3 and O_5. These metal ions, in low concentrations, have effects similar to radiation damage. They surely produce defect centers; and if crystallization occurs below the temperature of the upper transition point of the resulting material, they should tend to concentrate at domain walls. Again, one should look for concentration of Mo ions in these boundaries with the scanning electron microscope.

Finally, Buckley described one other mixed crystal system, which according to him is isomorphous with Rochelle salt. This is $(KNa)_x \cdot (KK)_{(1-x)} \cdot$ tartrate $\cdot$ $4H_2O$, with $0 < x < 0.5$[171]. No dielectric studies of this material are known to the present writer, and it is quite unclear how the structure can correspond to that of Rochelle salt since K ions surely cannot be accomodated at Na positions. We are looking into this system now.

Although it is established that radiation damage drives water molecules out of the structure, the process is no gentle one and certainly other structural disorders occur as well. Because the tartrate ions are separated from one another by water molecules and alkali ions, it is possible that relaxation of the disordered condition occurs, in time, except for the escape of waters along the channels, parallel to c; and in this partial recrystallization, if waters 7 or 8 have been driven out, waters 9 or 10 will tend to replace them.

Boutin, Frazer, and Jona[138] were chiefly concerned with evidence for structural changes which might result from X-ray or γ-ray damage. As indicator of such changes, the ratio of intensities of (050) to (060) neutron reflections was measured as a function of total radiation. [Thermal neutron energies are too low to produce measurable damage upon exposure equivalent to those necessary for a diffraction analysis; and Boutin et al. observed no change in (050)/(060) intensity ratio or (050) intensity alone after prolonged exposure to a thermal neutron beam.] The (050)/(060) intensity ratio was reduced by a factor of 5 after a 15-hr exposure to Co^{60} γ-rays of 2.5×10^5 rep/hr, and by a factor of 10 after twice that exposure. The total exposure was comparable to that required during a three-dimensional X-ray diffraction analysis. It was known that the damage from X-rays of ~10 to 20 kV energies and from the Co^{60} γ-rays was comparable; γ-radiation is preferable because exposure is uniform throughout the crystal volume.

The experiment demonstrated phenomenologically why odd (0k0) reflections were unobservable in X-ray diffraction patterns; radiation damage by the X-rays themselves obliterated evidence of the disappearance of the 2_1-axial symmetry parallel to b in the FE phase. This experiment did not demonstrate the X-rays were incapable of revealing the $x_A + x_B$ shifts along the FE axis which are given (roughly) in our Tables III, IV, and V and applied in Table VI to calculation of the spontaneous polarization. These $x_A + x_B$ values have nothing to do with (0k0) reflections, for k either even or odd. Neither do the X-ray results presented in this present paper demonstrate that X-rays cannot reveal the shifts of atoms along the a direction. It is unfortunate that neither our own group[137] (including the present writer!) nor Boutin et al.[138] were sensitive to the fact that the nonappearance of odd (0k0) X-ray reflections should not have prevented us from refining the FE structure in spacegroup $P2_1$ (2_1 parallel to a), with a polarized (single-domain) crystal. Since the evidence is amply available that the monoclinic symmetry is retained even after very extensive radiation damage, one can expect that the shifts along the x-direction, which destroy the rotational symmetry along both b and c directions, still will be evident in the X-ray data. One should even be able to see the effects of these shifts upon Bijvoet pairs ($I_{hk0}-I_{\bar{h}k0}$), using anomalous dispersion by the K atoms excited by CrKα (or even CuKα) radiation scattered by a polarized crystal.

Boutin et al.[138] do not speculate on specific structural changes which might be responsible for radiation-induced changes in I_{050}. Any unequivocal conclusion would have had to be based upon examination of a large number of changes in I_{hk0} values, and a survey of such changes was beyond the purpose of their study. The selection of (050) as the reflection to be monitored was natural, since one suspected that it was radiation damage which led to nonappearance of all odd (0k0) X-ray reflections from the monoclinic phase, and (050) is the strongest of these.

The neutron analyses whose results are presented in the Fourier projections of Figs. 23-27 were accomplished via many cycles of calculations of F_{hk0} values and adjustment of coordinates, to provide the best agreement between magnitudes of observed and calculated structure factors. Tables of F_{hk0} values will be published elsewhere, along with the corresponding nuclear coordinates. Agreement between observed and calculated F_{hk0}'s was satisfactory except for the weak F_{0k0} factors with k odd. F_{050} is the largest of these, and its magnitude is particularly sensitive to the y-coordinates of the nuclei in waters 9 and 10. But, as we have already indicated, there is something anomalous about the positions of these nuclei in crystals not subjected to X-ray and γ-ray damage. In Figs. 24(b), (c), and (d); 26(b) and (c); and 27(b) and (c), it can be seen that the peak for W_9 is low, suggesting that this oxygen is disordered in position or partially absent due to dehydration. The peak for D_9 is low, and that for D_9' very low indeed. It will also be recalled that the X-ray analysis of this phase assigned very high B-factors to W_9, W_{10}, and K_1. The largest $|y_A + y_B|$ values in Tables III, IV, and V appear for W_9, W_{10}, D_9, and D_9'.

The procedure applied by Boutin et al.[138] to the intensity ratio I_{050}/I_{060} should be extended to examination of all weak (hk0) reflections; and a study of the changes induced in these should permit deduction of the specific effect of radiation damage on all atoms. It is clear that the major effects will be upon the water molecules, the hydroxyl H's or D's, and the alkali metal ions, since shifts in these are involved in the two transitions. Confirmation of the predicted major effects on waters 9 and 10 must await such a study.

Moulton and Moulton[173] have exerted valiant effort to interpret electron paramagnetic resonance spectra from X-irradiated Rochelle salt, both hydrogenous and deuterated. Three sets of centers appeared to be present, as evidenced by: (a) an intense doublet which rapidly decays; (b) another doublet which grows and then decays; and (c) an 8-line spectrum which increases and is stable. The stable spectrum (c) seemed to be due to breaking of C_2-H_2 and C_3-H_3 bonds, and coupling to OH or H_2O hydrogens; the decaying spectra (a) and (b) were interpreted as arising from the water molecules. These interpretations would have been facilitated, perhaps, if the results of the neutron analysis had been analyzed and released earlier.

The Moultons presented another interesting observation: that when an irradiated crystal was subsequently dehydrated, it became opaque white, and crumbled when touched; but the EPR spectrum was not changed. This tends to support our hypothesis, that water molecules diffuse through the channels to domain walls (and then act to decouple the domains dielectrically), and when the resulting material is then dehydrated, water leaving the domain wall regions ceased to hold the domains in coherence. A crystal which was first stored in a dry atmosphere for two days and then irradiated showed only a broad resonance line which did not decay in the manner of the spectra from the normal crystals: demonstrating the healing influence of retained water molecules upon the structure.

The work initiated by the Moultons has been carried further by Abé and Suzuki[174]; these investigators agreed with the former EPR studies with respect to the stable lines (c), but assigned lines (a) to results of breaking of O—H bonds in hydroxyls 5 or 6 and lines (b) disruption of carboxyl C—O bonds. Iurin[172] had reported evolution of H_2, CO, CO_2, and CH_4 from irradiated crystals; these latter results should be reexamined, and there is no reason to restrict such decomposition studies to this particular tartrate salt.

The understanding of the structural mechanisms of the two transitions and polarization reversal in Rochelle salt remains as the prime problem. The main road to such understanding is open and well-paved: through three-dimensional neutron analyses of the three phases. Only after these are completed can one expect to apply resonance and relaxation studies in order to determine with confidence the supplementary information on internal fields and atomic dynamics which these methods can then provide. The neutron projection and X-ray analyses have been helpful not only in revealing the rough mechanism of polarization reversal but in showing what dangers lie in attempts at interpretation of resonance experiments without structural foundations.

C. Proposals for Experimental Studies

A variety of experiments, which are suggested by the structural knowledge now available, are in progress or planned for the immediate future. It is obvious, for example, that one should look at crystals which were radiation-damaged while in the FE temperature range, to determine whether or not resonance lines characteristic of the undamaged monoclinic phase persist. Resonance studies of partially substituted Rochelle salt is certainly suggested. Dehydration-rehydration cycling, with a view to possible preferential replacement of normal by heavy water molecules, has been mentioned in a previous section; such studies, which should be carried out as close as possible to 0°C, should extend to substituted as well as pure Rochelle salt. Is it possible that small monovalent cations (H^+, Li^+, even Na^+) can be electrodiffused into or even through the channels in which waters 9 and 10 lie; can OH^- ions move through them; why not try even Be^{+2} or Mg^{+2} + H ions (i.e., at low pH)? Can one MeOH molecule replace both waters 9 and 10? Was Buckley correct in his description of $Na_{0.5}K_{1.5}$ • tartrate • $4H_2O$ as a crystal isomorphous with Rochelle salt; or was he instead misled by partial replacement of Na^+ by H^+ and an incorrect analysis of the tartrate concentration?

It will be useful to examine the effects of small concentrations of foreign cations on dielectric properties, when crystallization is accomplished above the upper Curie temperature and then well below it; because in the former case the impurities will tend to serve as randomly distributed points of nucleation for domain walls when the temperature is lowered, and in the latter case at least some of the impurities should tend to concentrate in the domain walls during growth. The writer has already observed differences in dielectric behavior and domain structure in tris-glycine sulfate, grown with impurity ions both above and below the Curie point[175].

The greatest progress in understanding of the Rochelle salt problem will come from full three-dimensional neutron analyses of the structures of all three phases. Arrangements for accomplishment of such analyses are in progress. One anticipated result of such a program is that the structure of the LT phase differs significantly from that of the HT phase. It is quite probable that the presently-accepted symmetry of the LT phase is incorrect. Two possible symmetries were shown in Figs. 28(c) and (d), with (c) considered the more likely. These are both superstructures. No proper search for superstructure lines in neutron diffraction patterns has yet been made, and this should be done.

V. CONCLUSION AND ACKNOWLEDGMENT

The purpose of this paper was to demonstrate the importance of structural studies in elucidation of the mechanisms of FE transitions and dielectric properties. We have by now learned that understanding of one FE is of only limited value in development of knowledge of others. There is no possibility of a general structural theory, because the mechanisms vary so markedly from one FE family to the next. The writer could have illustrated the points he intended to make with a discussion of representatives of any one of a dozen different FE families. As already stated, the subject merits a treatise, not a single paper.

The writer is indeed grateful to his friend and colleague Professor Thomas Scott, of the Department of Physics of this university, for helpful discussions of magnetic resonance methods.

Encouragement and inspiration for this return to consideration of ferroelectric problems came to the writer during visits with his many friends and former associates in Japan, in September, 1969, on the occasions of the "Second International Meeting of Ferroelectricity" in Kyoto, and the "Kaya Conference" in Hokkaido. His gratitude for the hospitality and discussions provided to him in that country is extreme.

REFERENCES

[1]. E. Wainer and A. N. Salomon, Electr. Rept., 8 (1942); 9 and 10 (1943).

[2]. T. Ogawa, Busseiron Kenkyu No., 6, 1 (1947).

[3]. B. Wul and I. M. Goldman, Compt. Rend. Acad. Sci. USSR, 46, 139 (1945); 49, 177 (1945); 51, 21 (1946).

[4]. A. von Hippel, R. G. Breckenridge, F. G. Chesley and L. Tisza, Ind. Eng. Chem., 38, 1097 (1946); A. von Hippel, Z. Phys., 133, 158 (1952).

[5]. Reported in A. V. Schubnikov, I. S. Zheludev, V. P. Konstantinova, and I. M. Silvestrova, Étude des Textures Piézoélectriques, Paris, Dunod, 1958.

[6]. H. Blattner, B. T. Matthias, and W. Merz, Helv. Phys. Acta, 20, 225 (1947).

[7]. J. P. Remeika, J. Am. Chem. Soc., 76, 940 (1954); U. S. Patent 2,852,400 (16 September 1958).

[8]. Landolt-Börnstein: Numerical Data and Functional Relationships in Science and Technology, New Series, Group III, Volume 3: Ferro- and Antiferroelectric Substances, (T. Mitsui et al. eds.) Springer, New York, 1969.

[9]. F. Jona and G. Shirane, Ferroelectric Crystals, Macmillan, New York, 1962.

[10]. H. J. Martin, Die Ferroelektrika, Akademische Verlagsges., Geest and Portig K.G., Leipzig, 1964.

[11]. E. Fatuzzo and W. J. Merz, Ferroelectricity. North Holland, Amsterdam, 1967.

[12]. J. C. Burfoot, Ferroelectrics. Van Nostrand, London, Princeton, Toronto, 1967.

[13]. R. Pepinsky and Y. Makita, Dielectric Anomalies in Clathrate Crystals. Appendix H in R. Pepinsky, Final Report under Contract AF49(638)-1044, Air Force Office of Scientific Research, March, 1962.

[14]. G. Busch and P. Scherrer, Naturwiss., 23, 737 (1935).

[15]. G. Busch, Helv. Phys. Acta, 11, 269, (1938).

[16]. W. Bantle and Ch. Caflisch, Helv. Phys. Acta, 16, 235 (1943).

[17]. B. Zqicker and P. Scherrer, Helv. Phys. Acta, 17, 346 (1944).

[18]. A. N. Holden, B. T. Matthias, W. F. Merz, and J. P. Remeika, Phys. Rev., 98, 546 (1955).

[19]. R. Pepinsky and K. Vedam, Phys. Rev., 114, 1217 (1959).

[20]. R. Pepinsky, K. Vedam, Y. Okaya and S. Hoshino, Phys. Rev., 111, 1467 (1959).

[21]. R. Pepinsky, K. Vedam, and Y. Okaya, Phys. Rev., 110, 1309 (1958).

[22]. E. Fatuzzo and R. Nitsche, Phys. Rev., 117, 936 (1960).

[23]. B. T. Matthias and J. P. Remeika, Phys. Rev., 103, 262 (1956).

[24]. F. Jona and R. Pepinsky, Phys. Rev., 103, 1126 (1956).

[25]. R. Pepinsky and K. Vedam, unpublished; cited in S. Hoshino, Y. Okaya, and R. Pepinsky, Phys. Rev., 115, 323 (1959).

[26]. R. Pepinsky and F. Jona, Phys. Rev., 105, 344 (1957).

[27]. R. Pepinsky, Y. Okaya, K. Vedam, and S. Hoshino, Bull. Am. Phys. Soc., II, 3, 44 (1958); Y. Okaya, K. Vedam, and R. Pepinsky, Acta. Cryst., 11, 307 (1957).

[28]. S. Hoshino, K. Vedam, Y. Okaya, and R. Pepinsky, Phys. Rev., 112, 405 (1958).

[29]. M. Kay, Janie Lajzerowicz, Y. Okaya, and R. Pepinsky, Abstracts, Am. Cryst. Assoc., Cornell University, July, 1959; B. Singh and R. Pepinsky, Bull. Am. Phys. Soc., II, 7, 177 (1962).

[30]. B. Singh, Doctoral Thesis, Pennsylvania State University, 1962.

[31]. E. O. Schlemper and W. C. Hamilton, J. Phys. Chem., 44, 4498 (1966).

[32]. D. Eastman, F. Jona, and R. Pepinsky, unpublished (1956).

[33]. R. Pepinsky and K. Vadam, cited in R. Pepinsky, K. Vedam, and Y. Okaya, Abstracts, Am. Cryst. Assoc., Cornell University July 19-24, 1959.

[34]. Y. Takagi and Y. Makita, J. Phys. Soc. Japan, 13, 272 (1958).

[35]. W. J. Merz, Phys. Rev., 91, 513 (1953).

[36]. S. O. Morgan, Ann. N. Y. Acad. Sci., 40, 357 (1940).

[37]. A. L. Solomon, Phys. Rev., 104, 1191 (1956).

[38]. R. Pepinsky, to be published.

[39]. R. Pepinsky, F. Jona, and G. Shirane, Phys. Rev., 102, 1181 (1956).

[40]. D. E. Palin and H. M. Powell, Nature, 156, 334 (1945).

[41]. M. Hagan, Clathrate Inclusion Compounds, Reinhold, New York, 1962.

[42]. H. M. Powell, Chapter 7 in Non-stoichiometric Compounds, (L. Mandelcorn, ed.). Academic, New York, 1964.

[43]. J. S. Dryden, Trans. Faraday Soc., 49, 1333 (1953).

[44]. L. A. K. Stavely, Chapter 10 in Non-stoichiometric Compounds, (L. Mandelcorn, ed.). Academic, New York, 1964.

[45]. R. Pepinsky, p. 19 in Final Report, Contract No. AF49(638)-1044, Air Force Office of Scientific Research, Washington, D. C., March, 1962.

[46]. S. Roberts, J. Am. Ceram. Soc., 33, 63 (1950); G. A. Smolenskii, Zh. Tekhn. Fiz., 20, 137 (1950) (In Russian).

[47]. E. Sawaguchi, H. Maniwa, and S. Hoshino, Phys. Rev., 83, 1078 (1951); G. Shirane, E. Sawaguchi, and Y. Takagi, Phys. Rev., 84 476 (1951); F. Jona, G. Shirane, F. Mazzi, and R. Pepinsky, Phys. Rev., 105, 849 (1957).

[48]. A. Ya. Dantsiger and E. G. Fesenko, Sov. Phys.-Cryst., 8, 630 (1964).

[49]. R. Pepinsky, to be published.

[50]. R. Pepinsky, K. Vedam, and Y. Okaya, Abstracts, Am. Cryst. Assoc., Cornell University, July, 1959; Y. Makita and R. Pepinsky, Bull. Am. Phys. Soc., II, 7, 117 (1962).

[51]. K. Okada, Phys. Rev. Letters, 15, 252 (1965); K. Okada, Phys. Rev., 164, 683 (1967).

[52]. G. J. Goldsmith and J. G. White, J. Chem. Phys., 31, 1175 (1959); H. Futama, J. Phys. Soc., 17, 434 (1962): H. Futama, Y. Shiozaki, A. Chiba, E. Tanaka, T. Mitsui, and J. Furuichi, Phys. Letters, 25A, 8 (1967).

[53]. Cited in R. Pepinsky, K. Vedam, S. Hoshino, and Y. Okaya, Phys. Rev., 111, 1508 (1958).

[54]. R. Pepinsky and K. Vedam, cited in R. Pepinsky, K. Vedam and Y. Okaya, Abstracts, Am. Cryst. Assoc., Cornell University, July 19-24, 1959.

[55]. K. Vedam, R. Pepinsky, Joseph Lajzerowicz, Y. Okaya, and N. Stemple, Bull. Am. Phys. Soc., II, 4, 63 (1959).

[56]. R. Pepinsky, Some New X-ray and Neutron Studies of Hydrogen Bonding, in Hydrogen Bonding, (D. Hadzi, ed.) Pergamon, New York (1959), pp. 33-35.

[57]. R. Pepinsky, K. Vedam, Joseph Lajzerowicz, and Y. Okaya, Appendix N in R. Pepinsky, Final Report, Contract No. AF49(638)-1044 with Air Force Office of Scientific Research, Washington, D. C., March, 1962.

[58]. R. Pepinsky, pp. 13-16 in Final Report, Contract No. AF49(638)-1044 with Air Force Office of Scientific Research, Washington, D. C., March, 1962.

[59]. P. Baertschi, Helv. Phys. Acta, 16, 218 (1943); P. Baertschi, Helv. Phys. Acta, 18, 267 (1945); G. Busch, W. Känzig, and W. M. Meier, Helv. Phys. Acta, 26, 385 (1953); H. Gränicher, W. M. Meier, and W. Petter, Helv. Phys. Acta, 27, 216 (1954).

[60]. R. Ando, J. Phys. Soc. Japan, 17, 937 (1962).

[61]. R. Pepinsky and F. Jona, Abstracts, Fourteenth Pittsburgh Diffraction Conference, November, 1956; R. Pepinsky, F. Jona, K. Vedam, Y. Okaya, M. S. Ahmed, P. Gallitelli, S. Hoshino, T. Mitsui, J. Johns, D. Eastman, and F. Unterleitner, Acta Cryst., 10, 835 (1957); R. Pepinsky, K. Vedam, and Y. Okaya, Abstracts, Am. Cryst. Assoc., Cornell University, July, 1959; R. Pepinsky, K. Vedam, H. Futama, and Y. Makita, Abstracts, Am. Cryst. Assoc., Washington, D. C., January, 1960; R. Pepinsky, unpublished, 1957-1963.

[62]. Y. Takagi and Y. Makita, J. Phys. Soc. Japan, 13, 272 (1958).

[63]. J. Valasek, Phys. Rev., 15 (2), 537 (1920); 17 (2), 475 (1921); 19 (2), 478 (1922); 20 (2), 639 (1922); 24 (2), 560 (1924); Science, 65, 235 (1927).

[64]. B. T. Matthias and J. K. Hulm, Phys. Rev., 82, 108 (1951).

[65]. B. T. Matthias and J. P. Remeika, Phys. Rev., 107, 1729 (1957).

[66]. R. Pepinsky, Bull. Am. Phys. Soc., II, 7, 177 (1962).

[67]. R. Pepinsky, Paper S2-14, Program and Abstracts, Sixth Internat. Congress, Internat. Union Cryst., Rome, 1963.

[68]. H. Futama and R. Pepinsky, Bull. Am. Phys. Soc., II, 7, 177 (1962).

[69]. H. Futama and R. Pepinsky, J. Phys. Soc. Japan, 17, 725 (1962).

[70]. J. Kobayashi and N. Yamada, J. Phys. Soc. Japan, 17, 876 (1962).

[71]. K. Aizu, Phys. Rev., 133 A, 1584 (1964); K. Aizu, A. Kumada, H. Yumoto, and S. Ashida, Jr., Phys. Soc. Japan, 27, 511 (1969).

[72]. W. Känzig, Ferroelectrics and Antiferroelectrics, in Solid State Physics, Vol. IV, (F. Seitz and D. Turnbull, eds.), Academic, New York, 1957.

[73]. S. Sawada, S. Nomura, S. Fujii, and I. Yoshida, Phys. Rev. Letters, 1, 320 (1958).

[74]. B. T. Matthias, C. E. Miller, and J. P. Remeika, Phys. Rev., 104, 849 (1956).

[75]. D. E. O'Reilley and T. Tsang, J. Chem. Phys., 46, 1291, 1301 (1967); 50, 2274 (1969).

[76]. D. E. O'Reilly and T. Tsang, Phys. Rev., 160, 333 (1967).

[77]. W. C. Hamilton, J. Chem. Phys., 50, 2275 (1969).

[78]. R. Pepinsky, K. Vedam, S. Hoshino, and Y. Okaya, Phys. Rev., 111, 1508 (1958).

[79]. R. Pepinsky and K. Vedam, Phys. Rev., 117, 1502 (1960).

[80]. R. Blinc, Magnetic Resonance in Hydrogen-Bonded Ferroelectrics, in Advances in Magnetic Resonance, Vol. 3, (J. S. Waugh, ed.) Academic, New York, 1968.

[81]. A. Rigamonti, Phys. Rev. Letters, 19, 436 (1967).

[82]. G. P. O'Leary, Phys. Rev. Letters, 23, 782 (1969).

[83]. G. E. Bacon and R. S. Pease, Proc. Roy. Soc., A 220, 397 (1953).

[84]. G. E. Bacon and R. S. Pease, Proc. Roy. Soc., A 230, 359 (1955).

[85]. S. W. Peterson, H. A. Levy, and S. H. Simonsen, Jr., Chem. Phys., 21, 2084 (1953).

[86]. H. A. Levy, S. W. Peterson, and S. H. Simonsen, Phys. Rev., 93, 1120 (1954).

[87]. R. Pepinsky, Science, 117, 1 (1953).

[88]. R. O. Keeling, Jr., and R. Pepinsky, Zt. Krist., 106, 236 (1955).

[89]. L. Tenzer, B. C. Frazer, and R. Pepinsky, Acta Cryst., 11, 505 (1958).

[90]. B. C. Frazer, H. R. Danner, and R. Pepinsky, Phys. Rev., 100, 745 (1955).

[91]. G. Shirane, H. R. Danner, and R. Pepinsky, Phys. Rev., 105, 856 (1957).

[92]. G. Shirane, R. Pepinsky, and B. C. Frazer, Acta. Cryst., 9, 131 (1956).

[93]. F. Jona, G. Shirane, F. Mazzi, and R. Pepinsky, Phys. Rev., 105, 849 (1957).

[94]. M. I. Kay and B. C. Frazer, Acta Cryst., 14, 56 (1961).

[95]. M. I. Kay, B. C. Frazer, and R. Ueda, Acta Cryst., 15, 506 (1962).

[96]. I. Shibuya, J. Phys. Soc. Japan, 16, 490 (1961).

[97]. Y. Yamada, I. Shibuya, and S. Hoshino, J. Phys. Soc. Japan, 18, 1954 (1963).

[98]. I. Shibuya, Y. Iwata, N. Koyano, S. Fukui, S. Mitani, and M. Tokunaga, Ann. Rep. Research Reactor Inst. Kyoto University, 1, 246 (1968).

[99]. Y. Iwata, N. Koyano, and I. Shibuya, Ann. Rep. Research Reactor Inst. Kyoto University, 2, 13 (1969).

[100]. I. Shibuya, Y. Iwata, N. Koyano, S. Fukui, S. Mitani, and M. Tokunaga, Abstracts, Second Internat. Meeting on FE, Kyoto, September (1969), p. 218.

[101]. M. I. Kay and J. A. Gonzalo, Second Internat. Meeting on FE, Kyoto, September (1969).

[102]. S. Hoshino, Y. Okaya, and R. Pepinsky, Phys. Rev., 115, 323 (1959).

[103]. I. Shibuya et al., to be published (Private communication to R. Pepinsky, September, 1969).

[104]. F. N. Hainsworth, Dissertation, McMaster University, Hamilton, Ontario (Canada), (1965).

[105]. J. C. Taylor, M. H. Mueller, and R. L. Hitterman, Abstracts, Am. Cryst. Assoc. Meeting, Atlanta (1967).

[106]. A. Kikuchi, Y. Oka, and E. Sawaguchi, J. Phys. Soc. Japan, 23, 337 (1967).

[107]. I. D. Brown, Acta Cryst., 17, 654 (1964).

[108]. J. H. Van den Hende and H. Boutin, Acta Cryst., 17, 660 (1964).

[109]. V. M. Padmanabhan and R. Balasurbramanian, Acta. Cryst., 22, 532 (1967).

[110]. S. C. Abrahams, J. M. Reddy, and J. L. Bernstein, J. Chem. Phys. Solids, 27, 997 (1966).

[111]. S. C. Abrahams, W. C. Hamilton, and J. M. Reddy, J. Phys. Chem. Solids, 27, 1013 (1956).

[112]. S. C. Abrahams, H. J. Levinstein, and J. M. Reddy, J. Phys. Chem. Solids, 27, 1019 (1966).

[113]. S. C. Abrahams and J. L. Bernstein, J. Phys. Chem. Solids, 28, 1685 (1967).

[114]. S. C. Abrahams, W. L. Hamilton, and A. Sequeira, J. Phys. Chem. Solids, 28, 1693 (1967).

[115]. K. Okada, M. I. Kay, D. T. Cromer, and I. Almodovar, J. Chem. Phys., 44, 1648 (1966).

[116]. J. West, Zt. Krist., 74, 306 (1930).

[117]. J. C. Slater, J. Chem. Phys., 9, 16 (1941).

[118]. C. A. Beevers and W. Hughes, Proc. Roy. Soc. (London), A177, 251 (1941).

[119]. W. G. Cady, Piezoelectricity. McGraw-Hill, New York, 1946. Reprinted by Dover, New York, 1964.

[120]. I. -V. Kourtschatov, Le Champ Moléculaire dans les Diélectriques (Le sel de Seignette). Hermann, Paris, 1936. Original Russian text published in Moscow in 1933.

[121]. W.P. Mason, Piezoelectric Crystals and Their Application to Ultrasonics, D. van Nostrand, New York, 1950.

[122]. B. T. Matthias and W. J. Merz, Seignette-Elektrizität, in Beiträge zur Entwicklung der Physic (Festgabe zur 70. Geburtstag von Prof. Paul Scherrer), (H. Frauenfelder, O. Huber and P. Stähelin, eds.). Birkhäuser, Basel, 1960.

[123]. A. R. Ubbelohde and I. Woodward, Proc. Roy. Soc. (London) A 185, 448 (1946).

[124]. H. Mueller, Phys. Rev., 57, 829 (1940).

[125]. B. C. Frazer, M. McKeown, and R. Pepinsky, Phys. Rev., 94, 1435 (1954).

[126]. H. Mueller, Phys. Rev., 47, 175 (1935); E. Bradford, B. S. Thesis, Massachusetts Institute of Technology, 1934.

[127]. W. P. Mason, Phys. Rev., 55, 775 (1939).

[128]. J. Halblützel, Helv. Phys. Acta, 8, 499 (1935).

[129]. H. Mueller, Ann. N.Y. Acad. Sci., 40, 321 (1940).

[130]. J. Halblützel, Helv. Phys. Acta, 12, 489 (1939).

[131]. A. N. Holden and W. P. Mason, Phys. Rev., 57, 54 (1940).

[132]. B. Kourtschatov and M. Eremejev, Phys. Z. Sovjetunion, 1, 140 (1932). See also Reference 120.

[133]. A. Melmed, F. Jona, and R. Pepinsky, Techn. Report No. 22, Contract No. AF-33(616)-3133 with Air Force Office of Scientific Research; A. Melmed, M. A. Thesis, Pennsylvania State University, 1953.

[134]. H. Jaffe, Phys. Rev., 51, 43 (1937).

[135]. R. Pepinsky, Rev. Mod. Phys., 30, 100 (1958).

[136]. F. Mazzi, F. Jona, and R. Pepinsky, Bull. Am. Phys. Soc., 1, 131 (1956); Zt. Krist. 108, 359 (1959).

[137]. I. Krstanovic, Y. Okaya, and R. Pepinsky, Abstracts, Am. Cryst. Assoc., Washington, D. C., January 25-27, 1960. These results were apparently published in somewhat more detail by Krstanovic, in a paper listing the above three authors and appearing in Glas de l' Academie Serbes des Sciences, 249, 263 (1961); but the present writer has not seen this publication.

[138]. H. Boutin, B. C. Frazer, and F. Jona, J. Phys. Chem. Solids, 24, 1341 (1963).

[139]. L. Carroll, Alice In Wonderland. American Library, Signet Classic CD22, New York (1960).

[140]. R. Pepinsky, Abstracts, Tenth Pittsburgh Diffraction Conference, November, 1952; Abstracts, Eleventh Pittsburgh Diffraction Conference, November, 1953; Acta Cryst., 7, 690 (1954). See also discussions in: G. R. Ringo, Neutron Diffraction and Interference, p. 617, of pp. 552-642, in Encyclopedia of Physics Vol. XXXII, (S. Flügge, ed.) Springer, Berlin, 1957; also G. E. Bacon, Neutron Diffraction, Second Edition, pp. 244-45, Oxford, London, 1962.

[141]. W. J. Merz, Phys. Rev., 81, 1064 (1951).

[142]. R. C. Vernon and R. Pepinsky, unpublished; R. C. Vernon, Masters Thesis, Pennsylvania State College, 1953.

[143]. B. C. Frazer, J. Phys. Soc. Japan, 17, Suppl. B-II, 376 (1962).

[144]. M. Renninger, Zt. Krist., 97, 107 (1937). See also pp. 30-32 in H. Lipson and W. Cochran, The Determination of Crystal Structures, Third Edition, Cornell University Press, Ithaca, 1966.

[145]. J. L. Bjorkstam and J. H. Willmorth, Deuteron NMR and Dielectric Relaxation Study of Deuterated Rochelle Salt, pp. 728-738 in Magnetic Resonance and Relaxation, Proceedings of XIVth Colloque Ampére, Ljubljana, 1966, (R. Blinc, Ed.) North Holland, Amsterdam, 1967.

[146]. J. L. Bjorkstam, pp. 86-87 in Abstracts, Second International Meeting on FE, Kyoto, 1969.

[147]. R. Blinc, Magnetic Resonance in FE's, pp. 333-357 in Proceedings of International Meeting on Ferroelectricity

(V. Dvorák, A. Fousková, and P. Glogar, eds.) Vol. II, Prague, 1966. Inst. Phys. Czech. Acad. Sci., Prague, 1966.

[148]. R. Pepinsky Jr., Appl. Phys., 18, 601 (1947); R. Pepinsky, X-RAC and S-FAC: Electric Analogue Computers for X-ray Analysis, pp. 167-389 in Computing Methods and the Phase Problem in X-ray Crystal Analysis (R. Pepinsky, ed.), X-ray Crystal Analysis Laboratory, Pennsylvania State College, 1952.

[149]. B. C. Frazer and R. Pepinsky, to be published.

[150]. F. Unterleitner, Y. Okaya, K. Vedam, and R. Pepinsky, Abstracts, Am. Cryst. Assoc., Cornell University, July 19-24, 1959.

[151]. Y. Okaya, R. Pepinsky, and F. Unterleitner, Bull. Am. Phys. Soc.,(2),4, 62 (1959); R. Pepinsky, Y. Okaya, and F. Unterleitner, Acta Cryst., 13, 1071 (1960).

[152]. Y. Okaya and R. Pepinsky, New Developments in the Anomalous Dispersion Method for Structure Analysis, pp. 273-299 in Computing Methods and the Phase Problem in X-ray Crystal Analysis (R. Pepinsky, J. M. Robertson, and J. C. Speakman, eds.) Pergamon, New York, 1961.

[153]. R. Pepinsky, Record Chem. Progress, 17, 145 (1956).

[154]. J. M. Bijvoet, Koninkl. Ned. Akad. Wetenschap, 52, 313 (1949); J. M. Bijvoet, Endeavor, 14, 71 (1955).

[155]. Y. LeCorre, J. Phys. Radium, 18, 629 (1957).

[156]. H. R. Danner, B. C. Frazer, and R. Pepinsky, Bull. Am. Phys. Soc., II, 2, 127 (1957); Acta Cryst., 13, 1089 (1960).

[157]. B. C. Frazer and R. Pepinsky, Phys. Rev., 85, 479 (1952); Acta Cryst., 6, 273 (1953).

[158]. D. W. J. Cruickshank, Acta Cryst., 2, 65 (1949).

[159]. F. Stern and C. A. Beevers, Acta Cryst., 3, 341 (1950).

[160]. R. Blinc, M. Jamsek, J. Stepisnik, Z. Trontelj, and J. L. Bjorkstam, Solid State Comm., 7, 95 (1969).

[161]. R. Blinc, E. M. Peterson, and D. E. O'Reilly, Phys. Letters, 28A, 624 (1969).

[162]. I. Vigness, Phys. Rev., 48, 198 (1935).

[163]. K. Okada, J. Phys. Soc. Japan, 15, 363 (1960).

[164]. K. Okada, J. Phys. Soc. Japan, 16, 414 (1961).

[165]. K. Okada, J. Phys. Soc. Japan, 16, 1647 (1967).

[166]. K. Toyoda, A. Shimada, and T. Tanaka, J. Phys. Soc. Japan, 15, 536 (1959).

[167]. H. H. A. Drueger, W. R. Cook, Jr., C. C. Sartain, and H. P. Yockey, J. Appl. Phys., 34, 218 (1963).

[168]. Y. Makita and Y. Takagi, J. Phys. Soc. Japan, 13, 367 (1958).

[169]. H. E. Buckley, Mineral. Mag. 20, 159 (1924).

[170]. H. E. Buckley, Mineral. Mag. 21, 55 (1926).

[171]. H. E. Buckley, Crystal Growth. Wiley, New York (1951).

[172]. V. A. Iurin, Isv. akad. nauk SSSR, Ser. Fiz., 21, 329 (1957).

[173]. G. C. Moulton and W. G. Moulton, J. Chem. Phys., 35, 208 (1961).

[174]. R. Abé and I. Suzuki, Electron Paramagnetic Resonance in γ-Irradiated Rochelle Salt, pp. 358-363 in Proc. Internat. Meeting on FE, Vol. II, Prague, 1966 (V. Dvořák, A. Fouskova, and P. Glogar, eds.) Inst. Phys. Czech. Acad. Sci., Prague, 1966.

[175]. R. Pepinsky, unpublished.

Chapter 22

THERMODYNAMIC PHENOMENOLOGY OF FERROELECTRICITY IN SINGLE CRYSTAL AND CERAMIC SYSTEMS

L. E. Cross

Materials Research Laboratory and Department of Electrical Engineering
The Pennsylvania State University
University Park, Pennsylvania

I. INTRODUCTION

With the continuing development of the subject of ferroelectricity, it has become increasingly evident that, contrary to earlier beliefs, ferroelectric behavior is not a very rare and unusual phenomenon, but occurs in a wide diversity of compounds and solid solution systems.

In these circumstances, it becomes highly desirable to develop effective schemes for classifying and cataloging these diverse ferroelectrics, and many different bases have been proposed for such classifications. Probably the simplest, and certainly the only completely general classifications are based on symmetry considerations. From the symmetry also one can, for many ferroelectrics, develop simple free energy functions which will describe the static dielectric properties in both paraelectric and ferroelectric phases.

This type of phenomenological approach can be especially useful to the ceramist. From the initial classification one can tell at once those symmetries favorable to good ferroelectric behavior in a polycrystal, and using the phenomenological energy functions it is possible to take account of the very different mechanical and electrical boundary conditions which may exist on the crystallites in a ceramic ferroelectric.

In the following paper, three recent developments in the general phenomenology of ferroelectricity are considered, a systematic approach to using phenomenological information in quantifying the properties of ceramics is suggested, and a number of examples of the use of these techniques discussed.

II. OUTLINE OF THE PHENOMENOLOGICAL APPROACH

The phenomenological approach which we chose to use in seeking explanation of the dielectric behavior of ferroelectric crystals and ceramics is illustrated in the form of a simple "route map" in Fig. 1.

Let us assume a new ferroelectric crystal X. The most important information about X required for initiating the treatment is

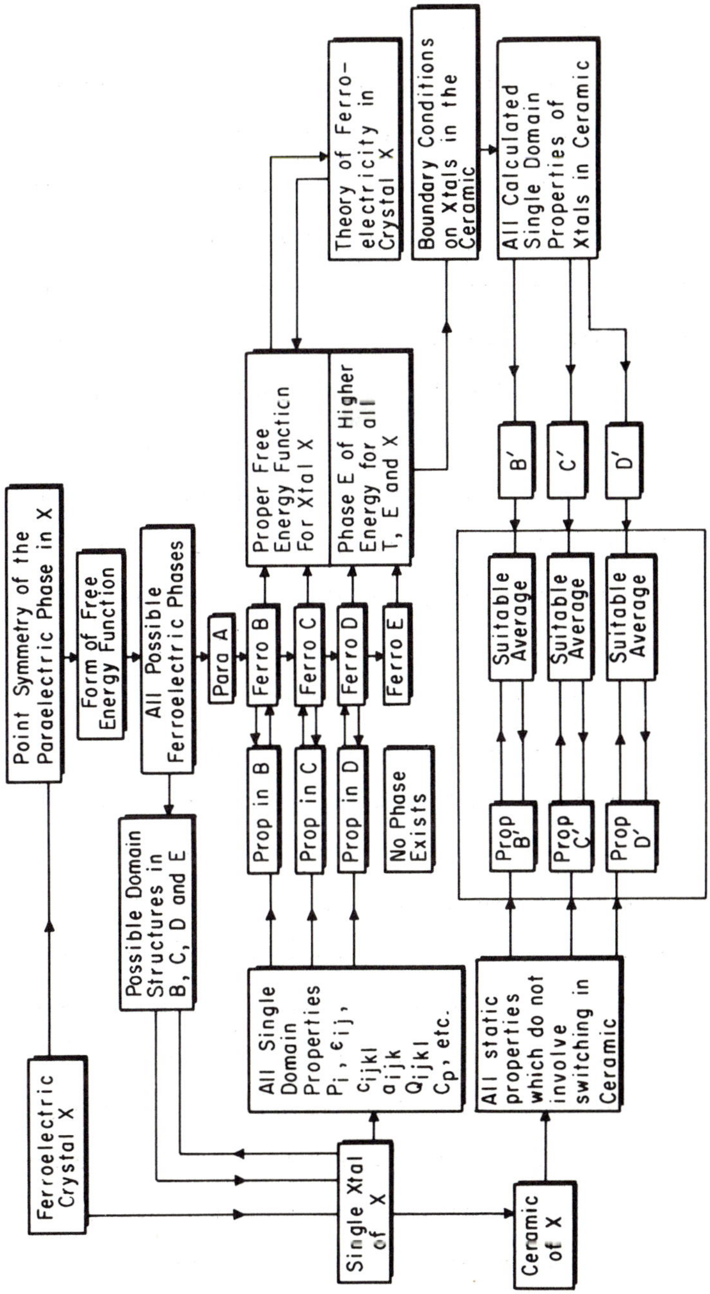

FIG. 1
Schematic diagram of the phenomenological thermodynamic approach to the study of ferroelectricity in crystals and ceramics.

the point symmetry of X in the paraelectric phase. From the point symmetry, using the recent formalism of Aizu[1] in Japan, we may predict all possible ferroelectric phases which crystal X can assume, and also the possible orientations of the polar vector P_s in these phases. If we choose to expand the thermodynamic potential describing the crystal in terms of the polarizations and strains measured from the paraelectric state, we may also determine from the symmetry, the allowed form for this energy function. Let us say that in our crystal X the point symmetry allows four possible ferroelectric phases which are designated B, C, D, E, and the paraelectric phase A. We can also predict in a quite general way from Aizu[2] something about the phase change from A to any one of the ferroelectric modifications, namely, whether it must be a first order change, or whether it may be either first or second order.

Now having made this derivation, we may go one step farther: using a simple criterion developed by Fousek and Janovec[3] we may derive in general all possible domain wall orientations for walls which could separate the possible polar vector directions in B, C, D, E.

Since all of this information is now tabulated for the different point groups, it is a simple accessible framework into which we know that crystal X _must_ fit. It is a rigorous but limited framework to go farther we must know more about crystal X. We must now grow a crystal of X, and measure some of its macroscopic properties, the P_i, ε_{ij}, $c_{ijk\ell}$, a_{ijk}, $Q_{ijk\ell}$ and so on, as a function of temperature, and determine some of these properties in each of the ferroelectric phases which X chooses to assume over the complete temperature range of interest.

Let us say that in this crystal only A, B, C, D phases appear. Now by intercomparing the properties predicted from the thermodynamic potential, and the actual properties, we may choose suitable magnitudes for the coefficients in the energy function to describe our actual crystal X. The method is frequently useful since for many crystals a sharply truncated power series (with only the lower order stiffness coefficients linearly temperature dependent) gives a rather good quantitative description of the behavior. The amount of experimental information required to generate the quantitative free energy function will vary, of course, with the symmetry of the system considered, but once the function has been quantified, it may be used to calculate and cross check with other macroscopic properties of the system. We could, for example, at this point realize why phase E does not occur over our measured range, and predict whether there are conditions of temperature, stress, or field which could make it the lowest energy, and, therefore, the stable phase.

Now we are also in a good position to turn to the theorist for a detailed microscopic explanation of the free energy, for if the microscopic theory predicts correctly the magnitudes of the terms in the phenomenological energy function, it will describe completely all facets of the static experimental behavior of the crystal. If the theory works in this sense, it may then be used to go on to predict time dependent intrinsic properties and compare with experiment; and if all works well, we have at this point a rather complete understanding of ferroelectricity in the single crystal X.

As a further stage in development we may now go on to look at a ceramic made from composition X. In considering any ferroelectric crystal we know from the complete energy function that the properties are very sensitive to the external variable, e.g., the permittivity at constant stress is not the same as permittivity at constant strain; permittivity at zero field is not the same as permittivity at a finite field E_o, etc. Fortunately, however, knowing the full energy function, we may predict new properties for any conceivable set of external variables; that is, any conceivable set of boundary conditions on our crystal.

The first, and certainly the most difficult, task in going from the single crystal to the ceramic system is in predicting effectively the correct boundary conditions on the crystallites in the ceramic. Only in a few special cases has this been done at present. If it can be done, we may then predict new properties, even perhaps new phases, designated B′, C′, D′ for the crystallite under the boundary conditions in the ceramic. From these predicted properties, by making suitable averages to take account of the orientations of crystallites in the ceramic, we may arrive at a set of predicted average properties which can then be compared with the measured ceramic properties.

It is in the last block of this diagram that the measure of success or failure of this macroscopic approach can be judged.

III. THE THERMODYNAMICS OF FERROELECTRICITY IN CRYSTALS

The phenomenological methods which we shall consider are basically quite simple extensions of conventional thermodynamic analysis to the macroscopic interrelationships in a general anisotropic nonlinear polarizable-deformable insulator. At this stage then it is perhaps advisable to briefly review the thermodynamic functions involved.

In thermodynamic equilibrium, any system may be described by a set of independent variables which describe the system as a whole, and a corresponding set of dependent variables which are uniquely determined by the independent ones. For the thermodynamic analysis of the polarizable, deformable insulator, these variable are: mechanical stress X, strain x, electric field E, polarization P, temperature T, and entropy S.

To derive a set of thermodynamic potentials we inquire how the internal energy U of the dielectric changes for an infinitesimal change in each dependent variable,

$$dU = dQ + EdP + Xdx \tag{1}$$

The change dU is made up of three parts: a heat flow dQ, work done against the electric field (EdP), and work done against the mechanical stress (Xdx). Assuming a reversible process dQ = TdS and Eq. (1) becomes

$$dU = TdS + EdP + Xdx \tag{2}$$

In Eq. (2) only the form is given, and clearly in any real system summations would be required to take account of the vector nature of E and P, and the tensor form of X and x.

Proceeding in a manner similar to that used in simple gas thermodynamics, a whole set of thermodynamic potentials may be developed from Eq. (2). For example, a Helmholtz free energy A given by

$$A = U - TS \tag{3}$$

so that an infinitesimal change dA is made up,

$$dA = -SdT + EdP + Xdx \tag{4}$$

An enthalpy H given by

$$H = U - Xx - EP \tag{5}$$

so that for an infinitesimal change

$$dH = TdS - PdE - xdX \tag{6}$$

an elastic Gibbs function G_1

$$G_1 = H - TS + EP \tag{7}$$

with an infinitesimal change dG_1 given by

$$dG_1 = -SdT - xdX + EdP \tag{8}$$

and a full Gibbs function G

$$G = H - TS \tag{9}$$

with incremental form

$$dG = -SdT - xdX - PdE \tag{10}$$

Since the treatments here are mostly concerned with isothermal behavior they use mostly the elastic Gibbs, Gibbs, and Helmholtz free energies.

To take a simple example, the elastic Gibbs function for a normal linear piezoelectric dielectric is given by

$$G_1 = G_{10}(T) - \frac{1}{2} s^P X^2 + \frac{1}{2} \chi^X P^2 - bXP \tag{11}$$

and the Helmholtz free energy takes the form

$$A = A_o(T) + \frac{1}{2} c^P x^2 + \frac{1}{2} \chi^x P^2 - axP \tag{12}$$

The partial derivatives

$$\frac{\partial G_1}{\partial P} = E = \chi^X P - bX \tag{13}$$

$$\frac{\partial G_1}{\partial x} = -x = -s^P X - bP \tag{14}$$

$$\frac{\partial A}{\partial P} = E = -ax + \chi^x P \tag{15}$$

$$\frac{\partial A}{\partial x} = X = c^P x - aP \tag{16}$$

give the interrelationship between dependent and independent variables, which now take proper account of the piezoelectric cross coupling between dielectric and mechanical behavior. Note that again only the form of the terms is given and in real materials summations would be required for the polarization, field, and stress and strain terms.

IV. THE PHENOMENOLOGY OF FERROELECTRICITY IN CRYSTALS

To generalize the phenomenological thermodynamic treatment to nonlinear dielectrics, all we need to do is to add higher order terms to describe the appropriate nonlinearity. Let us at the moment simplify the analysis by considering only phenomena at constant and zero stress for an isothermal system, and thus effectively eliminate all cross coupling terms.

In this case, following Aizu[1], the elastic Gibbs function takes the form

$$\Delta G = B_i P_i + B_{ij} P_i P_j + B_{ijk} P_i P_j P_k + B_{ijk\ell} P_i P_j P_k P_\ell + \cdots \qquad (17)$$

Clearly the B_{ij} are closely related to the dielectric inverse susceptibilities (dielectric stiffnesses), and the B_{ijk}, $B_{ijk\ell}$, etc., are higher order stiffness coefficients necessary to describe the dielectric nonlinearity. The B_i will only occur in polar crystals.

For any crystal, the B_i, B_{ij} and higher order terms will be limited by the point symmetry of the paraelectric phase, or if no paraelectric phase exists by a higher symmetry prototype from which the ferroelectric modification is derived. By considering all such functions* for the 31 point groups which may be prototypic, Aiza[4] shows that these can give rise to 88 species of ferroelectrics. Crystals are said to belong to the same species when they are the same with respect to their own point group, the point group of the prototype, and the relation between elements of the ferroelectric and prototypic groups. The species can be subdivided into two types: there are 55 species in which the P_s vector is

*It is shown that it is not necessary to expand the energy function as a series to be able to apply the symmetry restrictions.

completely reversible (species derived from the 22 nonpolar groups), and 33 species in which the P_s vector may be redirected but not completely reoriented.

Aizu[5] uses two quite compact notations to describe all ferroelectric species. Since the older notation appears the more useful for ceramics, it is given here, and the relation between old and newer notation is shown in Table II.

In the older notation, a typical Aizu species for the perovskite (point symmetry m3m) would be

Fm3m(3)D4

where F indicates a ferroelectric state,

m3m indicates the symmetry of the paraelectric or prototype phase,

(3) indicates the number of alternative polar orientations (twice this number gives the number of vector directions),

D indicates that the P directions are defined by the paraelectric symmetry element which follows. In this case it is the fourfold axis.

Thus the Fm3m(3)D4 is a compact notation for the tetragonal state in a perovskite like $BaTiO_3$. The symbol A in the place of D in Aizu's formalism leaves the P_s vector arbitrary but in a plane defined by the following symbol, i.e., perpendicular to a rotation axis, or in a mirror plane. The complete set of possible species for a perovskite m3m is shown in Table I. The first three states are the familiar Devonshire solutions for $BaTiO_3$, tetragonal, orthorhombic, and rhombohedral states, which 3, 6, and 4 alternative P_s directions respectively.

TABLE I

AIZU SYMBOLS FOR THE SPONTANEOUSLY POLARIZED STATES POSSIBLE IN A CRYSTAL WITH PARAELECTRIC (PROTOTYPE SYMMETRY) 13c

Polarization scheme	Point symmetry	Aizu symbol
$P_1^2 = P_2^2 = P_2^2 = 0$	m3m	Paraelectric
$P_1^2 \neq 0 \quad P_2^2 = P_3^2 = 0$	4mm	Fm3m (3) D4
$P_1^2 = P_2^2 \neq 0 \quad P_3^2 = 0$	2mm	Fm3m (6) D2
$P_1^2 = P_2^2 = P_3^2 \neq 0$	3m	Fm3m (4) D3
$P_p^2 \neq 0 \quad P_2^2 \neq 0 \quad P_3^2 = 0$ $P_1^2 \neq P_2^2$	2	Fm3m (12) A4
$P_1^2 = P_2^2 \neq 0 \quad P_3^2 \neq 0$ $P_3^2 \neq P_1^2$	m	Fm3m (12) A2
$P_1^2 \neq 0 \quad P_2^2 \neq 0 \quad P_3^2 \neq 0$ $P_1^2 \neq P_2^2 \neq P_3^2$	1	Fm3m (24) A1

It is interesting to notice that Aizu predicts three lower symmetry states also—states which would be particularly favorable in ceramic systems since they have 12, 12, and 24 alternative P_s orientations, i.e., up to 48 vector directions, so that however the crystallite were to lie in the polycrystal it must always be close to an "easy direction" of polarization.

A full list of the Aizu symbols for all ferroelectric species in both older and newer notation is given in Table II. It will be noted that in the older notation species in which the polarization is redirectable but not reversible are defined by a symbol (/2); the divisor indicates that the number above now refers to vector directions, and that complete reversibility is forbidden. In the newer notation, the ferroelectric species is simply categorized by its point symmetry. In most cases the relationship between ferroelectric and prototypic symmetry is obvious, but where this is not the case the additional symbols which indicate this relationship are described in footnotes to Table II (taken from Aizu[5]).

In a ceramic ferroelectric, the crystallites in a sample are oriented quite at random in normal processing. If a ferroelectric species with only one orientation were chosen, then whatever the orientation of a poling field on the ceramic, some crystallites will lie with this unique axis perpendicular to the field direction so that polarization reversal will be impossible. Similarly, it is obvious as suggested earlier, that the larger the number of equivalent polar directions in the individual crystallite, the more closely should the properties of the polycrystal approach those of the originating single crystal. Thus from the Aizu table of species it is possible to tell at a glance those symmetries which are useless in ceramic form, and those where the polycrystal at least has reasonable chance of being a successful ferroelectric. It is perhaps interesting to note that the perovskite structure (m3m) is probably the most widely used in ferroelectric ceramics, but even so, as far as is known no special effort has been made to realize the low symmetry states which might be the most attractive.

Turning now to the derivation of the order of the ferroelectric phase transition, Aizu[2] makes two additional postulations:

1. The degree of degeneracy of the principal values of the coefficients of the second order terms in the expansion of the elastic Gibbs function at T_c are the same as in the paraelectric state.

TABLE II

AIZU SYMBOLS FOR ALL FERROELECTRIC SPECIES IN BOTH OLDER AND NEWER NOTATIONS[a]

New symbol	Old symbol	New symbol	Old symbol	New symbol	Old symbol
$\bar{1}$F1	F$\bar{1}$(1)A1	4/*mmm*F*m*(*p*)	F4/*mmm*(4)A4	$\bar{6}$*m*2F*mm*2	F$\bar{6}$*m*2(3/2)D2
2F1	F2(2/2)A1	4/*mmm*F*mm*2(*s*)	F4/*mmm*(2)D2	$\bar{6}$*m*2F3*m*	F$\bar{6}$*m*2(1)D$\bar{6}$
*m*F1	F*m*(2/2)A1	4/*mmm*F4*mm*	F4/*mmm*(1)D4	6/*mmm*F1	F6/*mmm*(12)A1
2/*m*F1	F2/*m*(2)A1	3F1	F3(3/2)A1	6/*mmm*F*m*(*s*)	F6/*mmm*(6)A2
2/*m*F2	F2/*m*(1)D2	$\bar{3}$F1	F$\bar{3}$(3)A1	6/*mmm*F*m*(*p*)	F6/*mmm*(6)A6
2/*m*F*m*	F2/*m*(1)A2	$\bar{3}$F3	F$\bar{3}$(1)D$\bar{3}$	6/*mmm*F*mm*2(*s*)	F6/*mmm*(3)D2
222F1	F222(4/2)A1	32F1	F32(6/2)A1	6/*mmm*F6*mm*	F6/*mmm*(1)D6
222F2	F222(1)D2	32F2	F32(3/2)D2	23F1	F23(12/2)A1
*mm*2F1	F*mm*2(4/2)A1	32F3	F32(1)D3	23F2	F23(3)D2
*mm*2F*m*	F*mm*2(2/2)A*m*	3*m*F1	F3*m*(6/2)A1	23F3	F23(4/2)D3
*mmm*F1	F*mmm*(4)A1	3*m*F*m*	F3*m*(3/2)A*m*	*m*3F1	F*m*3(12)A1
*mmm*F*m*	F*mmm*(2)A*m*	$\bar{3}$*m*F1	F$\bar{3}$*m*(6)A1	*m*3F*m*	F*m*3(6)A*m*
*mmm*F*mm*2	F*mmm*(1)D*m*	$\bar{3}$*m*F2	F$\bar{3}$*m*(3)D*m*	*m*3F*mm*2	F*m*3(3)D*m*
4F1	F4(4/2)A1	$\bar{3}$*m*F*m*	F$\bar{3}$*m*(3)A*m*	*m*3F3	F*m*3(4)D3
$\bar{4}$F1	F$\bar{4}$(4/2)A1	$\bar{3}$*m*F3*m*	F$\bar{3}$*m*(1)D$\bar{3}$	432F1	F432(24/2)A1
$\bar{4}$F2	F$\bar{4}$(1)D$\bar{4}$	6F1	F6(6/2)A1	432F2(*s*)	F432(6)D2
4/*m*F1	F4/*m*(4)A1	$\bar{6}$F1	F$\bar{6}$(6/2)A1	432F4	F432(3)D4
4/*m*F*m*	F4/*m*(2)A4	$\bar{6}$F*m*	F$\bar{6}$(3/2)A$\bar{6}$	432F3	F432(4)D3
4/*m*F4	F4/*m*(1)D4	$\bar{6}$F3	F$\bar{6}$(1)D$\bar{6}$	$\bar{4}$3*m*F1	F$\bar{4}$3*m*(24/2)A1
422F1	F422(8/2)A1	6/*m*F1	F6/*m*(6)A1	$\bar{4}$3*m*F*m*	F$\bar{4}$3*m*(12/2)A*m*
422F2(*s*)	F422(2)D2	6/*m*F*m*	F6/*m*(3)A6	$\bar{4}$3*m*F*mm*2	F$\bar{4}$3*m*(3)D$\bar{4}$
422F4	F422(1)D4	6/*m*F6	F6/*m*(1)D6	$\bar{4}$3*m*F3*m*	F$\bar{4}$3*m*(4/2)D3
4*mm*F1	F4*mm*(8/2)A1	622F1	F622(12/2)A1	*m*3*m*F1	F*m*3*m*(24)A1
4*mm*F*m*	F4*mm*(4/2)A*m*	$\bar{6}$*m*2F*m*(*s*)	F622(3)D2	*m*3*m*F*m*(*s*)	F*m*3*m*(12)A2
$\bar{4}$2*m*F1	F$\bar{4}$2*m*(8/2)A1	622F6	F622(1)D6	*m*3*m*F*m*(*p*)	F*m*3*m*(12)A4

$42m$F2(s)	F$\bar{4}2m$(2)D2	$6mm$F1	F$6mm$(12/2)A1	$m3m$F$mm2(ps)$	F$m3m$(6)D2
$\bar{4}2m$Fm	F$\bar{4}2m$(4/2)Am	$6mm$Fm	F$6mm$(6/2)Am	$m3m$F$4mm$	F$m3m$(3)D4
$\bar{4}2m$F$mm2$	F$\bar{4}2m$(1)D$\bar{4}$	$\bar{6}m2$F1	F$\bar{6}m2$(12/2)A1	$m3m$F$3m$	F$m3m$(4)D3
$4/mmm$F1	F$4/mmm$(8)A1	$\bar{6}m2$F$m(p)$	F$\bar{6}m2$(6/2)Am		
$4/mmm$F$m(s)$	F$4/mmm$(4)A2	$\bar{6}m2$F$m(p)$	F$\bar{6}m2$(6/2)A$\bar{6}$		

[a]The letters "p" and "s" in species symbols are the initials of "principal" and "side," respectively. When the prototype belongs to a noncubic system, "p" means that the crystallographic unique axis of the ferroelectric or ferroelastic crystal is *along the crystallographic unique axis* of the prototype, and "s" means that the crystallographic unique axis of the ferroelectric or ferroelastic crystal is *perpendicular to the crystallographic unique axis* of the prototype. When the prototype belongs to the cubic system, "p" means that the unique axis or an important axis of the ferroelectric or ferroelastic crystal is *along one of the cubic principal axes* of the prototype, and "s" means that the unique axis or an important axis of the ferroelectric or ferroelastic crystal is *along a face diagonal of the cubic lattice* of the prototype.

"PG" and "FG" in sentences below are abbreviations of "the prototypic point group" and "the ferroelectric or ferroelastic point group."

422F2(s): The dyad axis of FG is along one of the dyad axes (except the tetrad axis) of PG.

$\bar{4}2m$F2(s): The dyad axis of FG is along one of the dyad axes (except the $\bar{4}$ axis) of PG.

$4/mmm$F$m(s)$: The mirror plane of FG is along one of the mirror planes parallel to the tetrad axis of PG.

$4/mmm$F$m(p)$: The mirror plane of FG is along the mirror plane perpendicular to the tetrad axis of PG.

$4/mmm$F$mm2(s)$: The dyad axis of FG is along one of the dyad axes (except the tetrad axis) of PG.

622F2(s): The dyad axis of FG is along one the dyad axes (except the hexad axis) of PG.

$\bar{6}m2$F$m(s)$: The mirror plane of FG is along one of the mirror planes parallel to the $\bar{6}$ axis of PG.

$\bar{6}m2$F$m(p)$: The mirror plane of FG is along the mirror plane perpendicular to the $\bar{6}$ axis of PG.

$6/mmm$F$m(s)$: The mirror plane of FG is along one of the mirror planes parallel to the hexad axis of PG.

$6/mmm$F$m(p)$: The mirror plane of FG is along the mirror plane perpendicular to the hexad axis of PG.

$6/mmm$F$mm2(s)$: The dyad axis of FG is along one of the dyad axes (except the hexad axis) of PG.

432F2(s): The dyad axis of FG is along one of the dyad axes (except the tetrad axes) of PG.

$m3m$F$m(s)$: The mirror plane of FG is along a mirror plane making an angle of 45° to a tetrad axis of PG.

$m3m$F$m(p)$: The mirror plane of FG is along a mirror plane perpendicular to a tetrad axis of PG.

$m3m$F$mm2(ps)$: One of the two mirror planes of FG is along a mirror plane perpendicular to a tetrad axis of PG, and the other is along a mirror plane making an angle of 45° to a tetrad axis of PG.

2. The form of the coefficients of tensors in the terms of order higher than 2 at T_c are the same as in the prototype.

   a. Coefficients which are nonzero from symmetry do not accidentally become zero at T_c.

   b. Elements unrelated by symmetry above T_c do not become accidentally equal at T_c.

On the basis of these assumptions, Aizu is able to predict which phase changes must be first order and which may be either first or second order.

From Aizu's work we have a compact description of the qualitative aspects of the behaviors of all ferroelectric crystal which are dictated by the crystal symmetry. Knowing just the paraelectric point symmetry we can now say what ferroelectric states are possible, and the orientations of the P_s vector which are possible in these phases.

V. DOMAIN WALLS IN AN INFINITE PERFECT FERROELECTRIC CRYSTAL

Taking the information generated by Aizu, Fousek, and Janovec[3] from the Academy of Sciences in Prague have developed a simple criterion for predicting possible domain wall orientations for walls which could separate the allowed P_s vector directions in an infinite perfect crystal.

The argument may be summarized as follows:

1. When the paraelectric crystal polarizes, it self-distorts; the resulting distortions can be predicted from the magni-

tude and direction of P_s, and the piezoelectric and electrostrictive coefficients in the paraelectric phase.

2. In the infinite perfect crystal, two domains having differently oriented P_s vectors may only join along planes where the two distortion systems mate exactly. These planes are then possible domain walls for that crystal.

To formulate the compatibility condition, Fousek and Janovec consider two domains with P_s vectors ${}^1P$ and ${}^2P$. The spontaneous deformation in these domains will then be

$$ {}^1e_{k\ell} = \sum_{i=1}^{3} b_{ik\ell}\, {}^1P_i + \sum_{i,j=1}^{3} Q_{ijk\ell}\, {}^1P_i\, {}^1P_j \tag{18} $$

$$ {}^2e_{k\ell} = \sum_{i=1}^{3} b_{ik\ell}\, {}^2P_i + \sum_{i,j=1}^{3} Q_{ijk\ell}\, {}^2P_i\, {}^2P_j \tag{19} $$

where $b_{ik\ell}$ and $Q_{ijk\ell}$ are piezoelectric and electrostriction coefficients, respectively.

If ds is an infinitesimal vector in the paraelectric phase, then the difference between the square of its length in the ferroelectric and in the paraelectric phases is given by

$$ d^m s^2 - ds^2 = 2 \sum_{k\ell=1}^{3} {}^m e_{k\ell}\, ds_k\, ds_\ell \tag{20} $$

For mechanical compatibility the deformation Eqs. (18) and (19) of a plane ($hk\ell$) must be equal (up to a rigid body motion). This requirement can be expressed as follows: Any vector ds in the plane $hk\ell$ changes its length equally with the appearance of ${}^1P$ or ${}^2P$, i.e.,

$$ d^1 s^2 - ds^2 = d^2 s^2 - ds^2 \tag{21} $$

for

$$hds_1 + kds_2 + \ell ds_3 = 0 \tag{22}$$

Using Eqs. (18), (19), and (20), Eq. (21) may be rewritten in terms of a distortion matrix $\Delta k\ell$ as

$$\sum_{k\ell=1}^{3} \Delta_{k\ell}\, ds_k\, ds_\ell = 0 \tag{23}$$

where

$$\Delta_{k\ell} = \sum_{i=1}^{3} b_{ik\ell}\left({}^2P_1 - {}^1P_1\right) + \sum_{i,j=1}^{3} Q_{ijk\ell}\left({}^2P_i\, {}^2P_j - {}^1P_i\, {}^1P_j\right) \tag{24}$$

Using a compact notation

$$\Delta_{k\ell} = \Delta_k, \qquad k = \ell$$

$$\Delta_{k\ell} = \Delta n/2, \qquad k \neq \ell$$

where n = 4,5,6 for $k\ell$ = 23, 13, and 12. Then Eq. (23) takes the form

$$\Delta_1 ds_1{}^2 + \Delta_2 ds_2{}^2 + \Delta_3 ds_3{}^2 + \Delta_4 ds_2 ds_3 + \Delta_5 ds_1 ds_3 + \Delta_6 ds_1 ds_2 = 0 \tag{25}$$

Then any plane Eq. (22) obeying Eq. (25) is a permissible wall.

For a given ferroelectric species, and two conjugate P_s vectors, all $\Delta\lambda$'s can be found and the permissible wall orientation determined. This evaluation has been carried out by Fousek[16] for all 55 species of the original Aizu classification, and a table of all permissible walls in these species, for all pairs of conjugate vectors is now available.

Quite aside from the great value of this general classification, some new, interesting, and unexpected results emerge. Just from the

TABLE III

TYPES OF FERROELECTRIC DOMAIN WALLS PREDICTED ON THE STRAIN CRITERION OF FOUSEK AND JANOVEC

W walls	W_∞	, a wall with arbitrary orientation
	W_f	, a crystallographically prominent wall with fixed orientation

	Type	Indexes depend on
S walls	S_1	direction of P_s
	S_2	b_{ijk} and/or Q_{ijkl}
	S_3	direction of P_s, b_{ijk} and/or Q_{ijkl}
	S_4	direction and magnitude of P_s, b_{ijk} and/or Q_{ijkl}
	S_5	magnitude of P_s, b_{ijk} and/or Q_{ijkl}

consideration of mechanical compatibility a completely new class of domain walls is predicted. The general classification is shown in Table III. The W walls are essentially conventional ferroelectric walls; the W_∞ has arbitrary orientation, and arises when the distortion produced by the conjugate P_s vectors are identical, as in the antiparallel walls in $BaTiO_3$. The suffix ∞ indicates that compatibility does not dictate any orientation, and all orientations are possible. The W_f walls are prominent crystallographic planes, like the 90° walls in $BaTiO_3$ or the 180° walls in KDP. Both categories correspond to the walls with which we are very familiar in many ferroelectric crystals, and the orientations predicted agree closely with those observed in reasonably perfect unstrained single crystals.

As well as these, however, Fousek and Janovec predict the existence of what they call S walls, S standing for strange. The S wall is not a crystallographically prominent plane, but one whose orientation depends on the magnitude and direction of P_s and on the b_{ijk} and $Q_{ijk\ell}$ coefficients, and several types of these S walls are predicted, as shown in the table. It would appear that in most ferroelectric crystals, conditions (apart from mechanical compatibility) are not favorable for S walls and they have not been observed. Recent measurements on NiI boracite, however, give a clear indication that such walls do appear in this system. The S wall is not then just an academic exercise, but a real entity in at least one ferroelectric, and it seems probable that as more people become aware of their possible existence and we look at more ferroelectric crystals in different symmetries, more cases will surely emerge.

VI. APPLICATION OF THE PHENOMENOLOGY TO BISMUTH TITANATE $Bi_4Ti_3O_{12}$

To illustrate the application of the Aizu and Fousek-Janovec developments, it is perhaps useful to take a specific example. In the study of bismuth titanate, work in the past had been hampered by a wrong assignment of symmetry and a misinterpretation of the polar state in the ferroelectric phase. A systematic application of the phenomenological methods helped in no small way to clear up these difficulties.

In bismuth titanate, Aurivillius[7] has shown that in the paraelectric phase (prototype) above 676°C the point symmetry is tetragonal, 4/mmm.

Following Aizu, the elastic Gibbs function for an unstressed crystal may be expanded in the form

$$\Delta G = B_i P_i + B_{ij} P_i P_j + B_{ijk} P_i P_j P_k + B_{ijkl} P_i P_j P_k P_l \cdots \tag{26}$$

where the coefficients B_{ijk} are limited by the paraelectric symmetry.

The condition that the *ith* component of polarization be nonzero may be derived by differentiating the energy function with respect to P_i and equating to zero (since $E_i = 0$). The reciprocal susceptibility χ_{ij} is given by the second differential with respect to P_i and P_j.

Taking the z(3) axis on the four-fold axis of the crystal, with mirror planes perpendicular to 1,2,3, the symmetry dictates the following conditions on the B_{ij},

1. Since the group is centrosymmetric all B_{ijk} with an odd number of indexes must be zero.
2. The only nonzero B_{ij} will be B_{11}, B_{22}, B_{33} and $B_{11} = B_{22}$.
3. The nonzero B_{ijkl} will be B_{1111}, B_{2222}, B_{3333}, B_{1133}, B_{2233}, B_{1122}, and $B_{1111} = B_{2222}$, $B_{1133} = B_{2233}$.
4. The first terms B_{ijklmn} will be B_{111111}, B_{222222}, B_{333333}.

Using the reduced notation 11 = 1, 22 = 2, 33 = 3, and absorbing all numerical constants into the new A_{ij}, the elastic Gibbs function at constant and zero stress may be written

$$\Delta G = A_1\left(P_x^2 + P_y^2\right) + A_3 P_z^2 + A_{11}\left(P_x^4 + P_y^4\right) + A_{33} P_z^4 + A_{13}\left(P_x^2 P_z^2 + P_y^2 P_z^2\right) + A_{12} P_x^2 P_y^2 + A_{111}\left(P_x^6 + P_y^6\right) + A_{333} P_z^6 \tag{27}$$

The conditions for stability are

$$\frac{\partial \Delta G}{\partial P_x} = 2A_1 P_x + 4A_{11} P_x^3 + 2A_{13} P_x P_z^2 + 2A_{12} P_x P_y^2 + 6A_{111} P_x^5 = 0 \tag{28}$$

$$\frac{\partial \Delta G}{\partial P_y} = 2A_1P_y + 4A_{11}P_y^3 + 2A_{13}P_yP_z^2 + 2A_{12}P_yP_x^2 + 6A_{111}P_y^5 = 0 \tag{29}$$

$$\frac{\partial \Delta G}{\partial P_z} = 2A_3P_z + 4A_{33}P_z^3 + 2A_{13}P_z\left(P_x^2 + P_y^2\right) + 6A_{333}P_z^5 = 0 \tag{30}$$

so that either

$$P_x = 0 \text{ or } 0 = 2A_1 + 4A_{11}P_x^2 + 2A_{13}P_z^2 + 2A_{12}P_y^2 + 6A_{111}P_x^4 \tag{31}$$

$$P_y = 0 \text{ or } 0 = 2A_1 + 4A_{11}P_y^2 + 2A_{13}P_z^2 + 2A_{12}P_x^2 + 6A_{111}P_y^4 \tag{32}$$

$$P_z = 0 \text{ or } 0 = 4A_{33}P_z^2 + 2A_{13}\left(P_x^2 + P_y^2\right) + 6A_{333}P_z^4 \tag{33}$$

There are eight possible solutions to these equations, one corresponding to the paraelectric state, and the seven possible ferroelectric states delineated in the following tabulation:

$P_x = P_y = P_z = 0$	Paraelectric		(34)
	Point symmetry of ferroelectric phase	Aizu Notation	
$P_x^2 \neq 0 \quad P_y^2 = P_z^2 = 0$	mm2	(2) D2	(35)
$P_z^2 \neq 0 \quad P_x^2 = P_y^2 = 0$	4mm	(1) D4	(36)
$P_x^2 = P_y^2 \neq 0 \quad P_z^2 = 0$	mm2	(2) D2	(37)
$P_x^2 \neq P_y^2 \neq 0 \quad P_z^2 = 0$	m	(4) A4	(38)
$P_x^2 \neq 0 \quad P_z^2 \neq 0 \quad P_y^2 = 0$	m	(4) A2	(39)
$P_x^2 = P_y^2 \neq 0 \quad P_z^2 \neq 0$	m	(4) A2	(40)
$P_x^2 \neq P_y^2 \neq P_z^2 \neq 0$	1	(8) A1	(41)

For each ferroelectric state, the point symmetry of the ferroelectric phase is given, together with the symbols describing this phase in the Aizu notation.

In bismuth titanate, only the paraelectric phase and F4/mmm(4)A2 appear stable[8]. In the ferroelectric state the point group is monoclinic, m, and the optical properties are quite consistent with this choice[9]. From the optical anisotropies, it is possible to deduce the behavior of P_x and P_z as functions of temperature, and to show that the optical effects are primarily due to quadratic electro-optic coefficients[10].

Making use of recent dielectric measurements by Fouskova, Cross, and Pohanka[11], it is possible to deduce numerical values for all the A_{ij} coefficients.

With A_1, A_{12}, and A_{13} linearly temperature dependent, it is possible to obtain good agreement with all the dielectric and polarization data throughout the temperature range from 20°C to 800°C.

Examination of the hierarchy of energies for all possible states shows that the monoclinic F4/mmm(4)A2 has the lowest energy over the complete range from 676°C to -180°C. Only the orthorhombic state F4/mmm(2)D2 is close but higher in energy over this whole range. It seems very probable that this close approach, and the parallelism in energy of these two states accounts for the easy "z" direction switching in $Bi_4Ti_3O_{12}$, and also for the peculiar inverse temperature dependence of the coercive field E_c for this switching.

To consider the possible domain structure in the ferroelectric phase, under the criterion of Fousek-Janovec, in specie F4/mmm(4)A2, there are eight possible vector orientations of the polar direction. Choosing orthogonal axes x,y,z, with z along the four-fold axis, and taking as the reference direction hhℓ, the seven alternative orientations of P_s are $[\bar{h}\bar{h}\bar{\ell}]$, $[\bar{h}h\ell]$, $[h\bar{h}\ell]$, $[hh\bar{\ell}]$, $[\bar{h}\bar{h}\ell]$, $[\bar{h}h\bar{\ell}]$, $[h\bar{h}\bar{\ell}]$. Considering these vectors in pairs, for example,

1. $[hh\ell]$ versus $[\bar{h}\bar{h}\bar{\ell}]$, in this case the distortion matrix $\Delta_i = 0$ so that any plane satisfies the zero strain criterion.
2. $[hh\ell]$ versus $[hh\bar{\ell}]$, however, $\Delta_1 = \Delta_2 = \Delta_3 = 0$, $\Delta_4 = 2Q_{44}\ h\ell = \Delta_5$, $\Delta_6 = 0$.

Therefore, plane x,y and the plane z, $\alpha = 45°$ to x and y are possible walls.

Proceeding systematically in this way through all possible pairs of conjugate vectors, the complete set of W walls shown in Fig. 2 may be derived. Comparing this set with the observed walls in $Bi_4Ti_3O_{12}$ gives good agreement, confirming beyond reasonable doubt the choice of ferroelectric system.

VII. POSSIBLE ALTERNATIVE ORIGINS FOR THE FERROELECTRIC STATE

It is interesting to examine in a rather general way what the significant properties of a "normal" ferroelectric are, and how they are reflected in the phenomenology. The Curie-Weiss behavior of the dielectric constant above T_c is a prominent feature of most ferroelectrics and clearly requires a linearly decreasing value of one of the B_{ij} coefficients in the free energy expansion. This coefficient must pass through zero in the vicinity of T_c, the Curie temperature. The higher order $B_{ijk...}$ coefficients are not necessarily very different from those in normal linear dielectrics; they become important in the ferroelectric simply because the low B_{ij} value allows much higher values of P than occur in normal solids,

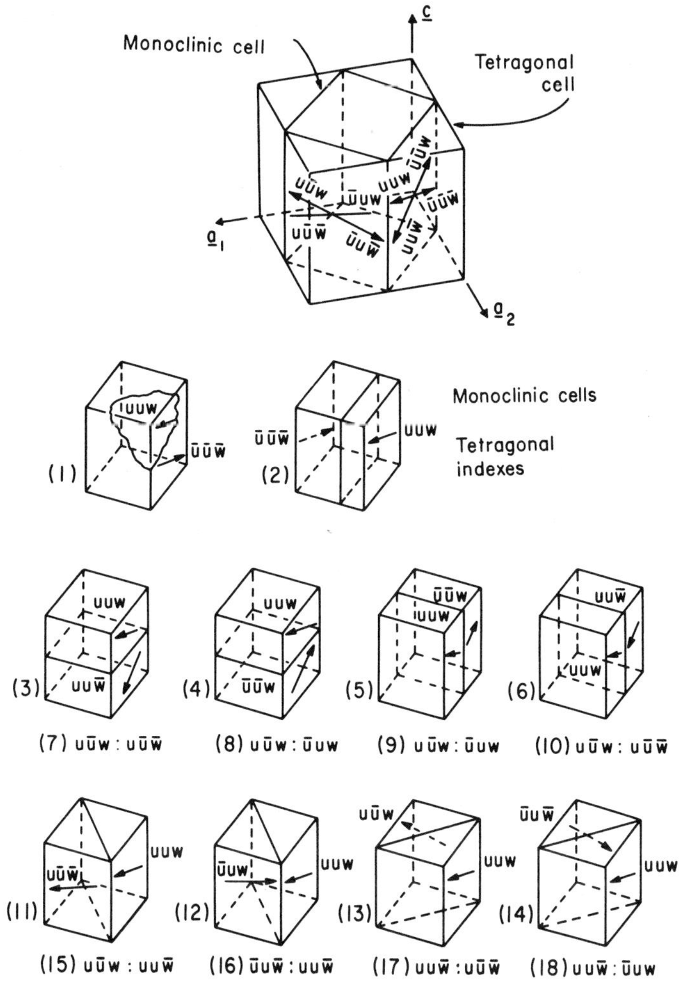

FIG. 2
Orientation of "W" domain walls in bismuth titanate predicted on the criterion of Fousek and Janovec.

particularly at temperatures near T_c. Below T_c it is these same higher order terms which limit the completely catastrophic polarization which would occur in a "linear solid" and thus dictate the magnitude of P_s.

If the ferroelectric state is generated by this dielectric type of instability, then clearly it is to be expected that in the vicinity of T_c there will be high and strongly nonlinear permittivities (low reciprocal susceptibilities) and indeed it is just these features which are exploited in TANDEL[12].

Recently, Borchardt and Bierstedt[13] reported a ferroelectric behavior in gadolinium molybdate, $Gd_2(MoO_4)_3$, which differs strongly from this expected pattern. The dielectric constant is low (≃ 10 to 11) and shows only a very weak anomaly at T_c. There is no evidence of nonlinearity even at temperatures above but very close to T_c. These puzzling features suggest that one look more closely at the phenomenology.

Gadolinium molybdate has point symmetry $\bar{4}2m$ above T_c, a point group which shows piezoelectricity. Taking into account only the lowest order terms, the Helmholtz free energy may be written[14]

$$\Delta A = \frac{1}{2} c_{66}^{P} x_6^2 - a_{36} P_3 x_6 + \frac{1}{2} \chi_{33}^{x} P_3^2 \tag{42}$$

where x_6 is the shear deformation about c, P_3 the polarization along c, c_{66}^{P} the elastic constant at constant polarization, a_{36} the piezoelectric constant and χ_{33}^{x} the inverse dielectric susceptibility at constant strain. The expression is essentially similar to that for KDP which has the same prototype symmetry.

$$\frac{\partial \Delta G}{\partial x_6} = -X_6 = c_{66}^{P} x_6 - a_{36} P_3 \tag{43}$$

$$\frac{\partial \Delta G}{\partial P_3} = E_3 = -a_{36} x_6 + \chi_{33}^{x} P_3 \tag{44}$$

where X_6 is the shear stress about c, and E_3 is the electric field in the direction of c.

For a spontaneous state $X_6 = 0$, $E_3 = 0$ and substituting from Eq. (44) into Eq. (43), the condition for a nonzero deformation becomes

$$c_{66}^{P} - a_{36}^{2}/\chi_{33}^{x} = 0 \tag{45}$$

As was indicated above, in conventional ferroelectrics χ_{33}^{x} is a linearly decreasing function of T, passing through zero in the vicinity of T_c. The catastrophic increase of x_6 and P_3 is limited by the higher order terms in P_3, and the dielectric properties are necessarily nonlinear for temperatures close to T_c.
That this simple phenomenological approach is not applicable in the case of gadolinium molybdate may be adduced from the following behavioral characteristics:

1. Measurements of P_s and of the spontaneous shear as a function of temperature below T_c show that a_{36} is not a function of temperature, but has the value of a "normal" piezoelectric constant.
2. Measurements of ε_c^x above the resonance frequency for a single crystal where the permittivity is intertia-clamped show no anomaly in ε_c^x, as in Fig. 3.
3. Measurements of the resonance frequency of mechanical vibration for a mode which includes a shear deformation x_6 shows a strong anomaly in frequency and therefore in the elastic constant c_{66} at the Curie temperature.

These results lead us to conclude that $Gd_2(MoO_4)_3$ is a new type of ferroelectric in which the polarization is a secondary

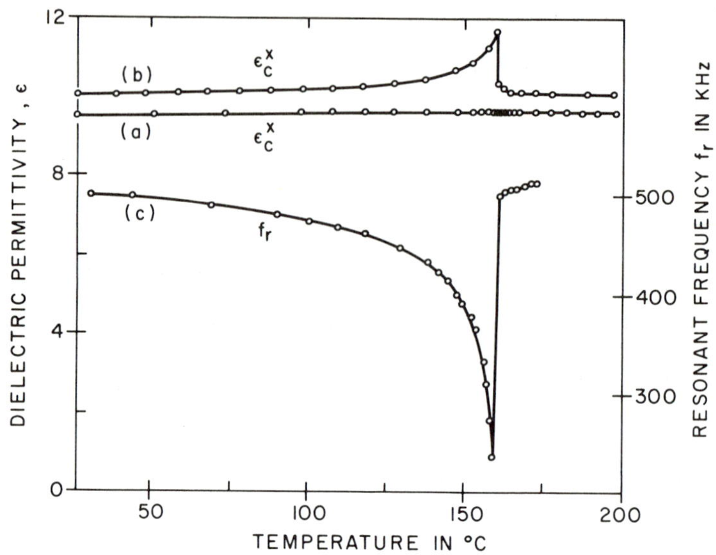

FIG. 3
Dielectric permittivity and mechanical response in gadolinium molybdate single crystal.

phenomenon, that at lower temperatures the crystal chooses to deform spontaneously by shear, and that the spontaneous shear forces a spontaneous polarization by quite normal piezoelectric coupling. Since $\pm x_6$, the spontaneous shear states, are equivalent, an electric field acting through the electromechanical coupling is able to reverse the spontaneous shear, giving rise to P_s reversal and ferroelectricity.

Very recently, Aizu[15] has extended the symmetry classification to include all types of mimetically twinned crystals. It would appear that the gadolinium molybdate is one rather interesting special case, encompassed within this general treatment. For this general classification Aizu considers all crystal in which there are two or more "orientation states" which are identical or enantiomorphous in symmetry, but which may be described as small "deformations" from a higher symmetry or prototypic state. The prototype symmetry reflecting the additional symmetry of all the rotation or inversion operations from all to all of the orientation states.

In the absence of external field or stress, the thermodynamic potential will be the same for all orientation states, but in the presence of field and stress, the potential ψ for an orientation state S may be expanded.

$$\psi = \psi_0 - \Sigma K_i E_i - \Sigma K_{ij} E_i E_j - \Sigma s_{ij} X_{ij} - \Sigma d_{ijk} E_i X_{jk} - \Sigma s_{ijkl} X_{ij} X_{kl} \ldots$$

and for another orientation state S′ of the same species would take the form

$$\psi' = \psi_0 - \Sigma K_i' E_i - \Sigma K_{ij}' E_i E_j - \Sigma s_{ij}' X_{ij} - \Sigma d_{ijkl}' E_i X_{kl} - \Sigma s_{ijkl}' x_{ij} X_{kl} \ldots$$

and clearly the difference in potential $\Delta\psi = \psi - \psi$ may be written in the form

$$\Delta\psi = -\Sigma \Delta K_i E_i - \Sigma \Delta K_{ij} E_i E_j - \Sigma \Delta s_{ij} X_{ij} - \Sigma \Delta d_{ijk} E_i X_{jk} - \Sigma \Delta s_{ijkl} X_{ij} X_{kl}$$

where $\Delta K_i = K_i - K_i'$, etc.

If $\Delta K_i \neq 0$ then the two orientation states are spontaneously polarized, and an electric field E may be chosen which will tend to drive the system from one to the other orientation state. If the difference from the prototype is small, a realizable field E may effect reorientation and the species would be ferroelectric.

If $\Delta s_{ij} \neq 0$ then the two orientation states are spontaneously strained, and a mechanical stress X_{ij} may be chosen which will tend

TABLE IV

THE 94 SPECIES OF FERROELECTRIC CRYSTAL TOGETHER WITH THEIR NUMBER OF STATES AND RELATIONSHIP TO FERROELECTRICITY[a]

Ferroelastic species	No. of states	Ferro-electric?	Ferroelastic species	No. of states	Ferro-electric?	Ferroelastic species	No. of states	Ferro-electric?
2F1	2	Yes	4/*mmm*F2/*m*(*p*)	4	—	6/*mmm*F*mmm*	3	—
*m*F1	2	Yes	4/*mmm*F*mmm*	2	—	23F1	12	Yes
2/*m*F$\bar{1}$	2	—	3F1	3	Yes	23F2	6	Yes
222F1	4	Yes	$\bar{3}$F$\bar{1}$	3	—	23F222	3	—
222F2	2	Yes	32F1	6	Yes	23F3	4	Yes
*mm*2F1	4	Yes	32F2	3	Yes	*m*3F$\bar{1}$	12	—
*mm*2F2	2	—	3*m*F1	6	Yes	*m*3F2/*m*	6	—
*mm*2F*m*	2	Yes	3*m*F*m*	3	Yes	*m*3F*mmm*	3	—
*mmm*F$\bar{1}$	4	—	$\bar{3}$*m*F$\bar{1}$	6	—	*m*3F$\bar{3}$	4	—
*mmm*F2/*m*	2	—	$\bar{3}$*m*F2/*m*	3	—	432F1	24	Yes
4F1	4	Yes	6F1	6	Yes	432F2(*s*)	12	Yes
4F2	2	—	6F2	3	—	432F2(*p*)	12	—
$\bar{4}$F1	2	Yes	$\bar{6}$F1	6	Yes	432F222(*ss*)	6	—
$\bar{4}$F2	2	Yes	$\bar{6}$F*m*	3	Yes	432F222(*pp*)	6	—
4/*m*F$\bar{1}$	4	—	6/*m*F$\bar{1}$	6	—	432F422	3	—
4/*m*F2/*m*	2	—	6/*m*F2/*m*	3	—	432F32	4	—
422F1	8	Yes	622F1	12	Yes	$\bar{4}$3*m*F1	24	Yes
422F2(*s*)	4	Yes	622F2(*s*)	6	Yes	$\bar{4}$3*m*F2	12	—
422F2(*p*)	4	—	622F2(*p*)	6	—	$\bar{4}$3*m*F*m*	12	Yes
422F222	2	—	622F222	3	—	$\bar{4}$3*m*F222	6	—

$4mm$F1	8	Yes	$6mm$F1	12	Yes	$\bar{4}3m$F$mm2$	6	Yes
$4mm$F2	4	—	$6mm$F2	6	—	$\bar{4}3m$F$\bar{4}2m$	3	—
$4mm$Fm	4	Yes	$6mm$Fm	6	Yes	$\bar{4}3m$F$3m$	4	Yes
$4mm$F$mm2$	2	—	$6mm$F$mm2$	3	—	$m3m$F$\bar{1}$	24	—
$\bar{4}2m$F1	8	Yes	$\bar{6}m2$F1	12	Yes	$m3m$F$2/m(s)$	12	—
$\bar{4}2m$F$2(s)$	4	Yes	$\bar{6}m2$F2	6	—	$m3m$F$2/m(p)$	12	—
$\bar{4}2m$F$2(p)$	4	—	$\bar{6}m2$F$m(s)$	6	Yes	$m3m$F$mmm(ss)$	6	—
$\bar{4}2m$Fm	4	Yes	$\bar{6}m2$F$m(p)$	6	Yes	$m3m$F$mmm(pp)$	6	—
$\bar{4}2m$F222	2	—	$\bar{6}m2$F$mm2$	3	Yes	$m3m$F$4/mmm$	3	—
$\bar{4}2m$F$mm2$	2	Yes	$6/mmm$F$\bar{1}$	12	—	$m3m$F$\bar{3}m$	4	—
$4/mmm$F$\bar{1}$	8	—	$6/mmm$F$2/m(s)$	6	—			
$4/mmm$F$2/m(s)$	4	—	$6/mmm$F$2/m(p)$	6	—			

TABLE IV—*Continued*

[a]422F2(*s*): See the notes in Table II.

422F2(*p*): The dyad axis of FG is along the tetrad axis of PG.

$\bar{4}$2*m*F2(*s*): See the notes in Table II.

$\bar{4}$2*m*F2(*p*): The dyad axis of FG is along the $\bar{4}$ axis of PG.

4/*mmm*F2/*m*(*s*): The dyad axis of FG is along one of the dyad axes (except the tetrad axis) of PG, or the mirror plane of FG is along one of the mirror planes parallel to the tetrad axis of PG.

4/*mmm*F2/*m*(*p*): The dyad axis of FG is along the tetrad axis of PG, or the mirror plane of FG is along the mirror plane perpendicular to the tetrad axis of PG.

622F2(*s*): See the notes in Table II.

622F2(*p*): The dyad axis of FG is along the hexad axis of PG.

$\bar{6}$*m*2F*m*(*s*): See the notes in Table II.

$\bar{6}$*m*2F*m*(*p*): See the notes in Table II.

6/*mmm*F2/*m*(*s*): The dyad axis of FG is along one of the dyad axes (except the hexad axis) of PG, or the mirror plane of FG is along one of the mirror planes parallel to the hexad axis of PG.

6/*mmm*F2/*m*(*p*): The dyad axis of FG is along the hexad axis of PG, or the mirror plane of FG is along the mirror plane perpendicular to the hexad axis of PG.

432F2(*s*): See the notes in Table II.

432F2(*p*): The dyad axis of FG is along one of the tetrad axes of PG.

432F222(*ss*): Two of the three dyad axes of FG are along dyad axes (other than the tetrad axes) of PG. (The remaining dyad axis of FG is along a tetrad axis of PG.)

432F222(*pp*): Two of the three dyad axes of FG are along tetrad axes of PG. (The remaining dyad axis of FG is also along a tetrad axis of PG.)

*m*3*m*F2/*m*(*s*): The dyad axis of FG is along one of the dyad axes (except the tetrad axes) of PG.

*m*3*m*F2/*m*(*p*): The dyad axis of FG is along one of the tetrad axes of PG.

*m*3*m*F*mmm*(*ss*): Two of the three mirror planes of FG are along mirror planes making an angle of 45° to tetrad axes of PG. (The remaining mirror plane of FG is along a mirror plane perpendicular to a tetrad axis of PG.)

*m*3*m*F*mmm*(*pp*): Two of the three mirror planes of FG are along mirror planes perpendicular to tetrad axes of PG. (The remaining mirror plane of FG is also along a mirror plane perpendicular to a tetrad axis of PG.)

TABLE V

EXAMPLES OF FERROIC CRYSTALS, SHOWING THE SPECIES, NUMBER OF STATES, ELECTRICAL, MECHANICAL, AND ELECTROMECHANICAL ORDERS

Substance	Temperature range (°C)	Species	No. of states	Electrical order	Mechanical order	Electro-mechanical order
$Ca_2Sr(C_2H_5COO)_6$	< 8.5	422F4	2	1st	2nd	1st
$LiH_3(SeO_3)_2$	< 110	2/mFm	2	1st	∞th	1st
KH_2PO_4	< -150	$\bar{4}$2mFmm2	2	1st	1st	1st
$ND_4D_2AsO_4$	< ca 30	$\bar{4}$2mF222	2	2nd	1st	1st
VO_2	< ca 70	4/mmmF2/m(s)	4	2nd (all)	1st (all)	1st (all)
$BaTiO_3$	rt	m3mF4mm	6	1st (all)	1st, ∞th	1st (all)
$NaBa_2Nb_5O_{15}$	< ca 300	4/mmmFmm2(p)	4	1st, 2nd	1st, ∞th	1st (all)
SiO_2 (quartz)	< 573	622F32	2	3rd	2nd	2nd
NaN_3	< ca 20	$\bar{3}$mF2/m	3	2nd (all)	1st (all)	1st (all)
Na_3AlF_6	< ca 550	m3mF2/m( )	12	2nd (all)	1st (all)	1st (all)
NH_4Cl	< -30.4	m3mF$\bar{4}$3m	2	3rd	∞th	2nd

to drive the crystal from one to the other orientation state. If this stress is within the elastic limit, then the stress:strain relation will become square hysteretic, and by analogy the species could be described as ferroelastic.

If $\Delta K_i = 0$ but $\Delta K_{ij} \neq 0$, then the orientation states differ in dielectric compliance, but not in polarization. Clearly, however, a high field will polarize these states differently, and thus a field may be chosen which will tend to drive the crystal from one to the other orientation state. If switching could be effected at a realizable field level, the term ferrobielectric is suggested to describe the species, the bi indicating the second order nature of the effect. Similarly it is obvious if $\Delta s_{ij} = 0$ but $\Delta s_{ijkl}X_{ij}X_{kl} \neq 0$ a similar mechanical case occurs, for which the term ferrobielastic is suggested.

Of particular interest would be the case in which only the Δd_{ijk} are nonzero. Here, of course, neither stress nor field alone will drive between the two orientation states, a case which is termed ferroelastoelectric!

In this way, Aizu classifies 212 species that he terms ferroic; of these, 88 (as have been shown previously) are ferroelectric. The 94 species which are fully ferroelastic are tabulated in Table IV. Clearly, since there is a coupling through the electromechanical constants, a number of the ferroelectric crystals are simultaneously ferroelastic, and these are indicated in the second column in the table. On the basis of this classification gadolinium molybdate would be termed a ferroelastic-ferroelectric. A number of examples of other different types of ferroic species are given in Table V (from Aizu).

VIII. THE PHENOMENOLOGY OF FERROELECTRICITY IN CERAMIC SYSTEMS

The development of phenomenological methods for the analysis of ceramic systems has, as yet, neither the rigor nor the generality of the corresponding single crystal treatments. First, we are hampered by the fact that for many of the compositions most interesting for use in, and most used in, ceramic systems, we do not yet have adequate single crystal data to generate the complete free energy function. Second, "cut and try" type body development has taken us, in many cases, to terribly complex systems for practical application, often multicomponent, polyphasic, deliberately heterogeneous systems where a multitude of dissimilar mechanisms may contribute to almost every measured property, and there is little hope of untangling directly the complex interactions.

At the moment we are constrained by these conditions to talk only about pure $BaTiO_3$ ceramics, and on these, the only property which has been extensively investigated and correlated is the weak field dielectric permittivity. Because of the nature of the domain walls in $BaTiO_3$ it seems probable that reversible wall motion does not contribute strongly to the real part of the permittivity at weak fields, though domain phenomena may contribute significantly to the dielectric loss.

If this basic assumption can be made, then the task of describing the weak field permittivity can be approached by first calculating the intrinsic, anisotropic permittivities of the crystallites under the constrained boundary conditions existing in the ceramic concerned, then averaging these properties for the random orientations of the crystallites in the ceramic.

Using a method of averaging developed by Bruggeman[16] the calculation was first carried out assuming that the crystallites behave essentially as free crystals. Marutake[17] in Japan then extended and improved the derivation, taking account of the fact that the crystallites constrain each other so that the full piezoelectric deformation of the free crystal may not be realized. Assuming spherical crystallites in an electrostrictive medium he recalculated the mean permittivity under the conditions of partial constraint, arriving at a more realistic value for the calculated behavior for a ceramic of normal grain size.

More recently, Dr. Buessem, Dr. Goswami and the author became interested in the grain size dependence of the weak field permittivity[18]. Based on the observation by Mrs. Hutchins at Sprague, that 90° domains do not occur in the fine grain $BaTiO_3$ when the grain size approached 1 μ, we were tempted to speculate that the dimension change at the Curie temperature, i.e., the developing tetragonality could not be relieved by twinning, so that the individual grains might be under stress. This argument is briefly summarized in Fig. 4.

Since the ceramic is formed at high temperature well above T_c where the symmetry is cubic, the sintering process will result in a closely interconnected structure which is largely strain free, each cubic grain being surrounded by a randomly oriented array of similar grains. Below T_c, the symmetry must change to tetragonal, but on average, because of the random arrangement in the crystallites no anisotropy can develop so that the mean environment must stay approximately cubic. For the individual grain, if tetragonality must develop, two types of effect may occur.

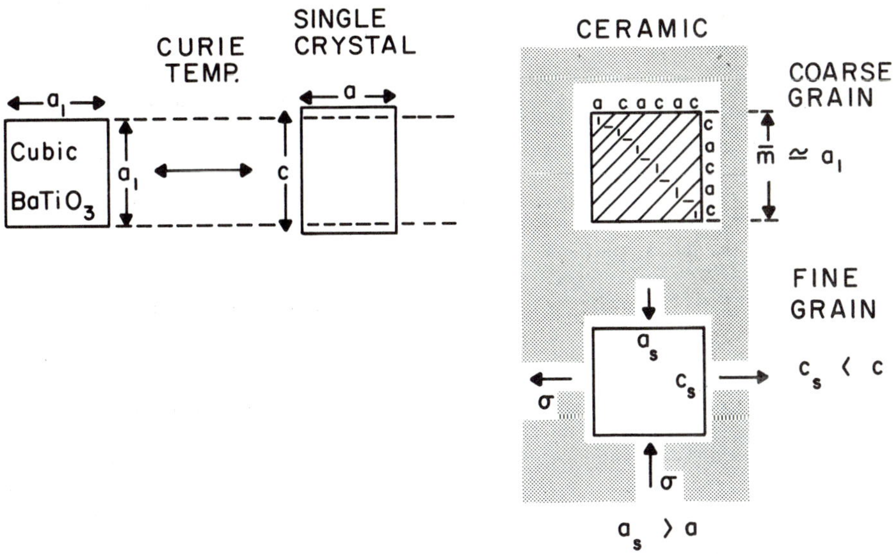

FIG. 4
Schematic diagram showing the origin of mechanical stress in a fine grain ceramic barium titanate sample.

1. If the grain size is large and 90° twins can occur, repeated twinning may average the dimensions of the grain so that little overall change takes place from the cubic, and no stress will develop.

2. If the grain size is small and 90° twins are excluded, then no dimensional accommodation is possible. On average, then, the grain will be subjected to a uniaxial compression along "c" and biaxial tensions along "a" directions, i.e., to a stress system tending to force it back towards the original cubic state.

Using the known thermodynamic functions it is easy to work out new permittivities for a range of the stress parameters, and to show that the type of stress system envisaged does in fact give rise to an enhancement of the permittivity of the type observed in the fine grain material.

One may note also that if the crystallites are under stress and one chooses a stress level of the correct order to account for the observed permittivity, the c/a ratio in the ceramic should be some 20% less than that in a corresponding unconstrained powder. This type of difference was, in fact, observed by Dr. Jonker and co-workers[19] at Phillips Laboratories.

It should be stressed that for a complete treatment of this problem the calculation of Marutake should be repeated for the new boundary conditions in the fine grain ceramic, and more refined methods used to average the resulting permittivities. However, in view of the drastic simplifications involved in the treatment of the very complex stress systems which must exist in real ceramics, the effort involved in the further calculation is barely justified.

One facet of this treatment which perhaps merits further consideration is the fact that it depends in quite a critical manner on the close approach of the orhorhombic-tetragonal phase change to room temperature, and on the resultant high values of ε_a, which are much modified by the stress system. Thus, on the basis of this hypothesis, strong grain size dependence in the form of enhancement of permittivity should be specific to $BaTiO_3$, and grain size dependence in this system should be markedly reduced by compositional modifications which would change the tetragonal-orthorhombic inversion temperature.

It must also be pointed out that the stress model is not the only macroscopic boundary condition modification which could account for the observed permittivity changes in fine grain $BaTiO_3$; as

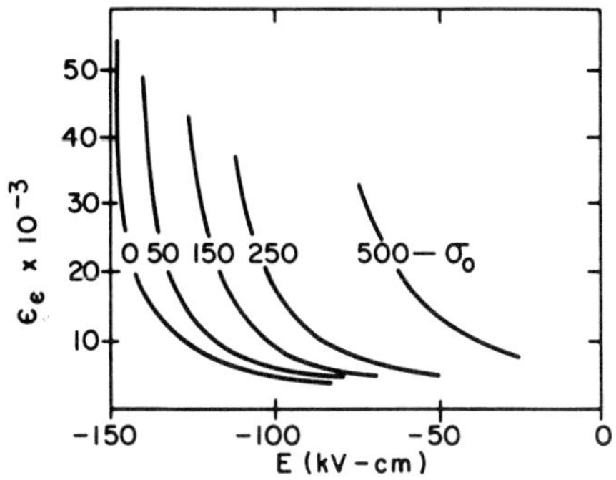

FIG. 5
Calculated arithmetic mean permittivity $\bar{\varepsilon}_\varepsilon$ as a function of reverse electric field at a series of internal stress values.

Goswami[20][21] has shown, other modifications can also enhance permittivity. For example, a reverse electric field, as shown in Fig. 5, or any number of combinations of field and stress, shown also in Fig. 5.

Since, however, there is a very plausible reason for the origin of the mechanical stress, and much less plausible arguments for the origin of a permanent electric field, the stress model looks to me much the more attractive.

Quite recently, Dr. Bradt at Pennsylvania State University has been extending the scope of the stress model, and examining the possibility of stress relief by slow changes in the twin structure as a possible explanation for some types of aging, and again though far from unequivocal, the explanation looks plausible and attractive.

In all the above, simple macroscopic considerations are being applied, and average properties determined from very large ensembles of unit cells, that is, from the bulk single crystal, are being used. Clearly, as the material is subdivided into smaller and smaller crystallites, there must come a size below which such treatments are unacceptable. Since ferroelectricity is a cooperative phenomenon, involving long range dipolar forces, there is good reason to fear that the breakdown of the macroscopic approach will occur at rather larger grain sizes than for many other simpler phenomena.

Unfortunately, it is very difficult to predict this size limit, which again will also depend quite critically on the boundary conditions. However, I would feel that below 0.5 μ grain size it is becoming doubtful that unperturbed macroscopic values will apply, and certainly by a size ∿ 0.1 μ one would expect the thermodynamic predictions to require drastic modification.

In very brief summary then, we have from the work of Aizu, and Fousek and Janovec, a conveniently tabulated frame of reference for all ferroelectric systems, and a ready set of tables from which to select new materials for ceramic study. From the quantitative energy function we have a shorthand description of all static intrinsic properties and the beginnings of a method from which to tackle the immensely complicated problem of a precise description of the ferroelectric ceramic.

REFERENCES

[1]. K. Aizu, Phys. Rev., 146, 423 (1966).

[2]. K. Aizu, Phys. Rev., 140, 590 (1965).

[3]. J. Fousek and V. Janovec, J. Appl. Phys., 40, 135 (1969).

[4]. K. Aizu, J. Phys. Soc. Japan, 23, 794 (1967).

[5]. K. Aizu, J. Phys. Soc. Japan, 27, 387 (1969).

[6]. J. Fousek, J. Phys. Soc. Japan (in press).

[7]. B. Aurivillius, Arkiv. Kemi, 1, 499 (1949).

[8]. S. E. Cummins and L. E. Cross, Appl. Phys. Let., 10, 14 (1967).

[9]. S. E. Cummins and L. E. Cross, J. Appl. Phys., 39, 2268 (1968).

[10]. L. E. Cross and R. Pohanka, J. Appl. Phys., 39, 3992 (1968).

[11]. A. Fouskova, L. E. Cross, and R. C. Pohanka, Electrochem. Soc. 134th National Meeting, Montreal, 1968.

[12]. A. Glanc, Z. Malek, J. Mastner, M. Novak, and J. Strajblova, J. Appl. Phys., 35, 1870 (1964).

[13]. H. J. Borchardt and P. E. Bierstedt, J. Appl. Phys., 38, 2057 (1967).

[14]. L. E. Cross, A. Fouskova, and S. E. Cummins, Phys. Rev. Let., 21, 812 (1968).

[15]. K. Aizu, Preprint Fall Mtg. Phys. Soc. Japan, 4, 28 (1969).

[16]. D. A. Bruggeman, Z. Physik, 92, 561 (1934).

[17]. M. Marutake, J. Phys. Soc. Japan, 11, 807 (1956).

[18]. W. R. Buessem, L. E. Cross, and A. K. Goswami, J. Am. Ceram. Soc., 49, 33 (1966).

[19]. G. H. Jonker and W. Noorlander, Science of Ceramics (G. H. Steward, Ed.), Academic Press, Inc., N. Y., 1962, p. 255.

[20]. A. K. Goswami, Ph.D. Thesis, The Pennsylvania State University, Sept. 1964.

[21]. A. K. Goswami, L. E. Cross, and W. R. Buessem, J. Phys. Soc. Japan, 24, 279 (1968).

Chapter 23

DYNAMICAL EFFECTS IN SOLID STATE PHASE TRANSFORMATIONS*

J. D. Axe

Brookhaven National Laboratory
Upton, New York

I. INTRODUCTION

It has been nearly ten years since Cochran and Anderson[1] suggested that the phase transformations in certain ferroelectrics might result from an instability in a normal vibrational mode of a lattice. It has been a very fruitful concept which has to a great extent revitalized the study of ferroelectric materials. However, the present discussion is concerned with some of the recent generalizations of these ideas. These generalizations were implied from the very beginning, but only recently have suitable experimental systems been found in which they may be studied.

---

*Work performed under the auspices of the U. S. Atomic Energy Commission.

It is necessary to define a generalized displacive phase transformation and this is done by reference to the concept of an order parameter, some quantity which is nonzero in one of the phases but vanishes in the other. We do this through the use of what may be called the phonon expansion given in Eq. (1)

$$\mathbf{u}_{\ell k}(t) = \sum_{\mu=(qj)} \xi_{k\mu} Q_{\mu}(t) \exp i\mathbf{q}\cdot\mathbf{X}_{\ell k} \tag{1}$$

which states that the displacement of the k*th* atom in the l*th* unit cell from some given reference position can be written as the sum of plane waves each with its own polarization vector ξ. Now there is no reason why such an expansion cannot also be used to describe static displacements, leading to static components of Q. Let a displacive transformation be one in which the order parameter can be chosen as the static component of such a normal phonon coordinate. In order to describe the structure which results when such a set of static displacements become incorporated into the lattice, we may say that a phonon has condensed out.

This discussion is concerned only with transitions which are nearly of second order, and this implies that there is a free energy function which can be expanded in a power series in the order parameter, and that this power series converges at least near the transformation temperature. In order to study the dynamics of such a phase transformation, it is necessary to look at the time dependent fluctuations in the order parameter. These fluctuations form some sort of excitation of the solid and, in the present case, these excitations are just the phonons themselves. Furthermore, we know that near a second order transformation, the amplitude of the fluctuations must diverge; since the amplitude of a harmonic oscillator is inversely proportional to its frequency we are led to the conclusion that the frequency of the phonons must tend towards zero at a second order phase transformation. The physics of the situation is now also clear because the frequency of a harmonic oscillator is a measure of the restoring force for the displacement and if this restoring force tends towards zero the displacements become free to incorporate themselves into the static structure.

How do these low frequency temperature dependent phonon modes, soft modes as they are called, come about? It is an anharmonic effect and, therefore, a difficult many body problem. The solution can be developed from the fact that the potential energy of a lattice can be expanded in a power series in a displacement starting with terms in u^2 then u^3 and so on. If the displacements are expressed in terms of normal coordinates using Eq. (1), then for simplicity neglecting the cubic term, the potential energy is given by Eq. (2)

$$V = \sum_{\mu} \frac{1}{2} \omega_{0\mu}^2 Q_\mu^2 + \sum V^{(4)}_{\mu_1\mu_2\mu_3\mu_4} Q_{\mu_1} Q_{\mu_2} Q_{\mu_3} Q_{\mu_4} + \ldots \tag{2}$$

which, although complicated, simplifies if we ask for the mean potential experienced by the μth normal mode.

$$\langle V \rangle_\mu = \frac{1}{2} \omega_{0\mu}^2 Q_\mu^2 + \sum_{\mu_1} V^{(4)}_{\mu_1\mu_1\mu\mu} \langle Q_{\mu_1} \rangle^2 Q_\mu^2 + \ldots \tag{3}$$

where the cross products between normal coordinates of different modes have disappeared because to, at least, lowest order, normal modes are independent. Equation (3) has a harmonic part $\frac{1}{2}\, \omega_{0\mu}^2 Q_\mu^2$

another part which is contributed by the fluctuations from all the other modes in the system. But both of these contributions are proportional to the square of the normal coordinate of the mode itself and, therefore, can be taken to define a quasi-harmonic frequency

$$\omega_{\mu}^{2}(T) = \omega_{0\mu}^{2} + \sum_{\mu_1} V^{(4)}_{\mu_1\mu_1\mu\mu} \left(2n(\mu_1)+1\right) + \ldots \tag{4}$$

using the well-known identity for the thermal mean square phonon amplitude.

In normal materials at moderately high temperatures, the harmonic part of Eq. (4) is always considerably greater than the anharmonic part which supplies a small temperature dependent correction. However, occasionally, materials are found in which a very delicate balance exists in which a negative harmonic contribution is compensated by a positive anharmonic contribution. Such a material would be unstable if the lattice could be held static but it is actually stabilized by the fluctuations of the other phonons. (Notice that there are zero point effects. The anharmonic term does not disappear at low temperatures.) Sometimes, however, the balance becomes more delicate still, and the zero point anharmonic contribution is not enough to stabilize the lattice configuration but may do so at some higher temperature where the fluctuations are greater. At high temperatures, the phonon occupation number is proportional to the absolute temperature and in such a case, we may approximate Eq. (4) as

$$\omega_{\mu}^{2}(T) = \alpha(T-T_0) \tag{5}$$

Such a material will undergo a second order displacive phase transformation at the temperature T_0. Equation (5) is the correct high temperature molecular field result even including cubic anharmonicity and gives a Curie-law susceptibility $\chi(T)$ which diverges at the temperature T_0 at which the lattice becomes unstable.

Because a very general definition of a displacive transformation has been given here, it may be seen that ferroelectrics comprise a very special subset of the general case. A ferroelectric must involve a condensation of a phonon which is both polar and of long wavelength because the ferroelectric state must have a macroscopic polarization. Clearly, in a material there are many more phonons without these special properties than there are with them, and they are a priori equally capable of becoming unstable, so that it might be expected that more examples of nonferroelectric displacive transformations than ferroelectric ones would be found. Ultimately this may perhaps be the case. However, whereas ferroelectrics have a divergent macroscopic property, namely, the dielectric response, to signal the occurrence of the transformation, phonon instabilities involving shorter wavelength phonons in general have no such marker and tend to be rather subtle. Only very recently has the first really clearcut example of such a generalized displacive phase transformation been definitely established. This work will be discussed in some detail since in some respects, it furnishes a better example of the dynamical effects associated with displacive transformations than any of the ferroelectric transformations that have thus far been studied.

II. STRONTIUM TITANATE TRANSFORMATION AT 110°K

The first indication of a phase transformation in strontium titanate came from the observation of discontinuous behavior of the elastic constants[2]. Subsequent X-ray and optical work[3] established that at about 110° stontium titanate transforms from a high temperature cubic perovskite form into a low temperature tetragonal form with a c/a ratio differing from unity by less than 1 part in 10^3. Because the transformation is such a slight one and also because, as we shall see, it involves only oxygen motion for which there is very little scattering power, there was no immediate X-ray crystallographic evidence concerning the nature of the internal atomic rearrangements. Some five years after the transformation was discovered, Unoki and Sakudo proposed a structure based upon their study of the splitting of the electron spin resonance spectrum of paramagnetic impurities[4]. The displacements which they proposed consist of rotation of the oxygen octahedron surrounding every titanium ion about a [100] axis. However, since the oxygen atoms are shared between octahedra, the sense of the rotation alternates on passing between adjacent unit cells. The new unit cell has twice the volume of the old original cubic one. The authors were quite aware of the dynamical implications of their model and stated that "it seems to suggest the occurrence of the instability of a phonon branch at large wave vector."

Once the atomic displacements are known, there exists a very practical test due to Cochran[5] for determining whether or not a displacive transformation is involved. For if the low temperature phase does indeed represent the condensation of a single normal phonon mode then the static displacements should have all of the symmetry properties of the condensed phonon, about which certain well-defined group theoretical statements can be made. If indeed a one to one correspondence is found between the symmetry properties of the static displacements and a single normal phonon mode, then the implication

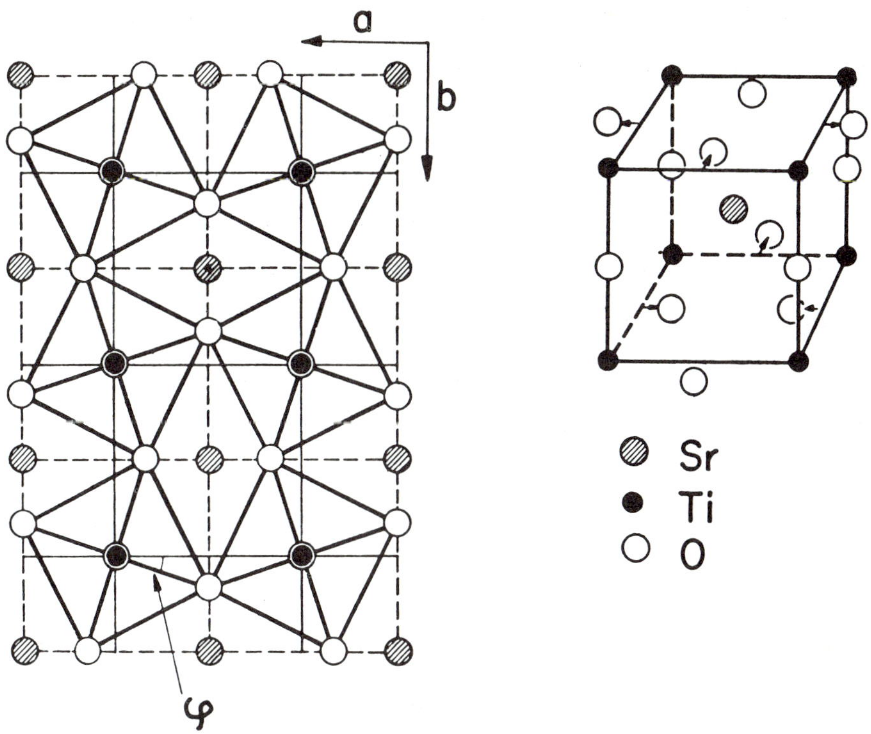

FIG. 1
Rotational displacements of the oxygen octahedra in $SrTiO_3$ which occur at the 110°K phase transformation.

is certainly strong that that mode is involved in the transformation and is thus soft.

Fleury, Scott, and Worlock[6] who were trying to understand the Raman scattering spectrum of strontium titanate took up Unoki and Sakudo's proposal and showed that the displacement suggested did indeed represent a single normal phonon mode and that the wave vector of the phonon was such that it lay at the [111] corner of the simple cubic Brillouin zone. This is a special point of very high symmetry in the reciprocal lattice and is known as the R point. (It is easy to see that the wave vector involved must be at the edge of the Brillouin zone for it is only for such wave vectors that the displacements reverse on passing from one unit cell to the next.)

The most striking aspect of the Raman spectrum is the appearance of many lines below the transformation temperature and the most interesting of these are shown in Fig. 2. Fleury et al. explained these features as follows: Above the transformation temperature, the soft phonons are triply degenerate (rotations about all three principal axes must be equivalent). These phonons are not directly visible in a light scattering experiment, however, because the wavelengths are much shorter than those of the exciting light. Below the transformation temperature, the unit cell has doubled and

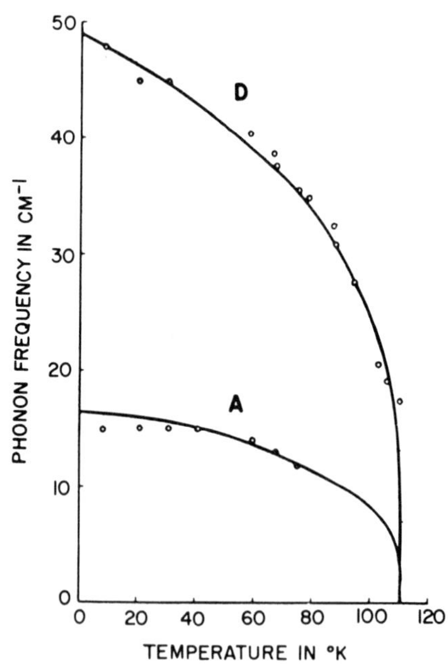

FIG. 2
Temperature dependence of the soft-phonon modes in tetragonal $SrTiO_3$. Smooth curves are of the form ω = const $\times(T_o-T)^n$ with T_o = 110°K, n = 0.31. After Fleury, Scott, and Worlock, Ref. [6].

the Brillouin zone, which is complementary, has halved in volume in such a way that the R point folds into the center of the zone. Since the phase of the displacements do not reverse in passing from one of the new larger unit cells to the next, it is properly counted as a long wavelength excitation and can, thus, now cause first order Raman scattering. Furthermore, the condensed distortion removes the cubic symmetry and lifts the threefold degeneracy of the modes thus accounting for two of the lines instead of one. The modes are stabilized by increasing distortion of the lattice, thus, accounting for the temperature dependence of the observed lines.

This proposal fit all of the known spectroscopic and structural data. What was lacking was a direct verification that there was a soft phonon in the cubic phase with a wave vector at the R point. The model was specific enough, however, that it could be easily tested by inelastic neutron scattering which is the only technique by which spectroscopy on such short wavelength excitations is possible. Independently, groups at both Brookhaven[7] and Chalk River [8] put together a complete picture. Figure 3 shows in more detail changes in the Brillouin zone implied by the Unoki-Sakudo model. The Brillouin zone of the simple cubic lattice shown by the dotted lines, enclose the reciprocal lattice point, Γ. By erecting perpendicular bisectors between this point and the new reciprocal lattice points, R, the Brillouin zone appropriate to the low temperature structure is obtained and can be recognized as a polyhedron of the body-centered type showing that the direct lattice of the new phase is face centered cubic. The slight spontaneous strains mentioned earlier which accompany the transformation are ignored here. Also shown in this figure is a cross section of the extended Brillouin zone showing reflections of the type [hℓℓ]. Here, the reciprocal lattice points of the simple cubic cell are shown in black , the superlattice R points by open circles. Concentrating on scattering about the superlattice points, from the previous discussion it would be expected that at high temperatures inelastic scattering due to the soft phonon modes would be seen but that as the temperature is lowered to the phase transition, the scattering should change to elastic Bragg scattering. Figure 4 shows how this evolution from inelastic to elastic scattering comes about (taken from the work of Shirane and Yamada)[7]. Some 100° above the transformation temperature there is a well resolved neutron group at about 5 meV corresponding to phonon creation and another symmetric with it about E = 0 but not shown due to phonon annhilation. At 5° above the transition temperature, these two phonon peaks have collapsed in towards the origin and are no longer completely resolved. This is the scattering from the soft phonon mode. (The sharper peak at E = 0 is due to a small amount of incoherent elastic scattering and is not relevant to the present discussion.) As the temperature is further lowered, below the phase transformation, a much stronger elastic Bragg peak grows out of the center of the phonon scattering marking the condensation of the static displacement. By studying the intensity of the elastic scattering at different superlattice points in the reciprocal lattice, Shirane and Yamada were able to verify that the static displacements did indeed correspond to a rotation of the oxygen octahedra and determined the rotational angle at 77°K to be 1.4° in good agreement with the spin resonance data. At temperatures, above the transition temperature, the square of the soft mode frequency is proportional to $T - T_0$, just the molecular field result. Below the

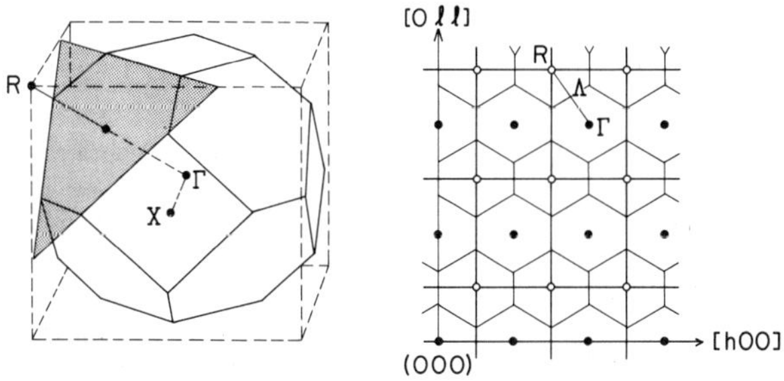

FIG. 3
Left: Brillouin zone of simple cubic perovskite lattice (dotted line), showing construction of new B.Z. in which R is also a reciprocal lattice point. Right: Cross section of B.Z. showing reflections of type [hℓℓ] studied by neutron scattering.

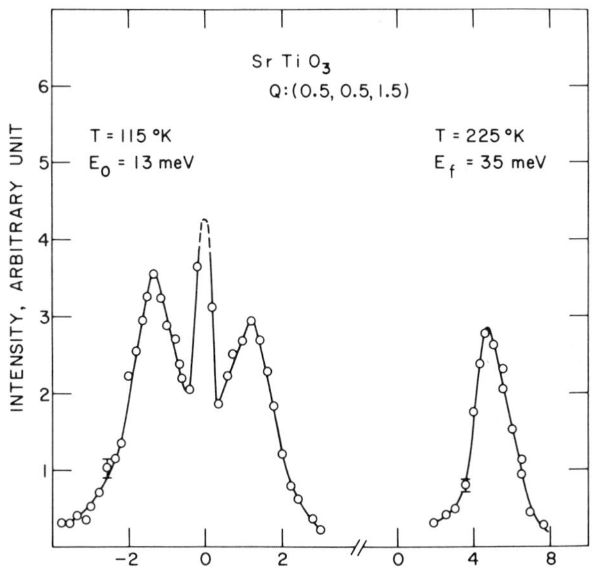

FIG. 4
Energy analysis of neutron scattering due to soft mode at B.Z. corner in $SrTiO_3$ at two temperatures. The sharp central peak corresponds to a small amount of incoherent scattering and is not of interest in the present context. After Shirane and Yamada, Ref. [7].

transition temperature, Fleury et al.[6] suggested $\omega^2 \propto (T - T_0)^{0.6}$. However, Pytte and Feder[9] have shown that a self-consistent molecular field model fits equally well. It appears to be an unanswered question at this time whether one can expect to observe a breakdown of the molecular field approximation at temperatures near T_0. This effect is well known in magnetic systems and is being extensively studied. In Fig. 5 the dispersion curves are shown for the two lowest transverse

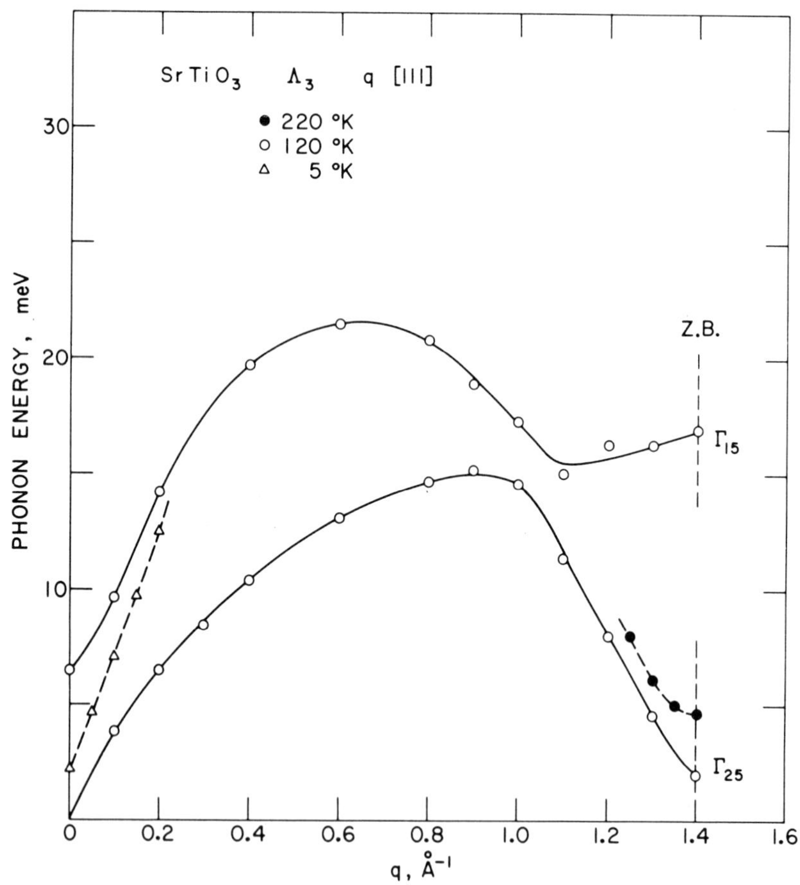

FIG. 5
Dispersion curve for the lowest transverse phonon branches in cubic $SrTiO_3$. After Shirane and Yamada, Ref. [7].

phonon branches with wave vectors along the [111] direction in strontium titanate. These curves can be best understood by realizing that they represent a normal acoustic branch starting at zero frequency at the center of the Brillouin zone and ending at the point marked Γ_{15} at the Brillouin zone edge. The optic mode, however, has quite an anomalous behavior. Its frequency is low and temperature dependent both at the center of the Brillouin zone and at the Brillouin zone edge and it rises to a maximum in the interior. Since the two branches have the same symmetry properties, they interact and repel one another in the region of the nominal crossing. Thus in strontium titanate the optical modes at the Brillouin zone center compete with those at the Brillouin zone edge to see which becomes unstable first. Had the zone center modes won, strontium titanate would have undergone a ferroelectric phase transformation, as it nearly does in any event.

III. THE LANTHANUM ALUMINATE TRANSFORMATION

Since the clarification of the mechanism of the strontium titanate transformation, several examples of similar transformations have been found in other systems. One which was studied by Shirane, Muller, and myself[10] occurs in lanthanum aluminate ($LaAlO_3$) which is also a cubic perovskite above 535° but transforms into a rhombohedral phase below that temperature. The dispersion curves for the [111] transverse phonons is very similar to that shown in Fig. 5 for strontium titanate insofar as there is a low frequency temperature dependent mode at the [111] zone boundary. Condensation of this mode clearly accounts for the transformation. There are, however, interesting differences between the transformation in strontium titanate and $LaAlO_3$ the first being that the same mode instability gives rise to two different structures, tetragonal in strontium titanate, and rhombohedral in $LaAlO_3$. Second, there are differences in the phonon line widths in the two cases, having to do with the relative magnitude of anharmonic damping. We will discuss this latter effect first.

The expression which determines the intensity of scattering due to a single phonon with momentum transfer $\hbar Q$ and energy transfer, $\hbar\omega$ is given by

$$\frac{d^2\sigma}{d\Omega d\omega} = \left|\frac{k'}{k}\right| \left|F_{in}(Q)\right|^2 \left(1 + n(\omega)\right) A(\omega) \tag{6}$$

where k′ and k are the momenta of the scattered and incident neutrons $1 + n(\omega)$ is the thermal amplitude of the phonons and $A(\omega)$, the spectral correlation function, is the measure of the power spectrum of these fluctuations. In lowest order anharmonic perturbation theory, $A(\omega)$ is given by

$$A(\omega) = \frac{1}{\pi} \left(\frac{\omega\Gamma}{(\omega_0^2 - \omega^2)^2 + (\omega\Gamma)^2}\right) \tag{7}$$

where ω_0 is the same quasi-harmonic frequency we have been discussing and Γ is a damping parameter which is itself a function of frequency and reflects the probability of possible scattering processes for the phonon. If Γ is small, the scattering cross section is a pair of Lorentzians centered about $\pm\omega_0$. However, with larger damping the peaks shift away from ω_0 toward the origin finally merging into a single overdamped peak centered about $\omega = 0$. The same behavior can be observed classically in a viscously damped pendulum. It is easily proved that damping does not affect the dynamical stability.

The instability condition for a damped or undamped oscillator is the same, namely $\omega_o^2 = 0$. As long as $\omega_o^2 > 0$ the system will recover if pushed away from its zero position. However, it is no longer true that one can determine the quasi-harmonic frequency from the peak of the scattering profile in the presence of heavy damping.

The degree of damping ordinarily encountered in lattice vibrations is so small that these effects are hardly visible except near displacive phase transformations (and this is one of the reasons for studying them). Although the damping effects are no larger than normal, the quasi-harmonic frequencies are extraordinarily low, and relatively, therefore, the damping effects can be very considerable. This is illustrated very clearly in the soft mode condensation in $LaAlO_3$, shown in Fig. 6. The scattering is symmetrical about E = 0 and only half is shown. The data were taken at three temperatures above that of the transition and the sharp peak at E = 0 is not true Bragg scattering but is rather elastic incoherent scattering and can be ignored. Just how well this data reflects the gradual overdamping of anharmonic oscillator can be seen from the curves drawn through the points which are least squares fits of Eqs. (6) and (7), treating ω_o and Γ as adjustable parameters. The values of ω_o obtained are indicated by arrows in the figure. The Γ turns out to be nearly temperature independent. The fit is not perfect but the deviations are about the same order as resolution corrections which could not be accurately made in this case. Although the quasi harmonic frequencies are not directly obtained from the peak scattering intensity it can be obtained indirectly because as may be shown from Eqs. (6) and (7),

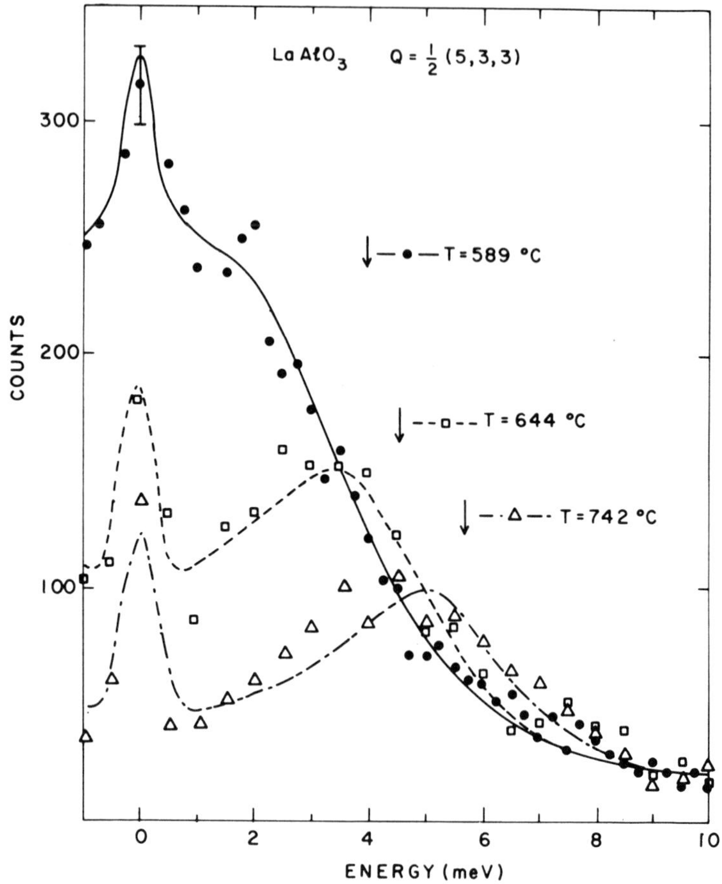

FIG. 6
Energy profile of quasi-elastic scattering about a super-lattice point in cubic $LaAlO_3$. The energy of the scattered neutrons was fixed at 25 meV. The narrow central component is due to elastic incoherent scattering from the sample and apparatus. The points represent experimental measurements and the lines best fitting curves of the form of Eq. (7).

the integrated intensity is independent of the damping and inversely proportion to ω_0^2. From the temperature dependence of the integrated intensity, we are able to show that ω_0^2 is proportional to $T - T_0$, again the molecular field result.

Consider now the differences in the structures of $SrTiO_3$ and $LaAlO_3$. Not only is low temperature $LaAlO_3$ rhombohedral, but the atomic displacements as determined by X-ray and ESR studies[11] consist of rotations of oxygen octahedra about [111] directions rather than [100] directions. How can these two different structures result from the same soft mode condensation? The answer is that this mode condensation at the R point consists of not a single mode but three triply-degenerate ones. Which linear combination of these modes best stabilizes the resulting structure is determined by the anharmonic forces present in the lattice in a way that can be very simply demonstrated[12]. Let Q_x represent the normal coordinate for oxygen rotation about the x axis, Q_y the rotation about y axis, and so on as shown in Fig. 7. Then by hypothesis, the difference in the free energy of the cubic and distorted phases may be represented by a power series expansion in the Q's as shown in Fig. 7. Only those combinations of Q's which have cubic symmetry

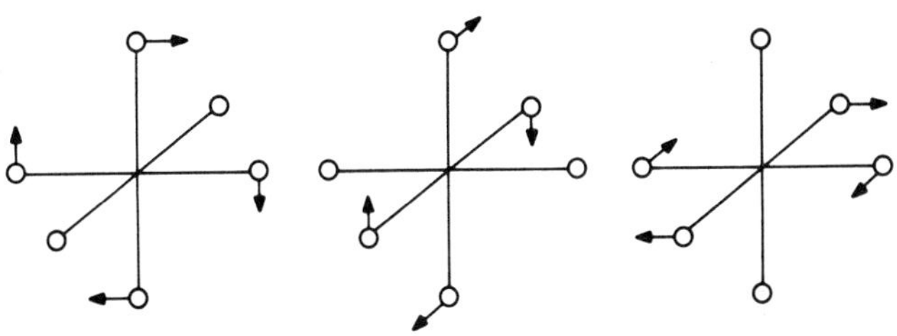

FIG. 7

Three order parameters representing oxygen rotation about [100] directions in cubic perovskites. The free energy due to such displacements can be represented in terms of a power series in Q.

$$F = \frac{1}{2} S^2 \left(Q_S^2 + Q_S^2 + Q_S^2\right) + \frac{1}{4} V_4 \left(Q_X^4 + Q_Y^4 + Q_Z^4\right) + \frac{1}{4} V_4' \left(Q_X^2 Q_Y^2 + Q_Y^2 Q_Z^2 + Q_Z^2 Q_X^2\right)$$

+ sixth and higher order terms.

can appear and there is one quadratic invariant, two fourth order invariants, and so on. The coefficient of the quadratic invariant is essentially the quasi-harmonic frequency which we now call ω_S. To determine the physically acceptable solutions, it is sufficient to minimize the energy with respect to $Q_{x,y,z}$ (see Fig. 8). Solution I, which is always of lowest energy provided ω_S^2 is greater than zero, is simply the undistorted cubic phase. When ω_S^2 is negative [see Eq. (5)] other solutions become possible (ω_S is the phonon frequency in the cubic phase; in the distorted phases $\omega^2 = \omega_S^2$ + renormalizing terms due to the displacements). There is first of all, a solution of type II in which one mode condenses out with amplitudes shown, giving rise to the strontium titanate structure. Another solution of type IV is possible in which all modes condense out at equal amplitudes. The net result is a rotation in about a [111] direction as occurs in $LaAlO_3$. Whether the type II or type IV solution is the most stable is determined by relative values of the quartic anharmonic coefficients as is shown in the lower portion of Fig. 8. For some ratios of V_4 to V_4' only first order transitions are possible. Solution III which corresponds to an oxygen rotation about a [110] axis gives rise to a saddle point, not a true minimum in the free energy. One thing that is left out of this simple treatment is the coupling of the internal displacements to macroscopic strains. This has been carried out by Slonczewski and Thomas [13] and predicts not only the correct spontaneous strains in the two cases, but also predicts elastic constant anomalies of the type which have been observed in $SrTiO_3$. In fact, it turns out that it is mainly the elastic term

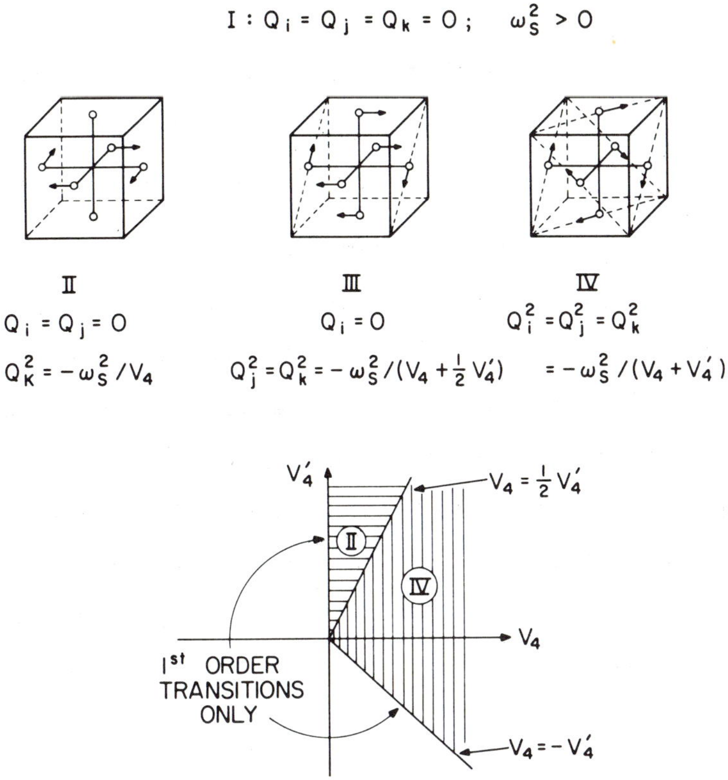

FIG. 8
Possible energetically favorable distortions. Solution I is stable above T_c. Which solution is most stable below T_c depends upon the ratio of the anharmonic constants V_4 and V_4'.

in the free energy which favors the tetragonal form over the rhombohedral form in $SrTiO_3$. It appears possible with these results to understand the switching from tetragonal to rhombohedral form under uniaxial stress that has recently been observed by Burke[14]. Not surprisingly, perhaps, Scott[15] has found evidence for similar transformations in several other rare earth aluminates and Minkiewicz and Shirane[16] have shown by a neutron study that $KMnF_3$ also shows such a transformation. The present discussion bears a considerable similarity to the treatment given by Devonshire[17] for a ferroelectric transformation in $BaTiO_3$. The similarity is not accidental and results from the fact that the symmetry of the Γ point and the R point of the cubic Brillouin zone are identical.

IV. THE α-β QUARTZ TRANSFORMATION

Figure 9 shows a projection upon the basal plane of the quartz structure and the arrows indicate the differences in the atomic position between the high temperature hexagonal β form and the low temperature trigonal α form. By a symmetry analysis, it may be shown that it is possible to almost exactly duplicate these displacements by a single normal phonon mode which result strongly suggests that a phonon instability is involved. Since the size of the unit cell does not change, the phonon would be of long wave-

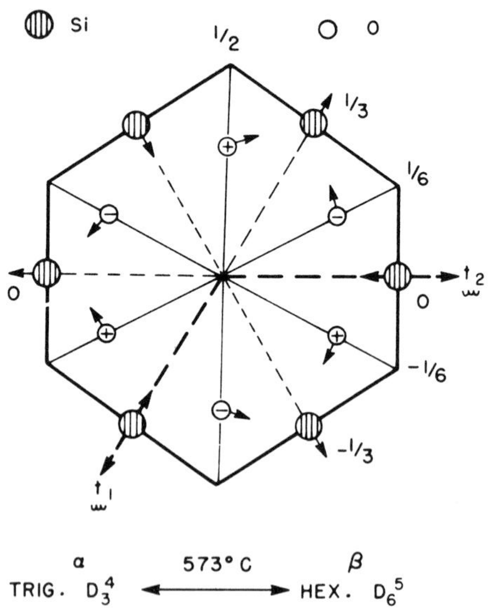

FIG. 9
A projection on the basal plane of the unit cell of β-quartz. The arrows indicate the shifts in atomic positions between β-SiO_2 and room temperature α-SiO_2.

length (zone center) but would be neither infrared nor Raman active in the high temperature phase, although it would be expected to cause Raman scattering in the low temperature phase. Cummings, O'Shea and Shapiro[18] have seen a soft temperature dependent mode which disappears abruptly at the transition temperature. Until this year, however, there had been no evidence of the existence of soft modes or critical behavior above this temperature. One advantage of neutron spectroscopy over conventional optical spectroscopy is that the selection rules are much less stringent. In fact, knowing the kind of displacements that can be expected, Shirane and the present author[19] were able to calculate, using Eq. (6), exactly where in the reciprocal lattice such scattering would be most favorable. Figure 10 shows an example of the scattering seen

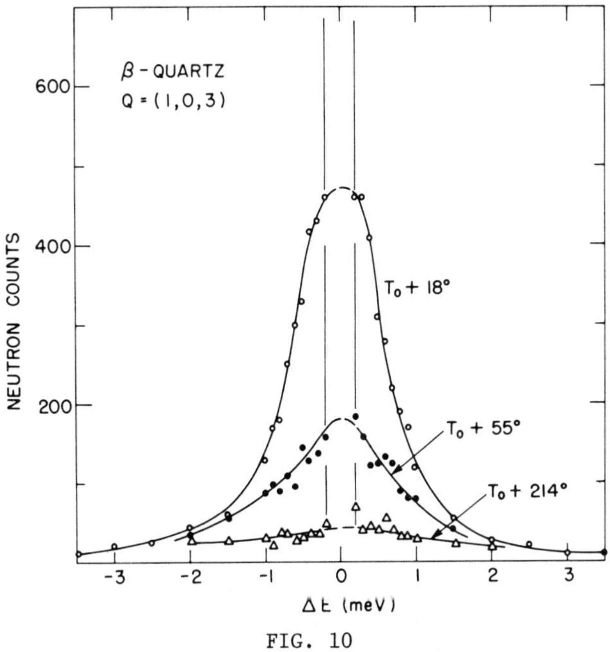

FIG. 10

Energy analysis of scattering about the (1,0,3) reflection in β-SiO at several temperatures above T_0. The strong central Bragg peak is superimposed upon weaker inelastic critical scattering.

about one such reciprocal lattice point in β-quartz just above the transformation temperature. We interpret this as scattering from an overdamped phonon similar to that observed in $LaAlO_3$. Here, however, the scattering exists exactly at the same place in the reciprocal lattice as does the Bragg scattering which accounts for the huge central component in the figure. Also, unlike $LaAlO_3$, we are unable to undamp this excitation with increasing temperature. We are, however, able to see a well resolved temperature dependent excitation as we move away from the center of the Brillouin zone in the basal plane, although the frequency remains quite low. The phonons traveling along the c axis appear to stay overdamped even up to very short wavelengths (Brillouin zone edge). The integrated intensity of the scattering shown in Fig. 10 follows a Curie law behavior with T_c about 10° below the transition temperature T_o showing that the transformation is slightly of first order.

By studying the intensity of the scattering (such as shown in Fig. 10) about many reciprocal lattice points, the atomic motion involved in the vibration may be deduced and in the present case, may be compared with the static displacements accompanying the phase transformation. Scattering was rather weak and the problem of background and Bragg scattering removal was severe. Nevertheless, by studying 13 phonon reflections, it was possible to define the displacements with moderate accuracy and to show that they do indeed strongly resemble the spontaneous displacements. A much more accurate mode determination was possible in the low temperature phase where the soft mode does become undamped.

Finally, a consistent explanation could be developed for the striking elastic anomaly (see Fig. 12) which accompanies the

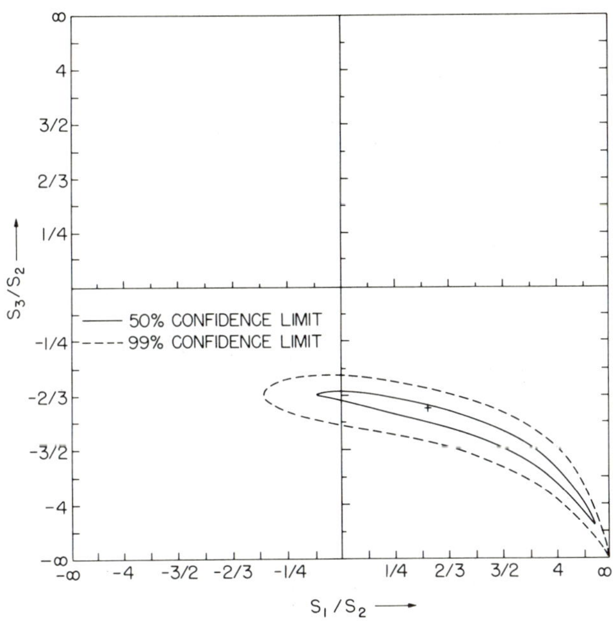

FIG. 11

Determination of the relative atomic displacements associated with the critical scattering in β-SiO . For a description of the coordinate system used to describe the displacements, see Ref. [19]. The spontaneous displacements associated with the β → α transformation are indicated by the cross.

phase transformation. The temperature dependent parts of these elastic constants obey special relations not required by the overall symmetry of the crystal[20] and these relations may be understood in the following way. Straining the lattice in certain directions can give rise to the virtual excitation of a pair of soft optical phonons traveling in opposite directions along the c axis. The strength of this interaction depends inversely upon the fourth power of the soft phonon frequency and thus increases rapidly near the transition temperature. The effect is an anharmonic one, however, and is basically small but because there are so many modes with low frequency along the c axis, the net effect can be quite large.

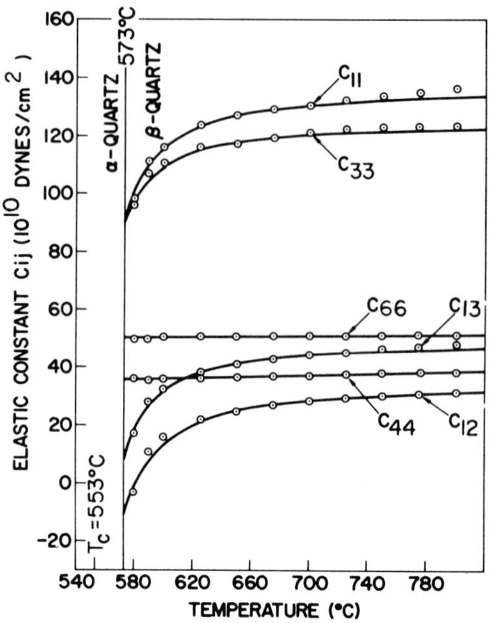

FIG. 12

The temperature dependence of the elastic constants of β-SiO . The solid lines are given by $c_{ij} = c_{ij}^{o} - c_{ij}/(T-T_o')$ with $\Delta c_{11} = \Delta c_{12}$; $(\Delta c_{11}\Delta c_{33}) = (\Delta c_{13})^2$; $\Delta c_{44} = \Delta c_{66} = 0$; and $T_o' = 533°C$. After Miller and Axe, Ref. [20].

Finally it would appear that phase transformations which are precipitated by vibrational instabilities of the lattice are quite general and by no means restricted to ferroelectrics. Judging by the rate at which they are presently being discovered, such systems may turn out to be rather common, but we have much to learn from them both from the point of view of phase stability and in the behavior of very heavily deamped excitations.

REFERENCES

[1]. W. Cochran, Advan. Phys., 9, 387 (1960); P. W. Anderson, in Fizika Dielectrikov (G. I. Skanovi, ed.) Akad. Nauk SSSR Moscow, 1960.

[2]. R. O. Bell and G. Rupprecht, Phys. Rev., 129, 90 (1963).

[3]. F. W. Lytle, J. Appl. Phys., 35, 2212 (1964).

[4]. H. Unoki and T. Sakudo, J. Phys. Soc. Japan, 23, 546 (1967).

[5]. W. Cochran and A. Zia, Phys. Status Solidi, 25, 273 (1968).

[6]. P. A. Fleury, J. F. Scott, and J. M. Worlock, Phys. Rev. Letters, 21, 16 (1968).

[7]. G. Shirane and Y. Yamada, Phys. Rev., 177, 858 (1969).

[8]. R. A. Cowley, W. J. L. Buyers, and G. Dolling, Solid State Commun., 7, 181 (1969).

[9]. E. Pytte and J. Feder, to be published.

[10]. J. D. Axe, G. Shirane, and K. A. Müller, Phys. Rev., 183, 820 (1969).

[11]. S. Geller and V. B. Bala, Acta Cryst., 9, 1019 (1956); K. A. Müller, W. Berlinger, and F. Waldner, Phys. Rev. Letters, 21, 814 (1968).

[12]. H. Thomas and K. A. Müller, Phys. Rev. Letters, 21, 1256 (1968).

[13]. J. C. Slonczewski and H. Thomas, to be published.

[14]. W. J. Burke and R. J. Pressley, to be published.

[15]. J. F. Scott, Phys. Rev., 183, 823 (1969).

[16]. V. J. Minkiewicz and G. Shirane, J. Phys. Soc. Japan, 26, 674 (1968).

[17]. A. F. Devonshire, Adv. in Phys., 3, 86 (1954).

[18]. S. M. Shapiro, D. C. O'Shea, and H. Z. Cummins, Phys. Rev. Letters, 19, 361 (1967).

[19]. J. D. Axe and G. Shirane, to be published in Phys. Rev.

[20]. P. B. Miller and J. D. Axe, Phys. Rev., 163, 924 (1967).

VIII. MAGNETIC CERAMICS

Chapter 24

THEORY OF ANTIFERROMAGNETISM AND FERRIMAGNETISM*

John B. Goodenough

Lincoln Laboratory
Massachusetts Institute of Technology
Lexington, Massachusetts

---

*This work was sponsored by the Department of the Air Force.

I. INTRODUCTION

Transition-metal oxides and fluorides have played an important role in the development of a theory of antiferromagnetism and ferrimagnetism. Magnetic susceptibilities like that of MnF_2 shown in Fig. 1 or that of the ferrospinels, which are similar to that for the ferrogarnet $Y_3Fe_5O_{12}$ shown in Fig. 2, led Néel[1] to propose his two-sublattice, collinear spin model of antiferromagnetism, and ferrimagnetism that has proven so useful for the interpretation of a large class of commercially important magnetic ceramics. Subsequently neutron-diffraction measurements provided direct confirmation of this model for these and many other transition-metal oxides and fluorides. The spin configurations found for the antiferromagnet MnO and for the ferrimagnetic spinel $Fe^{3+}[Ni^{2+}Fe^{3+}]O_4$ are shown in Figs. 3 and 4. In contrast to the ferromagnetic compounds, which

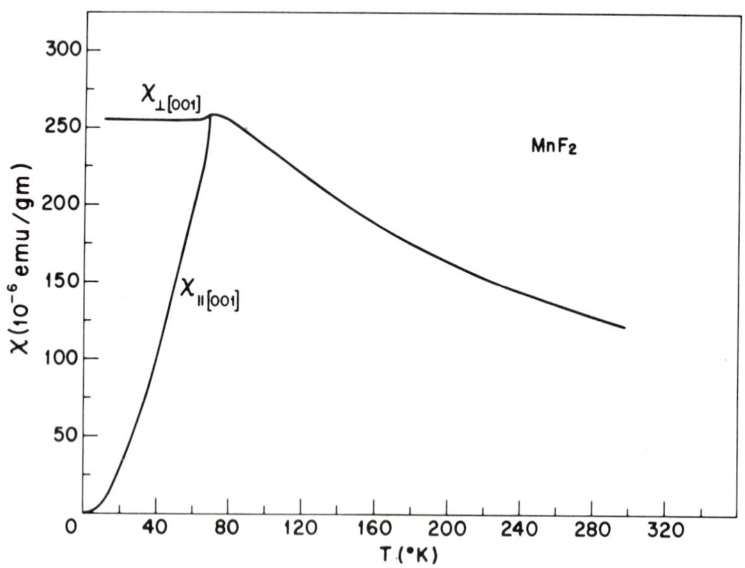

FIG. 1
Magnetic susceptibility versus temperature for single-crystal MnF_2. (After S. Foner, private communication.)

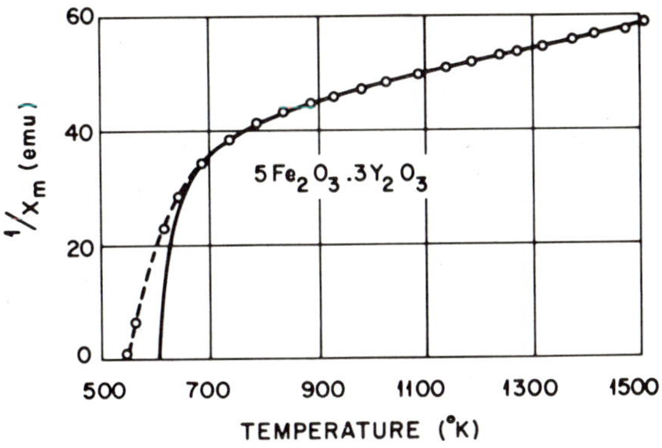

FIG. 2

Inverse paramagnetic susceptibility versus temperature for the yttrium ferrogarnet $Y_3Fe_5O_{12}$. Solid line is theoretical curve based on Néel model and empirically determined molecular-field constants. Deviation from experimental curve just above T_c is due to short-range order. [After R. Aléonard, J. Phys. Chem. Solids, 15, 167 (1960).]

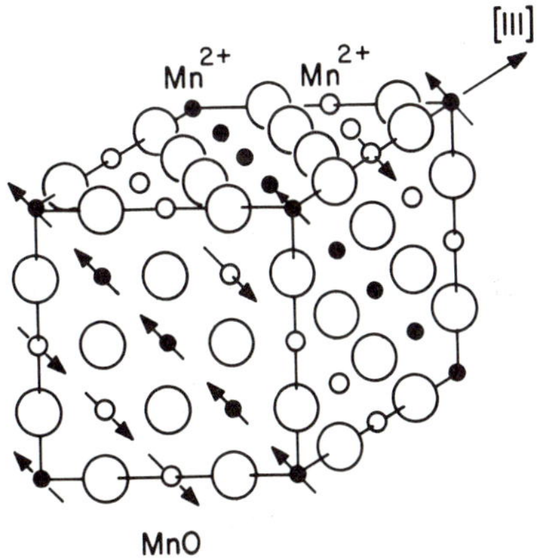

FIG. 3

Spin configuration found in antiferromagnetic MnO by neutron diffraction: C. G. Shull, W. A. Strauser, and E. O. Wollan, Phys. Rev., 83, 333 (1951), Spins lie in the (111) planes.

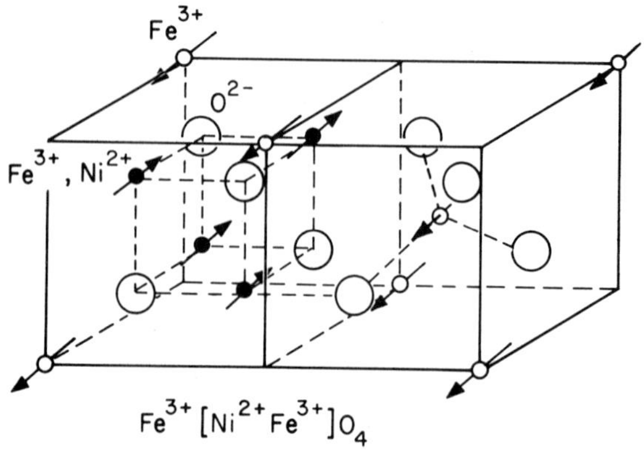

FIG. 4
Spin configuration found in ferrimagnetic $NiFe_2O_4$ by neutron diffraction: J. M. Hastings and L. M. Corliss, Revs. Modern Phys., 25, 114 (1953). One-quarter spinel unit cell. Spins are parallel to [111] axis.

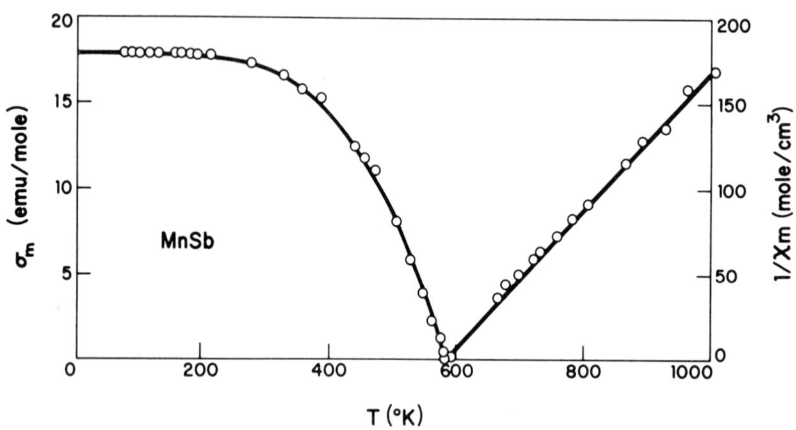

FIG. 5
Inverse paramagnetic susceptibility and magnetization versus temperature for ferromagnetic MnSb. [After T. Hirone, S. Maedar, I. Tsubokawa, and N. Tsuya, J. Phys. Soc. Japan, **11**, 1083 (1956).]

typically exhibit a temperature-dependent spontaneous magnetization and magnetic susceptibility similar to that of Fig. 5, antiferromagnetic MnO and ferrimagnetic $Fe^{3+}[Ni^{2+}Fe^{3+}]O_4$ have two interpenetrating, ferromagnetic subarrays that are coupled antiparallel to one another. In MnO the net magnetizations of these two subarrays just cancel each other, whereas, in $Fe^{3+}[Ni^{2+}Fe^{3+}]O_4$ only the iron moments cancel each other and there is a net magnetization per molecule due to the Ni^{2+} - ion contribution.

Subsequent studies on other oxides have shown that in many cases this simple collinear, two-sublattice model must be modified. In antiferromagnetic α Fe_2O_3, for example, there is a weak ferromagnetic component in the temperature interval $250 < T < 953°K$ in which the spins are ordered in the basal planes of the corundum structure, see Fig. 6. Similar canting has been observed in the orthoferrites having the orthorhombic perovskite structure, and this weak ferromagnetism presently shows great promise for computer applications. In antiferromagnetic MnO_2, on the other hand, it has been shown that competitive exchange interactions lead to the spiral spin configuration of Fig. 7. Much more complex spiral-spin configurations have been observed in chromium spinels and analyzed in terms of competitive exchange interactions. Figure 8 illustrates the ferrimagnetic-spiral spin configuration found in $Co^{2+}[Cr_2{}^{3+}]O_4$. An antiferromagnetic-spiral$\rightleftarrows$ferromagnetic phase change as a function of applied field H_a and temperature T has been found in $Ba_xSr_{1-x}Zn_2 \times Fe_{12}O_{22}$. This is known as metamagnetism. The perovskite $LaMnO_3$ was found to have the unusual Type A magnetic order shown in Fig. 9. Here Mn^{3+} ions, which interact in all three crystallographic directions

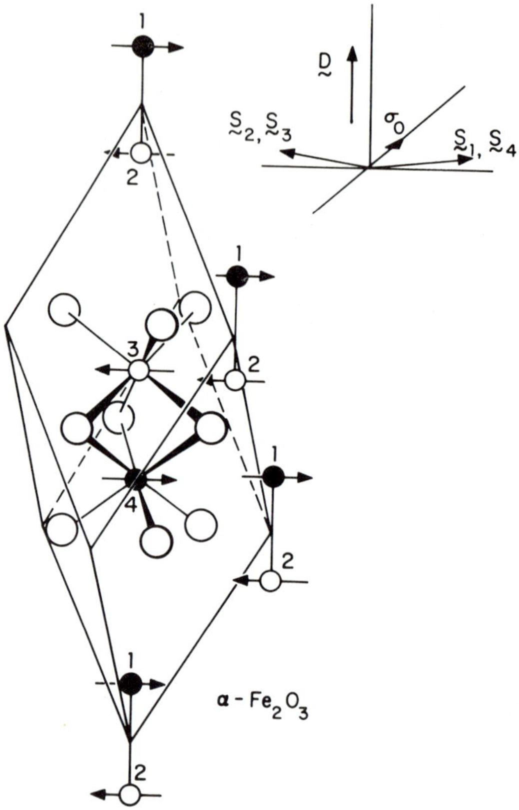

FIG. 6
Spin configuration found in α - Fe_2O_3. Weak ferromagnetism due to spin canting between basal planes in temperature interval 250 < T < 953°K, spins parallel to hexagonal c-axis if T < 250°K: C. G. Shull, W. A. Strauser, and E. O. Wollan, Phys. Rev., **83**, 333 (1951); F. J. Morin, Phys. Rev., **78**, 819 (1950).

through an intervening oxygen ion, are coupled ferromagnetically in an (001) plane and antiferromagnetically between (001) planes. Furthermore, mobile charge carriers in the perovskite system $La_{1-x}Ca_xMnO_{3+\delta}$ introduce a ferromagnetic component that increases with the carrier density until complete ferromagnetism is achieved. Metallic oxides, on the other hand, exhibit additional complexities. The compound V_2O_3 has a first-order semiconductor⇄metallic transition at the magnetic ordering temperature, whereas VO_2, which

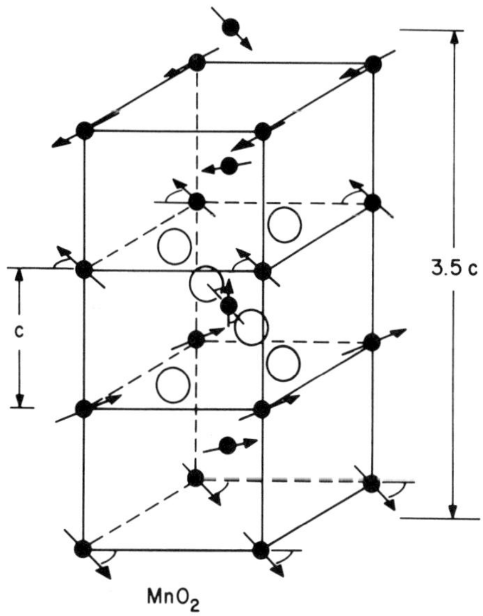

FIG. 7
Spiral spin configuration propagating parallel to c-axis found in MnO_2 by neutron diffraction: R. A. Erickson, Phys. Rev., **90**, 779 (1953); A. Yashimori, J. Phys. Soc. Japan, 14, 807 (1959). Turn angle between spins in successive (001) planes of rutile structure is $2\pi/7$.

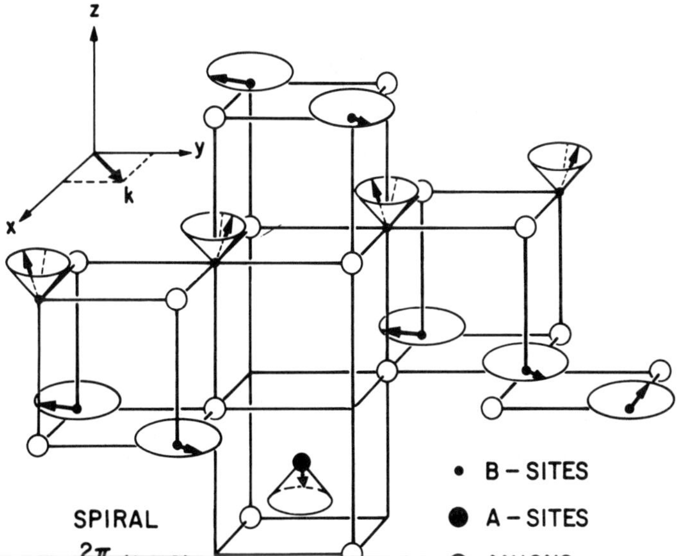

FIG. 8
Complex-spiral spin configuration found in the spinel $Co[Cr_2]O_4$ by neutron diffraction. K. Dwight and N. Menyuk, J. Appl. Phys., **40**, 1156 (1969). Spiral propagates along the [110] direction.

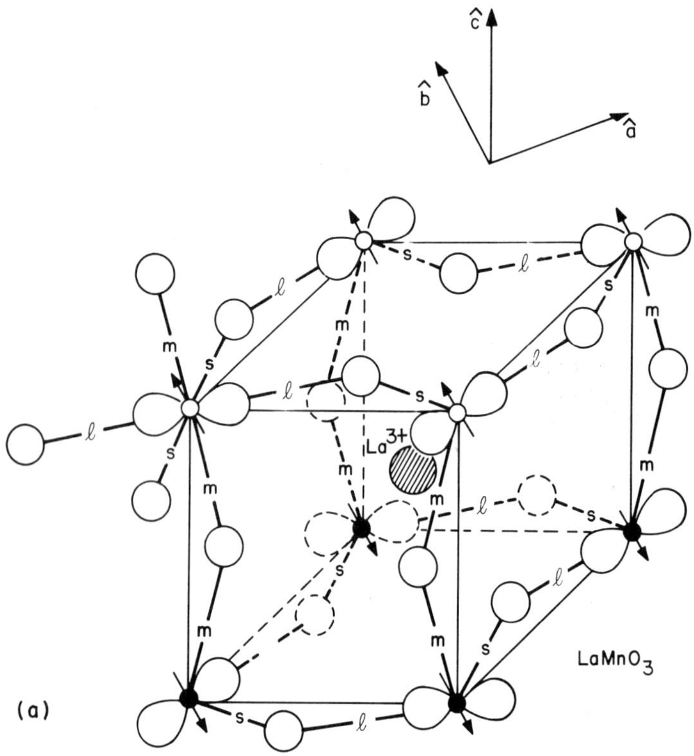

Fig. 9a

FIG. 9

(a) Magnetic order found in $LaMnO_3$ by neutron diffraction: E. O. Wollan and W. C. Koehler, Phys. Rev., **100**, 545 (1955). Jahn-Teller ordering of the occupied e_g orbitals also indicated: J. B. Goodenough, Phys. Rev., **100**, 564 (1955); J. Kanamori, J. Appl. Phys. Suppl., **31**, 14 S (1960). (b) Ferromagnetic moment versus composition for the system $La_{1-x}Ca_xMnO_3$.

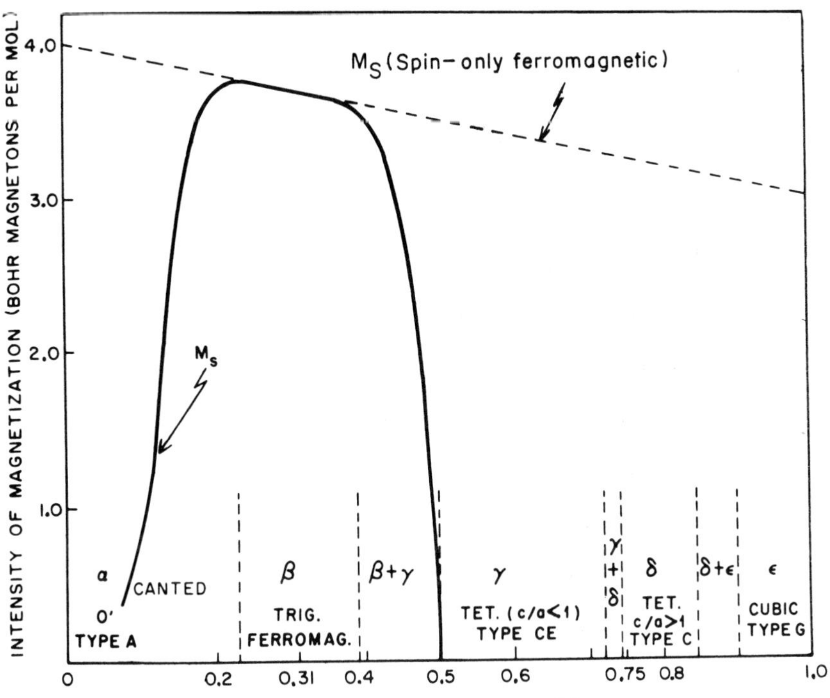

INTENSITY OF MAGNETIZATION (BOHR MAGNETONS PER MOL)
4.0
3.0
2.0
1.0
M_S (Spin-only ferromagnetic)
M_s
α
CANTED
O'
TYPE A
β
TRIG.
FERROMAG.
β+γ
γ
TET. (c/a<1)
TYPE CE
γ
+
δ
δ
TET.
c/a>1
TYPE C
δ+ε
ε
CUBIC
TYPE G
0
0.2
0.31
0.4
0.5
0.6
0.75
0.8
1.0
(b)
x = FRACTION OF Mn WHICH IS Mn^{4+}

Fig. 9b

exhibits a similar first-order transition, has no magnetic order in the low-temperature phase. $SrRuO_3$ is a metallic ferromagnet whereas isostructural $CaRuO_3$ is a metallic antiferromagnet. Metallic $SrCrO_3$ has a temperature-independent Pauli paramagnetism, whereas isostructural $CaCrO_3$ is antiferromagnetic with a weak ferromagnetic component. Metallic CrO_2, on the other hand, behaves like a classic ferromagnet. To develop the theoretical framework for these diverse phenomena requires the interplay of several concepts.

II. NÉEL TWO-SUBLATTICE MODEL

A. The Model

Néel[1] developed a phenomenological model of antiferromagnetism and ferrimagnetism that is applicable to a particular class of magnetic insulators. Insulators belonging to this class contain two identifiable ferromagnetic subarrays that are coupled antiparallel to one another, and the antiferromagnetic coupling between subarrays is much stronger than any antiferromagnetic coupling within an array.

In an insulator, each magnetic ion carries a localized spin, and the leading term in the spin-spin interaction energy is an isotropic superexchange term

$$\mathcal{H}_{ex} = -\sum_{ij} J_{ij}\mathbf{S}_i \cdot \mathbf{S}_j \tag{1}$$

which, in the Weiss molecular-field formalism, is approximated by

$$\mathcal{H}_{ex} = -\sum_{n=1} \mathbf{M}_n \cdot \mathbf{H}_{wn}, \qquad \mathbf{H}_{wn} \equiv \sum_{m=1}^{\nu} W_{nm}\mathbf{M}_m \tag{2}$$

where $\mathbf{M}_n$ is the magnetization of the nth subarray, and $\mathbf{H}_{wn}$ is the Weiss molecular field acting at a spin in the nth subarray. In the

Néel two-sublattice model, $\nu = 2$ and $W_{12} < 0$. Néel used the notation

$$-W_{12} = n > 0, \; W_{11} \equiv n\alpha, \; W_{22} \equiv n\beta \tag{3}$$

This type of theory may be generalized to larger numbers of subarrays (larger ν), but the two-sublattice model illustrates the important features of the theory.

B. Paramagnetism

The magnetic susceptibility $\chi_m = M/H_a$ for a collection of uncoupled, localized spins in an applied field H_a is given by the Curie law

$$\chi_m = C/T, \; C = N\mu^2/3k \tag{4}$$

where N is Avogadro's number and k is Boltzmann's constant. The effective localized atomic moment is

$$\mu_{eff} = \begin{cases} g[J(J+1)]^{1/2}\mu_B & \text{for } \Delta \gg kT \\ [4S(S+1) + L(L+1)]^{1/2}\mu_B & \text{for } \Delta \ll kT \end{cases}$$

$$= [3kC/N]^{1/2} = [8C_{mol}]^{1/2}\mu_B \tag{5}$$

where Δ is the multiplet splitting of the localized-electron levels, g is the spectroscopic splitting factor, $\mathbf{J} = (\mathbf{L} + \mathbf{S})$ is the total angular-momentum quantum number, and μ_B is the Bohr magneton. The effective orbital and spin angular-momentum quantum numbers **L** and **S** as well as **g** are obtained from crystal-field theory. This means that the atomic moment of each magnetic ion can be calculated, and the sum over all $\mu_{eff}^2/3k$ gives the total Curie constant C. In the special case L = 0, there g = 2.0 and

$$\mu_{eff} \text{ (spin-only)} = 2[S(S+1)]^{\frac{1}{2}}\mu_B = [8C_{mol}]^{\frac{1}{2}}\mu_B \tag{6}$$

In the Weiss molecular-field theory, the applied field $\mathbf{H}_a$ is replaced by the total field $(\mathbf{H}_a + \mathbf{H}_{wn})$ acting on a spin of the nth array, which gives the set of equations

$$M_n T - C_n \left(H_a + \sum_{m=1}^{\nu} W_{nm} M_m \right) = 0. \tag{7}$$

The total moment induced by $\mathbf{H}_a$ is

$$M = \sum_{m=1}^{\nu} M_m, \tag{8}$$

and for a two-sublattice model ($\nu = 2$), elimination of the M_n from Eqs. (7) and (8) gives

$$\frac{1}{\chi_m} = \frac{1}{C}\left(T - \theta_a - \frac{\theta_b^2}{T - \theta} \right) \tag{9}$$

where $C = C_1 + C_2$. In the Néel notation, where

$$C_1 = C\lambda, \ C_2 = C\mu, \text{ and } \lambda + \mu = 1 \tag{10}$$

and n, α, β are defined by Eq. (3), the parametric temperatures entering Eq. (9) are

$$\theta_a = -nC(2\lambda\mu - \alpha\lambda^2 - \beta\mu^2) \tag{11}$$

$$\theta = nC\lambda\mu(2 + \alpha + \beta) \tag{12}$$

$$\theta_b = nC(\lambda\mu)^{\frac{1}{2}}(\lambda(1+\alpha) - \mu(1+\beta)) \tag{13}$$

1. Antiferromagnets

For the special case where the two sublattices are equivalent, $C_1 = C_2$ (or $\lambda = \mu$) and $\alpha = \beta$, which makes $\theta_b = 0$. Then Eq. (9) reduces to the form of a **Curie-Weiss** law, the sign of θ_a reflecting the sign of the dominant molecular-field parameter $W_{12} = -n$. This is illustrated in Fig. 10.

2. Ferrimagnets

In the more general case, where $\theta_b \neq 0$, $1/\chi_m$ versus T is curved, approaching a Curie-Weiss law only asymptotically at higher temperatures $(T - \theta) \gg \theta_b^2$. This is illustrated in Fig. 11, which is

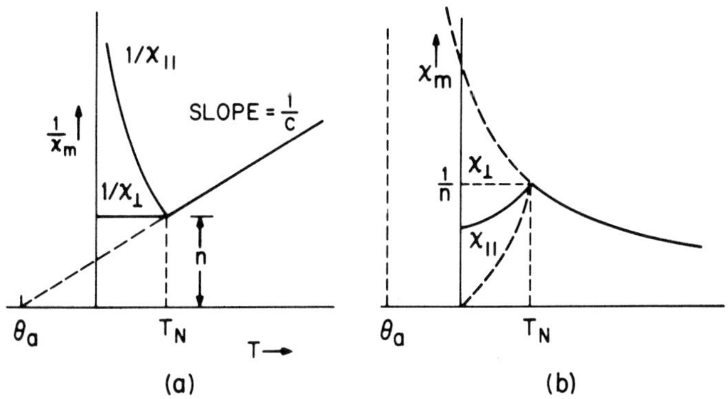

FIG. 10
Temperature dependence of (a) reciprocal susceptibility and (b) susceptibility for Néel antiferromagnets, in molecular-field approximation.

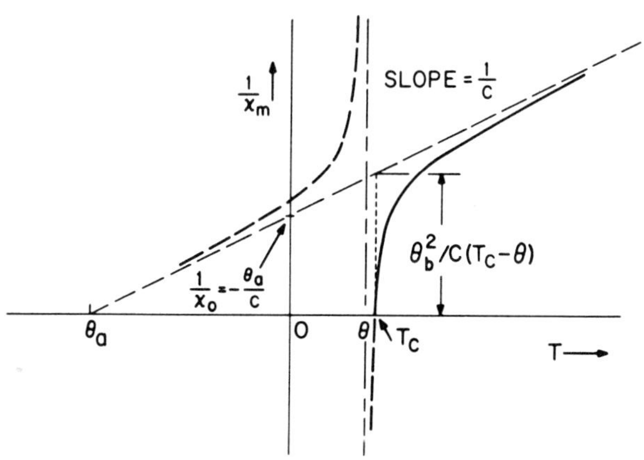

FIG. 11
Reciprocal susceptibility versus temperature for a Néel ferrimagnet, in molecular-field approximation.

to be compared with Fig. 2. The asymptotic Curie temperature θ_a is negative, again reflecting the sign of the dominant molecular-field parameter $W_{12} = -n$.

Theoretically the relative coupling parameters n, α, β of Eq. (3) may be obtained from the parametric temperatures θ_a, θ, θ_b determined graphically as shown in Fig. 11. In practice it is sometimes important to introduce a temperature-dependent Weiss constant

$$-W_{12} = n_0 \, (1 + wT) \tag{14}$$

where $w \simeq 10^{-4}$ is proportional to the coefficient of thermal expansion. Then the measured parameters are related to the theoretical parameters of Eq. (9) as follows:

$$\begin{aligned} C^* &= C/\left(1 - w\theta_a\right) & \qquad \theta_a^* &= \left(\theta_a + 2w\theta_b^2\right)/\left(1 - w\theta_a\right) \\ \theta^* &= \theta/\left(1 + w\theta\right) & \qquad \left(\theta_b\right)^* &= \theta_b^2\left(1 - 3w\theta\right)/\left(1 - w\theta_a\right) \end{aligned} \tag{15}$$

where the asterisk denotes the measured parameter.

Although the significance of such parameters is limited by the inadequacies of the Weiss field approximation, nevertheless this simple theory provides the essential qualitative features. Comparison with Fig. 5 shows that it is possible to distinguish ferromagnetism from antiferromagnetism from ferrimagnetism with a knowledge of the paramagnetic $1/\chi_m$ versus T curves above.

C. Néel and Curie Temperatures

To determine the order ⇄ disorder transition temperature, it is only necessary to set equal to zero the determinant for the coefficients of M_n in Eq. (7) with $H_a = 0$, since this determines the critical temperature T_c at which the magnetization has a nontrivial solution in zero external field. This gives an equation of the νth degree in T_c, and the largest real root corresponds to the most stable magnetically ordered phase having collinear atomic moments. For $\nu = 2$, solution of the determinantal equation gives the Curie temperature

$$T_c = \frac{1}{2} nC\left[\left((\lambda\alpha - \mu\beta)^2 + 4\lambda\mu\right)^{\frac{1}{2}} + \lambda\alpha + \mu\beta\right] \tag{16}$$

For an antiferromagnet with both sublattices identical, this reduces to

$$T_N = \frac{1}{2} nC(1 - \varepsilon) \tag{17}$$

where $\alpha = \beta = -\varepsilon$ and the critical temperature is called the Néel temperature. A measure of the relative intrasublattice-to-intersublattice coupling ε for antiferromagnets is then given by a comparison of the ratio

$$|\theta_a|/T_N = (1 + \varepsilon)/(1 - \varepsilon) \tag{18}$$

subject of course to the limitations of the Weiss field approximation. From Eq. (18) it might appear that the ratio $|\theta_a|/T_N$ can be made arbitrarily large. Actually this is not the case, since too large an ε leads to a different type of magnetic order, even under the constraint of collinear spins. In a body-centered-cubic array of magnetic atoms, for example, the two simple-cubic subarrays order antiferromagnetically, to give ordering of the first kind, only if the nearest-neighbor interactions predominate. However, if $\varepsilon > \varepsilon_c = \frac{1}{2}$, then a collinear-spin model calls for ordering of the second kind, in which there is antiferromagnetic ordering within each of the simple-cubic subarrays. As shown below, the ground-state spin configuration for $\varepsilon > \varepsilon_c = \frac{1}{2}$ is actually an antiferromagnetic spiral. This means that for a body-centered-cubic magnetic array, the maximum value of $|\theta_a|/T_N$ is 3. If a ratio much in excess of this value is found, then it may be suspected that the assumption of localized spins has broken down. If, in addition, the Curie constant C is larger than that calculated from a crystal-field model, then the Weiss field theory is not applicable, and a band theory of antiferromagnetism must be used.

D. Susceptibility of an Antiferromagnet below T_N

In an antiferromagnet, θ_b of Eq. (13) is $\theta_b = 0$, and from Eqs. (9), (11), and (17) the magnetic susceptibility at $T = T_N$ is

$$\chi_m(T_N) = \frac{1}{n} \tag{19}$$

as indicated in Fig. 10. This expression is only slightly modified by the inclusion of magnetic anisotropy, unless there is a crystallographic Jahn-Teller distortion below T_N induced by spin-orbit coupling and a collinear ordering of the spins. Even without such a distortion, below T_N it is necessary to distinguish $\chi_{||}$ from $\chi_{\perp}$, the susceptibilities for $\mathbf{H}_a$ parallel and perpendicular to the unique axis defined by $\mathbf{M}_n$.

An $\mathbf{H}_a \perp \mathbf{M}_n$ below T_N turns the sublattice moments $\mathbf{M}_1$ and $\mathbf{M}_2$ by an angle ϕ from the original axis to give a net moment $\mathbf{M}||\mathbf{H}_a$ of magnitude

$$M = (M_1 + M_2) \sin \phi \simeq 2M_1\phi \tag{20}$$

Since the anisotropy field is generally much smaller than the exchange field, at equilibrium H_a is equal and opposite to the parallel component of the Weiss field, which is $W_{12}M_1 \sin 2\phi \simeq -nM$. Therefore,

$$\chi_{\perp} = M/H_a = 1/n = \chi_{\perp}(T_N) \tag{21}$$

is constant for all $T \leq T_N$, as shown in Figs. 1 and 10.

At T = 0°K, where there is perfect alignment of the spins, an $\mathbf{H}_a || \mathbf{M}_n$ provides no torque $\mathbf{H}_a \times \mathbf{M}_n$ to turn the sublattice moments, so that $\chi_{||}$ (T = 0°K) = 0. An analytic expression for $\chi_{||}$ (T) can be obtained by expanding in powers of H_a, and retaining the first-order term only, the temperature-dependent **expressions** for the sublattice magnetizations

$$M^{\pm} = M_0 B_j(y), \text{ where } y = M_0\left(H_a \pm H_w\right)/NkT \tag{22}$$

and J is the atomic total-angular-momentum quantum number. Since the Brillouin function is defined as

$$B_J(y) = \frac{J + (1/2)}{J} \coth\left(\frac{J + (1/2)}{J} y\right) - \frac{1}{2J} \coth\left(\frac{1}{2J} y\right) \tag{23}$$

this procedure gives

$$\chi_{||} = \frac{M^+ - M^-}{H_a} = \frac{c\Lambda}{T - \theta_a \Lambda}, \text{ where } \Lambda = \frac{3JB'_J(y_0)}{J + 1} \tag{24}$$

and $y = M_0 H_w/NkT$. At high temperatures ($T \geq T_N$) the Brillouin function may be expanded as

$$B_J(y) = \frac{J + 1}{3J} y - 0(y^3) \cdots \qquad (y << 1) \tag{25}$$

At low temperatures

$$B_J(y) \approx 1 - \frac{1}{J} \exp(-y/J) \qquad (y >> 1) \tag{26}$$

Since

$$B_J'(y_0)/T \to 0 \text{ as } T \to 0 \text{ and } B_J'(y_0) \to (J + 1)/3J$$

as

$$y_0 \to 0 \text{ (or } T \geq T_N)$$

$\chi_{||}$ of Eq. (24) vanishes at T = 0°K and reduces to the Curie-Weiss law ($\Lambda = 1$) for $T \geq T_N$. The single-crystal anisotropy in the susceptibility below T_N permits a convenient determination of the unique magnetic axis, see Fig. 1.

For polycrystalline samples in an applied field strength $H_a \ll H_c$, the measured susceptibility χ_{mp} is a mean value for all the crystallites. At T = 0°K, a powder sample of a uniaxial antiferromagnet like Cr_2O_3 has

$$\chi_{mp}(0) = \frac{2}{3}\chi_{\perp}(0) + \frac{1}{3}\chi_{||}(0) \simeq \frac{2}{3}\chi_{mp}(T_N) \qquad (27)$$

as indicated in Fig. 10. For most antiferromagnetic crystals, the Néel temperature T_N is well defined by a sharp maximum in the χ_{mp} versus T curve. However, two-dimensional antiferromagnets may present an exception. Were there no crystalline anisotropy, two-dimensional antiferromagnets would have no long-range order, only regions of short-range order that increase their volume toward infinity as T → 0. In the presence of small anisotropy fields, long-range order is established, but the ordering temperature may occur only on the low-temperature side of a broad χ_{mp} versus T maximum and without introduction of a noticeable accident in the curve, as illustrated by $BaMnF_4$ in Fig. 12(a).

If $\mathbf{H}_a$ is applied parallel to $\mathbf{M}_n$, there is a critical field strength H_c above which the spins flop nearly 90° to give a $\chi_\perp$. Since the energy per unit volume required to flop the spins from an easy direction to a hard direction of magnetization is given by the anisotropy constant K, the critical field is obtained from the energy balance

$$-\frac{1}{2}\chi_{\parallel}H_c^{\ 2} = \frac{1}{2}\chi_\perp H_c^{\ 2} + K \tag{28}$$

This expression, together with Eq. (21), gives

$$H_c = [2nK/(1 - n\chi_{\parallel})]^{\frac{1}{2}} \tag{29}$$

At low temperatures, where $\chi_{\parallel} \rightarrow 0$, measurement of $\chi_\perp$ and H_c provides a good estimate of the anisotropy constant K. Where T_N is large, the product nK is generally too large for observation of the spin-flop field strength H_c in conventional laboratory fields. However, H_c is readily observed in antiferromagnets having a small $T_N (T_N < 30°K)$ and a small anisotropy energy ($K < 10^5 ergs/cm^3$). Figure 12(b) displays a spin flop at $H_c = 10.4$ kOe in $BaMnF_4$ at 1.4°K.

Above H_c, a $\chi_{mp} = \chi_\perp$ is found in polycrystalline samples, so that for low T where $\chi_{\parallel} \simeq 0$

$$\chi_{mp}\ (H_a > H_c)/\chi_{mp}\ (H_a << H_c) \simeq 3/2 \tag{30}$$

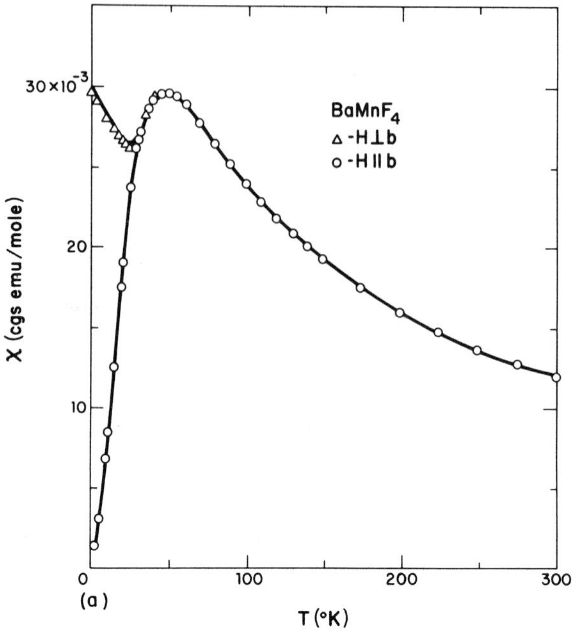

Fig. 12a

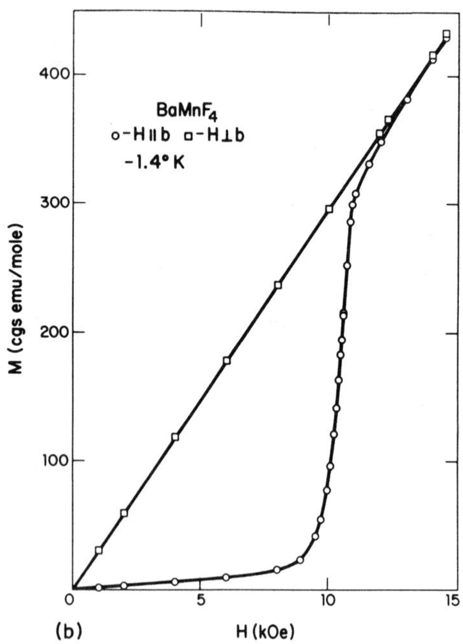

Fig. 12b

and it is possible to observe a break in the χ_{mp} versus H_a curve in an interval ΔH_a below H_c, as illustrated in Fig. 12(c) for the hexagonal insulator $CsCoF_3$.

E. Temperature Dependence of the Magnetization in Ferrimagnets

Below the Curie temperature of a collinear ferrimagnet, there is a spontaneous magnetization

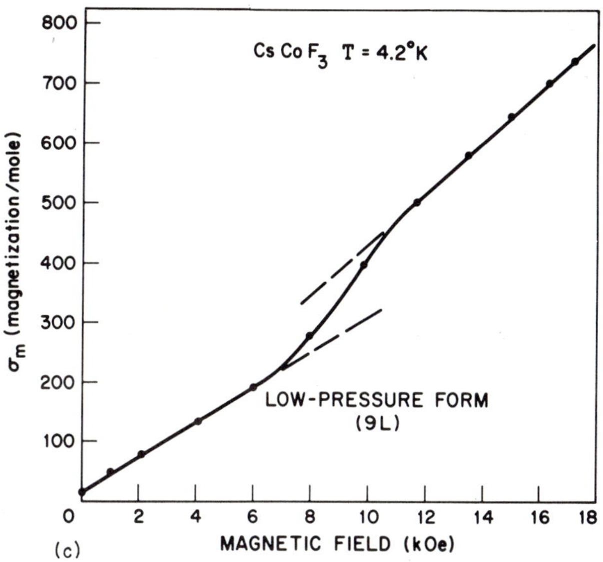

FIG. 12c
(a) Magnetic susceptibility versus temperature and (b) magnetization versus applied field for $H_a \parallel \mathbf{b}$ and $H_a \perp \mathbf{b}$ in single-crystal $BaMnF_4$: L. Holmes, M. Eibshütz, and H. J. Guggenheim, Solid State Communications, 7, 973 (1969). (c) Magnetization versus applied field for polycrystalline $CsCoF_3$ at 4.2°K: J. M. Longo, J. A. Kafalas, J. R. O'Connor, J. B. Goodenough, J. Appl. Phys., 41, 935 (1970).

$$M_s = |M_1 - M_2| \tag{31}$$

Although the individual M_n have temperature dependences

$$M_n(T) = M_{on}B_{Jn}(y_n), \text{ where } y_n = M_{on}(H_a + H_{wn})/NkT, \tag{32}$$

nevertheless the temperature dependence of M_s is not restricted to the Brillouin-type shape as is illustrated in Fig. 13.

In order to determine the principal characteristics of the M_s versus T curve to be anticipated for various values of α and β in Eq. (3), Néel[1] proceeded in three steps:

a. The energy of a two-sublattice structure is, in the approximation of isotropic exchange interactions only,

$$W = -\frac{1}{2}n\,(\alpha M_1{}^2 + \beta M_2{}^2 - 2\mathbf{M}_1 \cdot \mathbf{M}_2) \tag{33}$$

which is a minimum for a collinear-spin configuration having $\mathbf{M}_1$ antiparallel to $\mathbf{M}_2$. Nevertheless, in the whole α-β plane there are four distinguishable situations.

<u>Case I</u>: $T_c \leq 0$ so that the material is always paramagnetic: From Eq. (16), the boundary $T_c = 0$ is given by the curve $\alpha\beta = 1$ in Fig. 14.

<u>Case II</u>: M_{02} is less than its maximum possible value μM, but $M_{01} = \lambda M$ is its maximum value: Then $M_{02} = -\lambda M/\beta$ minimizes W, which is consistent provided $\beta < -\lambda/\mu$. This defines the boundary CE of Fig. 14.

<u>Case III</u>: $M_{02} = \mu M$ and $M_{01} < \lambda M$: Similar to Case II, but $M_{01} = -\mu M/\alpha$ minimizes W and is consistent

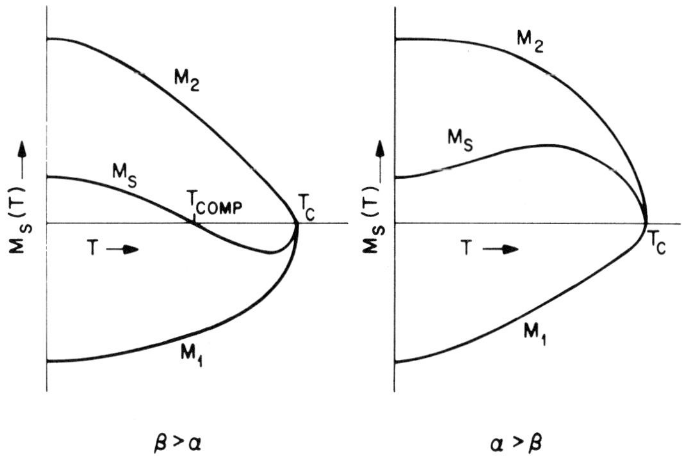

FIG. 13
Schematic M_s versus T curves for Néel ferrimagnets. $M_s = M_2 - M_1$ is the resultant magnetization. T_{comp} is the compensation temperature.

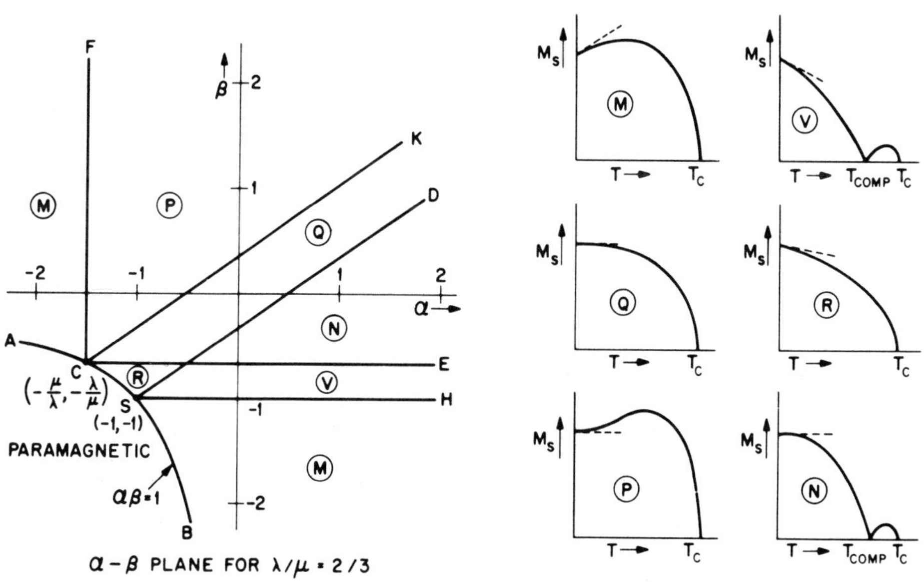

FIG. 14
Types of magnetization curves corresponding to various Néel regions in the α - β plane.

provided $\alpha < -\mu/\lambda$. This defines the boundary CF in Fig. 14.

Case IV: $M_{01} = \lambda M$ and $M_{02} = \mu M$: This solution minimizes W in the region FCE. Further, if the sublattices are so chosen that $\lambda < \mu$ and $M_s > 0$ for $\mathbf{M}_s \cdot \mathbf{M}_1 > 0$, then in region BCE (Case III)

$$M_s(0) = M_{01} - M_{02} = \lambda M(1 + 1/\beta) \qquad (34)$$

From this it follows that $\mathbf{M}_s(0)$ is $\parallel \mathbf{M}_{01}$ in region HSB, is $\parallel \mathbf{M}_{02}$ in region HSA.

b. From Eq. (25), (31), and (32) it is possible to calculate M_s and to show that it changes sign along the line defined by

$$\lambda(\alpha + 1) - \mu(\beta + 1) = 0 \qquad (35)$$

which is the line SD of Fig. 14. Just below T_c, $\mathbf{M}_s \parallel \mathbf{M}_1$ in the region BSD, $\mathbf{M}_s \parallel \mathbf{M}_2$ in the region DSA. It follows that there should be a compensation point T_{comp} (see Fig. 13 for definition) for samples falling in region DSH of the α-β plane.

c. In the regions ACF and BCE, only one sublattice is saturated. This sublattice experiences the larger H_w, and from Eq. (26) $B_J(y)$ is less sensitive to changes in T for larger H_w. Therefore, the unsaturated M_n decreases with T faster than the saturated M_n, and from T = 0°K an $\mathbf{M}_s \parallel$ saturated $\mathbf{M}_n$ (regions ACF and BCH) increases initially with T, but an $\mathbf{M}_s \parallel$ unsaturated $\mathbf{M}_n$ (region ECSH) decreases initially with T. On the other hand, if both sublattices are saturated (region FCE), then at T = 0°K $dM_1/dT = dM_2/dT = 0$. However, just above T = 0°K there is a small, but finite variation. Substitution of Eq. (26) into Eq. (31) gives the line CK of Fig. 14 defined by

$$\lambda\alpha + \mu = \mu\beta + \lambda \tag{36}$$

as the boundary between $dM_s/dT > 0$ (region FCK) and $dM_s/dT < 0$ (region KCL).

This reasoning is summarized in the schematic M_s versus T curves shown in Fig. 14 for various regions of the α-β plane. In Fig. 15 is shown the α-β plane for the special case $\lambda = \mu = 0.5$. For ideal antiferromagnets the condition $\alpha = \beta$ (the line SD) is also fulfilled, so that $M_s = 0$ for all temperatures.

The principal predictions of the Néel two-sublattice theory of ferrimagnetism are: (1) a nonlinear behavior of the $1/\chi_m$ versus T curve that is specified by the parameters appearing in Fig. 11; (2) a saturation magnetization at T = 0°K given by $M_s(0) = |M_{10} - M_{20}|$, where $M_{i0} = N_i g_i \mu_B S_i$ (if there are more than one type of magnetic ion on a sublattice, it is of course necessary to replace $N_i g_i S_i$ by $\Sigma_m N_{im} g_{im} S_{im}$), (3) the shape of the M_s versus T curve, including the possibility of a compensation point; (4) the possibility of a large $g_{eff} = (M_1 - M_2)/[(M_1/g_1) - (M_2/g_2)]$ in the vicinity of a compensation point; (5) two magnetic resonance modes, corresponding to in-phase and antiphase resonance of the two sublattices; and (6) identification of the chemical phases wherein ferrimagnetism versus antiferromagnetism may be anticipated, given a $W_{12} < 0$ between two identifiable

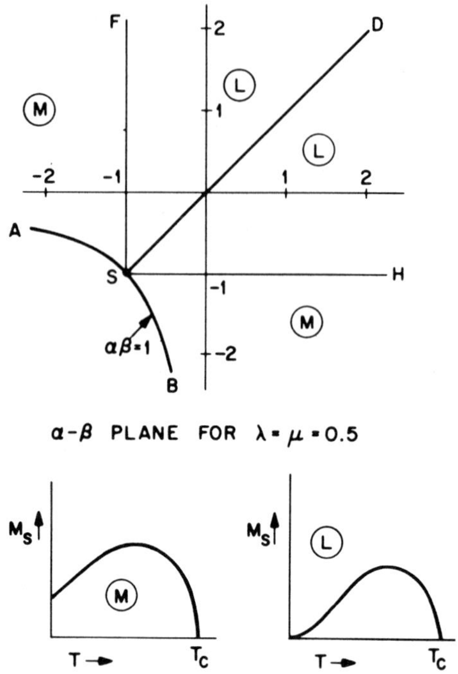

FIG. 15
Possible magnetization curves for $\lambda = \mu = 0.5$, according to Néel model. Along the line SD, $M_s = 0$ for all temperatures.

magnetic sublattices. All of these qualitative features have been found in nature, particularly in the ferrospinels, ferrogarnets, and magnetoplumbites, which have such important practical applications. Figure 16 shows the magnetization data for the spinel system $Li_{0.5}Fe_{2.5-x}Cr_xO_4$, which has a compensation temperature in the interval $x \geq 1.0$. However, the Néel model is a molecular-field model and is subject to the same criticisms as the Weiss field theory for ferromagnets. It is necessary to apply spin-wave theory to obtain a correct temperature dependence for M_s near $T = 0°K$, and just above T_c short-range order alters the $1/\chi_m$ versus T curve as indicated in Fig. 2.

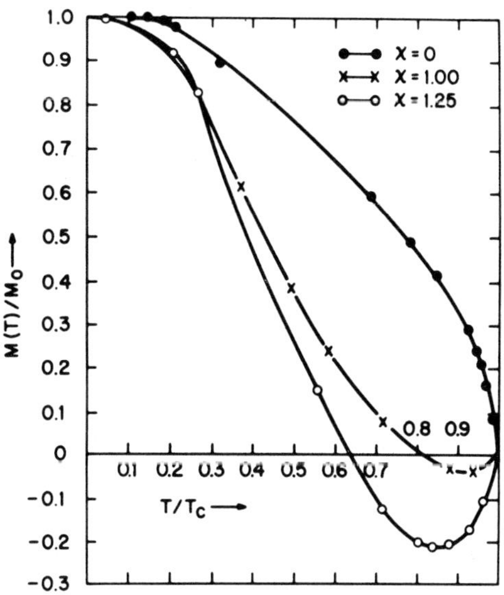

FIG. 16
Reduced magnetization versus temperature data for the system $Li_{0.5}Fe_{2.5-x}Cr_xO_4$, which exhibits a compensation temperature in the range $1 \leq x \leq 2$. [After E. W. Gorter, Philips Research Repts., **9**, 403 (1954).]

III. CANTED-SPIN CONFIGURATIONS

Many antiferromagnetic compounds exhibit a specific magnetization σ_g in high external fields H_a that, for $T < T_N$, varies as

$$\sigma_g = \sigma_o + \chi_g H_a \tag{37}$$

where χ_g is the magnetic susceptibility per gram. Such a weak ferromagnetism appears in the low-field ($H_a \ll H_c$) limit in $CsCoF_3$, for example, as may be seen in Fig. 12. The weak ($10^{-1}$to 10^{-5} times the sublattice magnetization) ferromagnetic component σ_o is intrinsic to the compound and is caused by a small canting of the sublattice magnetization vectors $\mathbf{M}_1$ and $\mathbf{M}_2$ that reduces the 180° angle between them by 2ϕ. Since $M_1 = M_2$, the resultant moment between them is

$$M_s = (M_1 + M_2) \sin \phi \simeq 2M_1\phi \qquad (38)$$

This canting may be caused by three different mechanisms: local anisotropy fields that are puckered, antisymmetric superexchange, and/or the presence of mobile electrons.

1. Local Anisotropy Fields

The compound NiF_2, which like MnO_2 has the rutile structure shown in Fig. 7, illustrates the first of these mechanisms. The two Ni^{2+} ions in a unit cell have orthorhombic environments that are rotated with respect to one another by 90° about the c-axis. Since dipole-dipole forces keep the Ni^{2+} -ion spins in the basal planes and the body-center spins are coupled antiparallel to the corner spins, the local anisotropy caused by the distortion of the Ni^{2+} octahedral sites from cubic to orthorhombic symmetry can cant the spins. If the [110] [$\bar{1}$10], and [001] axes are, respectively, the x-, y-, and z-axes, then the spin Hamiltonian is

$$-\Sigma_{ij} J_{ij} s_i \cdot s_j - K \sum_i^{(1)} (S_{ix}^2 - S_{iy}^2) + K \sum_j^{(2)} (S_{jx}^2 - S_{jy}^2) \qquad (39)$$

where only the isotropic, symmetric part of the spin-spin exchange interaction is included, just as in the molecular-field theory. (The antisymmetric-exchange contribution is zero in NiF_2.) If the spins within each subarray remain parallel, only the antiferromagnetic parameter

$$J_a = -8J_{12} > 0 \qquad (40)$$

is important for estimating the strength of the spin canting, which is, therefore, calculated by minimizing the energy

$$J_a \cos \theta - 2K \sin 2\psi \sin \theta \qquad (41)$$

where $\theta = 180° - 2\phi$ and ψ is the angle between $\mathbf{S}_1 + \mathbf{S}_2$ and the x-axis. This gives

$$2\psi = \left(n + \frac{1}{2}\right)\pi, \quad \tan\theta = -\left(2K/n\right) \sin 2\psi \tag{42}$$

which means that the weak ferromagnetic component is directed along a [100] or [010] axis with magnitude

$$M_s = (M_1 + M_2) \sin \frac{1}{2} (\pi - \theta) \simeq (M_1 + M_2)K/J_a \tag{43}$$

In NiF_2, a $K/J_a \simeq 0.02$ is found.

2. Antisymmetric Superexchange

The most general form of the bilinear spin-spin coupling between two localized spins S_i and S_j on neighboring atoms is

$$V_{ij} = \mathbf{S}_i \cdot \mathbf{K}_s \cdot \mathbf{S}_j + \mathbf{S}_i \cdot \mathbf{K}_A \cdot \mathbf{S}_j \tag{44}$$

where $\mathbf{K}_s$ and $\mathbf{K}_A$ represent symmetric and antisymmetric tensors, respectively. The first term remains invariant under the interchange of two spins, whereas the second term changes sign. The symmetrical part can always be expressed in the form

$$K_\xi S_{i\xi} S_{j\xi} + K_\eta S_{i\eta} S_{j\eta} + K_\zeta S_{i\zeta} S_{j\eta} \tag{45}$$

by a proper choice of the coordinate axes ξ, η, ζ (the principal axes). This, in turn, can be separated into an isotropic and an anisotropic term as

$$-2J_{ij} \mathbf{S}_i \cdot \mathbf{S}_j - 2\mathbf{S}_i \cdot \Gamma_{ij} \cdot \mathbf{S}_j \tag{46}$$

where Γ_{ij} represents a symmetrical tensor. Dzialoshinskii[2] was the first to point out that the antisymmetric term, which may be written as

$$-2\mathbf{D}_{ij}\cdot\mathbf{S}_i\times\mathbf{S}_j \tag{47}$$

is frequently allowed by the symmetry requirement that, for a constant $\mathbf{D}_{ij}$, the magnetic energy of Eq. (47) be invariant under the symmetry operations of the crystallographic unit cell. In the corundum structure, for example, the space group D_{3d}^6 - $R\bar{3}c$ contains the symmetry elements

$$E,\ 2C_3,\ 3U_2,\ I,\ 2S_6,\ 3\sigma_d$$

Because of the threefold rotation C_3, the vector **D** for any pair along the c-axis of Fig. 6 must be parallel to the trigonal axis. As there are inversion centers I at the midpoints of the lines connecting ions 1 and 4 and 2 and 3, there is no antisymmetrical coupling between ions 1 and 4 or 2 and 3. The couplings between 1 and 2 and 1 and 3 must be of opposite sign to those between 3 and 4 and 2 and 4 because of the glide plane σ_d. Therefore, the antisymmetric couplings among spins in a unit cell are

$$\mathbf{D}\cdot\{(\mathbf{S}_1\times\mathbf{S}_2)-(\mathbf{S}_3\times\mathbf{S}_4)\}+\mathbf{D}'\cdot\{(\mathbf{S}_1\times\mathbf{S}_3)-(\mathbf{S}_2\times\mathbf{S}_4)\} \tag{48}$$

Neutron-diffraction experiments on α - Fe_2O_3 and Cr_2O_3 have shown that the antiferromagnetic spin arrangements are:

$$\alpha\text{ - }Fe_2O_3:\quad S_1 \parallel S_4 \text{ and } S_2 \parallel S_3 \text{ anti} \parallel S_1$$

$$Cr_2O_3:\quad S_1 \parallel S_3 \text{ and } S_2 \parallel S_4 \text{ anti} \parallel S_1$$

Therefore, Eq. (48) vanishes for Cr_2O_3, but not for α - Fe_2O_3 having a spin component in the basal planes. In the temperature interval $250 < T < 953°K$, the spins of α - Fe_2O_3 lie in the basal plane, and there is a weak ferromagnetic component. Below a spin-flip temperature $T_s \simeq 250°K$ the spins are parallel to the c-axis, and Eq. (48) vanishes, thus eliminating the small ferromagnetic moment σ_0.

3. Canting by Mobile Charge Carriers

If a magnetic ion is present in equivalent lattice positions in two different valence states, as in the perovskite system

$$La_{1-x}Ca_xMn_{1-x}^{3+}Mn_x^{4+}O_3,$$

then the electron transfer represented by

$$M^{m+} + M^{(m+1)+} \rightarrow M^{(m+1)+} + M^{m+} \tag{49}$$

couples an initial state and a final state of equal energy. Therefore, admixture of an electron-transfer excited state into the crystal-field ground state gives a stabilization in first-order perturbation theory:

$$V_{ij}^D = ct_{ij}^D = cb_{ij}^D \cos(\theta/2) \tag{50}$$

where c is the concentration of mobile carriers and θ is the angle between spins $\mathbf{S}_i$ and $\mathbf{S}_j$, which by the rules of spinor operators transform as follows their α- and β-spin components on rotating the spin an angle θ:

$$\begin{aligned} \alpha &= \alpha' \cos(\theta/2) + \beta' \sin(\theta/2) \\ \beta &= -\alpha' \sin(\theta/2) + \beta' \cos(\theta/2) \end{aligned} \tag{51}$$

Since the mobile electron transfers without a change of spin, the matrix element t_{ij} couples the neighboring spins parallel through the operation of either Pauli's exclusion principle or Hund's rule within atoms (here assumed infinitely strong). This gives ferromagnetic coupling with a cos (θ/2) dependence, and the spin-independent part of t_{ij} is the transfer integral

$$b_{ij} = (\psi_i, H_1\psi_j) \simeq \varepsilon_j(\psi_i, \psi_j) \qquad (52)$$

where H_1 is the perturbation of the electron potential at site j by the proximity of the cation at site i. Here ε_j is a one-electron energy, and (ψ_i, ψ_j) is the overlap integral for the interacting cationic wave functions on neighboring atoms. In tight-binding theory, the bandwidth is ~b, the b_{ij} for near-neighbor transfer. If b is so small that the hopping time $\tau \simeq \hbar/b$ is long compared to the period of an optical vibration mode, then the electron becomes trapped in a potential well created by a relaxation of the near-neighbor atoms. Such an entity, localized electron plus associated atomic displacements, is called a small polaron, and its mobility is described by diffusion theory:

$$\mu_p = (eD_0/kT) \exp (-\varepsilon_A/kT) \qquad (53)$$

where ε_A is the activation energy for an electron hop. In this limit

$$b^D \simeq b_0 \exp (-\varepsilon_a/kT) \qquad (54)$$

but Eq. (50) is still a correct expression for the ferromagnetic coupling, which is known as double exchange, provided $\varepsilon_A < kT$.

This interaction is in addition to the superexchange interaction between the stationary electrons in half-filled orbitals. With only isotropic, symmetric superexchange, the total coupling energy is

$$H_{ex} = - \sum_{ij} \left(J_{ij} S^2 \cos \theta_{ij} + cb_{ij}^D \cos (\theta_{ij}/2) \right) \quad (55)$$

Minimization with respect to θ gives a cant angle

$$\cos (\theta_0/2) = cb^D/4|J|S^2 \quad (56)$$

where J is here the net superexchange integral between the two sublattices. In this case the ferromagnetic component may be large, the magnetic order becoming purely ferromagnetic for $c > c_f$, where

$$c_f = 4|J|S^2/b^D \quad (57)$$

In the system $La_{1-x}Ca_xMnO_3$, the end member $LaMnO_3$ is antiferromagnetic, and in the range $0.1 \leq x \leq 0.25$ there is a ferromagnetic component that increases monotonically with x until it reaches the saturation value for $(1 - x)Mn^{3+} + xMn^{4+}$ ions, viz $[4(1 - x) + 3x] \times \mu_B$/Mn atoms. Below T_c, an $\varepsilon_A = 0$ is stabilized in this system.

IV. SPIRAL-SPIN CONFIGURATIONS

Even in the absence of local anisotropy fields, antisymmetric exchange, or mobile charge carriers to cant the sublattice magnetizations of an antiferromagnet, the restriction to collinear-spin configurations may be inadequate. For example, the magnetic energy density for a spiral spin configuration propagating along the [001] axis of a body-centered-cubic array of magnetic atoms is, in the absence of crystalline anisotropy,

$$\varepsilon_M = - M_s^2 \left(4W_1 \cos \theta + (W_2 + 4W_3) \cos 2\theta - 2W_3\right) \quad (58)$$

where W_1, W_2, W_3 are the interaction parameters between first-, second-, and third-nearest neighbors. The spins in any (001) plane are parallel, but they turn an angle θ on going from one plane to the next. Minimization of ε_M with respect to θ gives antiferromagnetic ordering of the first kind (ferromagnetic simple-cubic subarrays coupled antiparallel) provided $W_1 \geq (W_2 + 4W_3)$, but otherwise a spiral with turn angle

$$\cos \theta_0 = - W_1/(W_2 + 4W_3) \tag{59}$$

Since the collinear-spin model shows a breakdown to ordering of the second kind for $W_1 < W_2$, the spiral-spin configuration is more stable than ordering of the second kind. Thus this simple exercise demonstrates that competitive exchange interactions may produce a simple-spiral spin configuration having a periodicity that is independent of the lattice periodicity. However, in general the crystalline-anisotropy energy stabilizes spirals that are commensurate with a lattice periodicity. In MnO_2, for example, an antiferromagnetic spiral of wavelength 3.5 c propagates along the c-axis of the body-centered-tetragonal rutile structure, see Fig. 7.

The fact that a spiral spin configuration is more stable than any collinear-spin configuration does not prove that it is the ground-state spin configuration. A problem of considerable interest is, therefore, a rigorous (all 10^{21} spins permitted to have variable orientations) determination of ground-state spin configurations at $T = 0°K$ given an isotropic, symmetric exchange energy. Although this problem has not been solved in general, rigorous solutions have been obtained for a large class of compounds[3]. In particular it has been shown that, with such an exchange Hamiltonian, the ground state for Bravais and close-packed-hexagonal lattices is always a simple spiral. Collinear antiferromagnetism and ferromagnetism represent special cases corresponding, respectively, to $\theta = \pi$ and $\theta = 0$. In these crystals, spin configurations that are more complex than a simple spiral require additional terms in the Hamiltonian, as has been discussed for the canted-spin configuration. Where the crystallographic structure is more complex, competitive exchange interactions may produce extremely complicated spin configurations, as is illustrated in Fig. 8 for the spinel $Co[Cr_2]O_4$.

$Ba_2Zn_2Fe_{12}O_{22}$ is a ferrimagnetic, hexagonal oxide with spins oriented in the basal planes. The anisotropy in the basal plane is low. In the Sr-rich compounds of the system $Ba_{2-x}Sr_xZn_2Fe_{12}O_{22}$, the zero-field state is a spiral spin configuration propagating along the c-axis. The angular-dependent part of the exchange energy is due to interactions along the c-axis:

$$\varepsilon_M = - M_s^2 \, (W_1 \cos \theta + W_2 \cos 2\theta) \qquad (60)$$

where W_1 and W_2 are the net coupling parameters between first- and second-neighbor planes. In this structure, $W_1 > 0$ and $W_2 < 0$. Minimization of ε_M with respect to θ gives ferromagnetism if $|4W_2| > |W_1|$, but otherwise a spiral with turn angle

$$\cos \theta_0 = -W_1/4W_2 \qquad (61)$$

If a field $\mathbf{H}_a$ is applied perpendicular to the c-axis, the angle ϕ between $\mathbf{H}_a$ and a ferromagnetic basal-plane layer is a function of the coordinate z along the c-axis. Enz[4] solved the variational problem

$$\delta\int\left(\varepsilon_M - H_aM_s \cos\phi\,(z)\right) dz = 0 \qquad (62)$$

to obtain the deformed helix described by

$$\phi(z) \simeq \theta_0 z/a + \left(H_a/5M_s\theta_0^4|W_2|\right) \sin\phi \qquad (63)$$

where a is the c-axis spacing between basal planes. This has a relative magnetization

$$M/M_s = H_a/10\ M_s\theta_0^4|W_2|$$

obtained from averaging cos $\phi(z)$ over one period. Therefore, the initial susceptibility of the helix is

$$\chi_{\perp c} = M_s/H_a = 1/10\ \theta_0{}^4|W_2| \qquad (64)$$

Furthermore, if H_a exceeds a critical value $H_c \simeq \frac{1}{2} W_2\theta_0{}^4$, a marked discontinuity in the magnetization occurs. In the range $H_c < H_a < 2H_c$, the ferromagnetic component oscillates with z, but for $H_a > 2H_c$ there is only a ferromagnetic component and the magnetization is the same as for the collinear ferrimagnet $Ba_2Zn_2Fe_{12}O_{22}$. In the Sr-rich compounds an $H_c \simeq 1$ koe is found. Such a metamagnetic behavior is typical of spiral spin configurations having $W_1 > 0$ and $W_2 < 0$.

V. SUPEREXCHANGE

A. Microscopic Theory

In the limit of weak interactions between outer electrons on neighboring atoms, it is possible to use crystal-field theory to describe the one-atom properties of those electrons. Crystal-field theory rests on the assumption that the outer electrons are localized at discrete atomic positions for a time long compared to the period of an optical-mode vibration. In classic crystal-field theory, the neighboring atoms were considered point charges and the various atomic terms were identified by symmetry notations. For a free atom, which has spherical symmetry, the symmetry notations corresponding to total orbital angular momentum L = 0, 1, 2, 3 ... carry the spectroscopic identifications S, P, D, F .... For a single outer d electron in a cubic crystalline field, the atomic, orbitally fivefold-degenerate ^2D level is split into orbitally threefold-degenerate $^2T_{2g}$ and twofold-degenerate ^2E_g levels. The e_g orbitals of an octahedral-site cation are the $(2z^2 - x^2 - y^2)$ and $(x^2 - y^2)$ orbitals, which are directed toward near-neighbor anions; the t_{2g} orbitals are the yz, zx, xy orbitals directed in between the anions. Modern crystal-field theory includes covalent mixing with the near-neighbor anionic orbitals. Since the symmetrized anionic orbitals only mix with cationic orbitals having the same symmetry properties, introduction of covalent mixing does not change the symmetry notation. For an octahedral-site cation, the wave functions of e_g and t_{2g} symmetry have the form

$$\psi_e = N_e \left(f_e - \lambda_s\phi_s - \lambda_\sigma\phi_\sigma\right) \tag{65}$$

$$\psi_t = N_t \left(f_t - \lambda_\pi\phi_\pi + \lambda_c\phi_c\right) \tag{66}$$

where N_e, N_t are normalization constants; f_e, f_t are the free-atom wave functions; ϕ_s, ϕ_σ, ϕ_π are the symmetrized anion s, p_σ and p_π functions and ϕ_c represents symmetrized s and/or p wave functions at any near-neighbor cations sharing common octahedral-site edges or faces; λ_s, λ_σ, λ_π, λ_c are the corresponding covalent-mixing parameters. Admixed into the ground-state terms are all the appropriate single-ion excited states. From these terms it is possible to calculate an atomic ground-state moment $\mu_A = gS\mu_B$, where S is the effective localized spin and g is the spectroscopic splitting factor for the atom in a crystal. It is also possible to calculate the magnetic anisotropy energy from the dipole-dipole interactions and from any anisotropy in the g-factor. Finally, it is possible to estimate the multiplet splittings Δ due to spin-orbit coupling, and hence to calculate the Curie constant C.

Superexchange considers explicitly the interactions between localized spins on neighboring cations by admixing into the one-ion terms electron-transfer excited states. For localized electrons, the electron transfer represented by

$$M^{m+} + M^{m+} \rightarrow M^{(m + 1)+} + M^{(m - 1)+} \tag{67}$$

requires an electrostatic energy $U > 4$ eV, where

$$U = \left(\left|\psi_i(1)\right|^2, V\left|\psi_i(2)\right|^2\right), \quad V = \left(e^2/r_{12}\right) \exp\left(-\xi r_{12}\right) \tag{68}$$

and $\xi = \xi\,(b_{ij})$ is a screening parameter that increases with increasing transfer integral b_{ij}. If an electron can be transferred to an unoccupied orbital, this energy may be reduced by the intra-atomic exchange stabilization to give

$$U' = U - \Delta_{ex} \tag{69}$$

Because there is an energy difference between the initial and final states, the stabilization energy due to admixture of the electron-transfer excited state is given by second-order perturbation theory as

$$\Delta\varepsilon = - t_{ij}^{2}/U \tag{70}$$

To abstract the spin-dependent part from the matrix element t_{ij}, use is made of Eq. (51). As in the case of double exchange, electron transfer conserves the spin angular momentum. Therefore, if this transfer is from a half-filled orbital to a half-filled orbital, the Pauli exclusion principle only permits transfer between electrons of antiparallel spin. This means that, from Eq. (51),

$$\Delta\varepsilon \text{ (half-half)} = - \left(b_{ij}^{2}/U\right) \sin^2 (\theta/2) = \left(2b_{ij}^{2}/4S^2U\right) \mathbf{S}_i \cdot \mathbf{S}_j - \text{const.} \tag{71}$$

Thus the spin-dependent part has the form of the isotropic exchange interaction with

$$J_{ij}^{S} \text{ (half-half)} = - 2b_{ij}^{2}/4S^2U \tag{72}$$

On the other hand, if this transfer is from a full to a half-filled orbital or from a half-filled orbital to an empty orbital, then the transferred electron may have either spin, but that with spin parallel to the two localized net spins is stabilized by Δ_{ex}:

$$J_{ij}^{S} \text{ (full-half)} = J_{ij}^{S} \text{ (half-empty)} = -\left(2b_{ij}^{2}/4S^2\right)\Delta\left(1/U\right) = +2b_{ij}^{2}\Delta_{ex}/U^2 \tag{73}$$

This formalism thus provides not only contact with the phenomenological theories, but also rules for predicting the sign of the magnetic interactions: ferromagnetic ($J_{ij} > 0$) or antiferromagnetic ($J_{ij} < 0$). (Direct exchange gives a $J > 0$, but in ceramics $J \ll |J^S|$.)

The antiferromagnetic order found in the perovskite $LaMnO_3$, see Fig. 9, demonstrates strikingly the operation of these rules. The octahedral-site Mn^{3+} ion has four outer d electrons in the high-spin configuration $^5E_g(t_{2g}^3e_g^1)$. Since the orbital angular momentum is quenched by the crystalline fields, the atomic moment is $M_A = gS\mu_B = 4\mu_B$. This E_g configuration is orbitally twofold degenerate and, since it contains an accidental degeneracy without associated orbital angular momentum, there is no spin-orbit coupling to remove this degeneracy. However, a distortion of the octahedral interstice to tetragonal or orthorhombic symmetry would remove the degeneracy, thereby stabilizing the occupied state by an energy proportional to the atomic displacements. Since the elastic energy expended varies as the square of these displacements, a spontaneous distortion to lower symmetry occurs below some finite temperature T_t. This is known as a Jahn-Teller distortion. In $LaMnO_3$, an anomaly in χ_m^{-1} versus T (due to a reduction in the paramagnetic Curie temperature) marks $T_t \simeq 850°K$. At room temperature the orthorhombic perovskite structure has $c/\sqrt{2} < a < b$, corresponding to a pseudotetragonal $(c/a < 1) \rightleftarrows$ pseudocubic transformation at T_t. In the pseudotetragonal cell, the O^{2-} ions are shifted as shown in Fig. 9, indicating that in the basal planes half-filled σ-bonding orbitals interact with empty σ-bonding orbitals to give a dominant ferromagnetic interaction. Along the c-axis, on the other hand, each O^{2-} ion is equally spaced from its two near-neighbor cations, and antiferromagnetic interactions between half-filled orbitals are dominant. Thus it is possible to understand why two thirds of the $Mn^{3+} - O^{2-} - Mn^{3+}$ interactions are ferromagnetic and one third are antiferromagnetic. Also the presence of ferromagnetic interactions explains why the χ_m^{-1} versus T plot intercepts the positive temperature axis ($\theta_a > 0$) even though the compound is antiferromagnetic. Above T_t, where there is no static Jahn-Teller distortion, the degenerate e_g orbitals are one quarter filled and the dominant electron transfer is from a half-filled to an empty orbital in all three directions. This increases the net ferromagnetic interaction, so that θ_a is larger above T_t.

B. Exchange Striction

Thus far we have neglected the crystallographic distortions that generally accompany magnetic ordering, since these reflect relatively smaller energies. However, these distortions provide important information and, in some instances, modify significantly the predictions that are made if these energies are neglected. There are two general types of strictive energies: magnetostriction and exchange striction. Magnetostriction reflects the direction of the spin vectors and is therefore important for magnetic anisotropy. Spin-orbit coupling and dipole-dipole forces are the principal origin of magnetostriction. Exchange striction, on the other hand, reflects the symmetry of the magnetic order. Long-range magnetic order increases the binding energy between those atoms having a negative V_{ij}, decreases the binding energy between those atoms having a positive V_{ij}.

Since every force is met by a reactive force, exchange striction also influences the form of the spin-spin interaction energy V_{ij}. Thus addition of the elastic energy to the isotropic-exchange term gives an energy

$$h_{ij} = -J_{ij}(R)\mathbf{S}_i\cdot\mathbf{S}_j + (1/2)k_{e\ell}(R - R_0)^2 + \gamma_{ij}^0(\mathbf{S}_i\cdot\mathbf{S}_j)^2 \qquad (74)$$

where the biquadratic, isotropic exchange term is usually neglected because $\gamma_{ij}/J^0_{ij} < 0.01$. Here R is the separation between coupled atoms, and R_0 is the equilibrium separation in the absence of magnetic order. Minimization of h_{ij} with respect to R gives a biquadratic, isotropic exchange term with coefficient

$$\gamma_{ij} = \gamma^0_{ij} + \left(1/2k_{e\ell}\right)\left(\partial J_{ij}(R)/\partial R\right)^2 \tag{75}$$

and this coefficient may be large enough to make significant inclusion in V_{ij} of the biquadratic term.

In MnO, for example, the Mn^{2+} ion is in a high=spin ${}^6A_{1g}\left(t^3_{2g}e^2_g\right)$ state, so that all the d orbitals are half-filled and, therefore, all the interactions are antiferromagnetic via Eq. (72). The observed magnetic order, see Fig. 3, shows that the $Mn^{2+} - O^{2-} - Mn^{2+}$ interactions dominate the $Mn^{2+} - Mn^{2+}$ interactions, but a $|\theta_a|/T_N \simeq 5.0$, which is the maximum anticipated for a face-centered cubic array of magnetic atoms, shows that the $Mn^{2+} - Mn^{2+}$ interactions are strongly competitive. In fact, magnetic ordering of the first kind may be stabilized only because a spontaneous distortion to rhombohedral ($\alpha > 60°$) symmetry below T_N significantly increases the $Mn^{2+} - Mn^{2+}$ interaction between antiferromagnetically coupled ions in neighboring (111) planes while decreasing the interaction within (111) planes. In this compound, $\gamma_{ij} \simeq 0.05$. In general, the biquadratic exchange term increases the curvature of the M_n versus T curves over the Brillouin function, and the magnetic-ordering temperature becomes first-order if γ_{ij} is large enough.

From Eqs. (51) and (72) or (73), the strength of an exchange interaction is proportional to the square of an overlap integral

(ψ_i, ψ_j), which in turn depends sensitively upon R in cation-cation interactions, but upon the square of a covalent-mixing parameter (λ_i^2) in cation-anion-cation interactions. Since the λ_i are relatively insensitive to R, a larger exchange striction is associated with cation-cation than with cation-anion-cation interactions of comparable energy. If a given coupling parameter J_{ij} consists of an antiferromagnetic cation-cation component and a ferromagnetic cation-anion-cation component, for example, this difference in exchange striction causes an increase in R with stabilization of ferromagnetic order! This is well illustrated by ferromagnetic CrO_2, which has the rutile structure of MnO_2 shown in Fig. 7. In this compound, a half-filled orbital directed along the c-axis introduces an antiferromagnetic $Cr^{4+} - Cr^{4+}$ interaction between cations sharing common octahedral-site edges. Two other nearly degenerate d orbitals per cation are one-quarter filled (hence couple ferromagnetically) and interact with their counterpart on near-neighbor and next-near-neighbor cations via the p_π orbitals of an intermediary anion. (In fact this π bonding is strong enough to transform these d orbitals from localized to band orbitals, see next section, which is why the compound is metallic [5].) Since the ferromagnetic interaction dominates and is relatively insensitive to R, there is an anomalous expansion of the c-parameter as T decreases through the ferromagnetic Curie temperature.

Energy changes due to exchange striction may also give rise to first-order antiferromagnetic $\rightleftarrows$ ferromagnetic transitions as the relative magnitudes of two competing exchange interactions are unevenly changed by thermal expansion of the lattice. Although spiral $\rightleftarrows$ ferromagnetic transitions may be second-order, collinear-spin antiferromagnetic $\rightleftarrows$ ferromagnetic transitions are first-order.

VI. METALLIC OXIDES

To this point we have argued that for localized outer electrons, the magnitude of the atomic moments and the magnetic anisotropy can be calculated from crystal-field theory together with dipole-dipole interactions. This includes any Jahn-Teller distortions associated with collinear-spin ordering plus spin-orbit coupling. We have also argued that the spin-spin interaction energy is

$$H_{ex}^{S} + H_{ex}^{D} = -\sum_{ij}\Big\{J_{ij}(\mathbf{S}_i\cdot\mathbf{S}_j) + \gamma_{ij}(\mathbf{S}_i\cdot\mathbf{S}_j)^2 + \mathbf{D}_{ij}\cdot\mathbf{S}_j\times\mathbf{S}_j + \mathbf{S}_i\cdot\Gamma_{ij}\cdot\mathbf{S}_j + cb_{ij}\cos(\theta/2)\Big\} \tag{76}$$

where the parameters J_{ij}, which are calculated from perturbation theory, are given by Eqs. (72) or (73) and γ_{ij} by Eq. (75). A Dzialoshinskii vector $\mathbf{D}_{ij} \neq 0$ requires introduction of spin-orbit coupling into the superexchange theory[6]. The symmetric tensor Γ_{ij} depends upon spin-orbit coupling in higher order and is generally negligible. The total spin Hamiltonian to be used in any practical problem must also include the crystalline anisotropy energy and the Zeeman interaction with the total internal magnetic field. Because the assumption of localized outer electrons is correct for the outer 4f and 5f electrons, this formalism is always adequate for the rare-earth and actinide atoms or ions. In fluorides and in many first-row transition-element oxides this assumption is also adequate for outer d electrons at the cations, and this is why the theories discussed here have been so useful. However, there are other transition-metal oxides, which must also be classed as ceramics, in which this assumption breaks down[7].

There are two ways by which the localized-electron hypothesis may be broken down in oxides: by strong cation-cation interactions and by strong cation-anion-cation interactions. Superconducting TiO has the rocksalt structure with 15% cation and anion vacancies simultaneously present. The vacancies are ordered if annealed below 990°C or randomly distributed if quenched from above 990°C. The metallic properties of this compound are due to strong $Ti^{2+} - Ti^{2+}$ interactions across the shared octahedral-site edges. Note that the formal valence Ti^{2+} represents here, as in the case of localized d electrons, the number of d electrons per cation, which in this case is 2. As in more normal metals, TiO has a weak, temperature-independent Pauli paramagnetism and a measurable Fermi surface. The compound ReO_3, which has the cubic perovskite structure (see Fig. 9) without the large A cations at the body-center position, is also metallic with a measurable Fermi surface and a temperature-independent Pauli paramagnetism. In this case the $Re^{6+} - Re^{6+}$ separation across a cube face is $\simeq 5.5\text{Å}$, which is too large for strong cation-cation interactions, and the metallic conductivity is due to large Re-O covalent mixing, which makes the cation-anion-cation transfer integral $b \simeq N_\pi \varepsilon_\pi \lambda_\pi^2$ large enough to break down the conditions for localized electrons. This means that any adequate theory of the transition-metal oxides must include the transition from those conditions where localized-electron theory is applicable to those where conventional band theory is applicable. Some essential physical features that such a theory must contain are illustrated by the special case of half-filled interacting orbitals.

Figure 17(a) shows schematically the energy levels versus near-neighbor interaction parameter b for N cations containing one

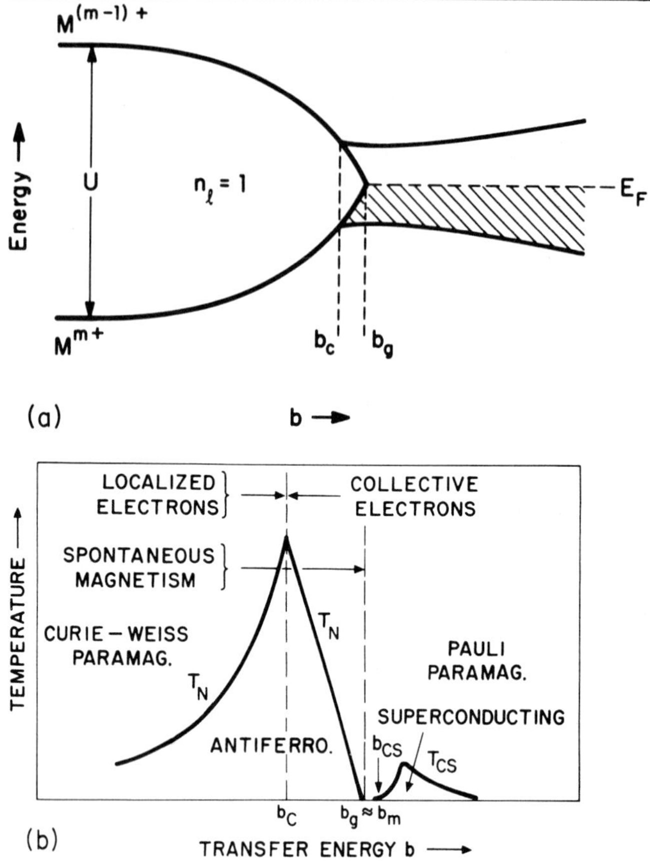

FIG. 17
(a) Energy levels versus near-neighbor interaction parameter b and (b) the T - b phase diagram for N cations containing one half-filled interacting orbital.

half-filled interacting orbital. In the no-interaction limit $b = 0$, two N-fold degenerate localized levels are separated by the electrostatic energy U. In the strong-interaction limit (large b), the 2N states are broadened into a half-filled band of itinerant-electron states, as in metallic sodium. The energy U of Eq. (68) decreases with increasing b because of the screening parameter $\xi(b)$, and in the strong-interaction limit $U < b$. The assumption $U = 0$ is implicit in conventional band theory. In certain special geometries, a $U \neq 0$ may immediately open an energy gap about the Fermi energy, but in general it initially introduces only a minimum in the density-of-states versus energy curve at E_F.

Since the Néel temperature T_N increases with the strength of the interaction parameter:

$$T_N \simeq J_{12} \simeq b^2/U \quad (77)$$

it must increase with b throughout the localized-electron domain where perturbation theory is applicable. This is shown graphically in Fig. 17(b). Clearly T_N must fall to zero before the strong-interaction limit is reached, and it is reasonable to anticipate that $T_N \to 0$ as $b \to b_m \simeq b_g$, where b_g is the value of b at which the energy U just creates an energy gap. This depends, of course, upon the particular crystallographic structure. Since superconductivity requires a Fermi surface, the superconducting $\rightleftarrows$ normal conducting transition temperature T_{cs} must go to zero at a $b_{cs} > b_g$.

The compound VO_x, where $0.8 \leq x \leq 1.3$, has the rocksalt structure with cation and anion vacancies simultaneously present, as in TiO_x. However, in VO_x these vacancies are always disordered, so that there can be no splitting of the d bands as a result of the translational symmetry of the structure. At an octahedral-site V^{2+} ion, the three t_{2g} orbitals are half-filled, and it can be argued[7] that the isolated anion vacancies trap two electrons, isolated cation vacancies two holes, and paired cation-anion vacancies trap an exciton. This means that for all x the Fermi level stays approximately at the center of the degenerate bands formed from the t_{2g} orbitals. This compound is n-type and essentially metallic for $x < 1$, p-type and a semiconductor $(d\varepsilon_g/dx > 1)$ for $x > 1$.[8] This, together with a semiempirical $R_c \lessapprox R$ in VO_x[9] and a $dR/dx > 0$, suggests that the band splitting is due to a $db/dx < 0$ with $b \simeq b_g$ at VO. This makes the magnetic susceptibility, which is shown in Fig. 18, particularly instructive. It contains two terms: a temperature-independent (Pauli) term dominant at low values of x and a Curie-Weiss term due to unpaired spins trapped at a fraction of the cation vacancies. Although the density of unpaired spins increases sharply with x, any long-range magnetic ordering, even at largest values of x, can only occur at very low temperatures[10]. It appears that here $b_g > b_m$.

The compound VO_2 has the rutile structure of Fig. 7 above $T_t \simeq 343°K$. Below T_t, the chains of vanadium atoms along the c-axis contain V-V pairs, indicative of homopolar bonding by the single outer d electron per V^{4+} ion, and a tilting of each V-V bond axis from the c-axis to make one short V-O bond per vanadium, as in a ferroelectric-type distortion. A weakly temperature-dependent susceptibility drops abruptly at T_t to a weak, temperature-independent value, and there is no magnetic order below T_t. These observations are consistent with homopolar V-V bonding below T_t. Although the antiferroelectric component of the distortion may determine T_t, a doubling of the periodicity along the c-axis splits a half-filled, V-sublattice band in two. V-V pairing concentrates the electrons in homopolar bonds where they are spin-paired[9][11].

The compound V_2O_3 has the corundum structure (see Fig. 6) above $T_t \simeq 150°K$ and a monoclinic structure below T_t. Although the susceptibility drops abruptly at T_t as in VO_2, V_2O_3 is antiferromagnetic below T_t with $\mu_V{}^{3+} \simeq 1.2\mu_B$ (reduced moment due to disproportionation

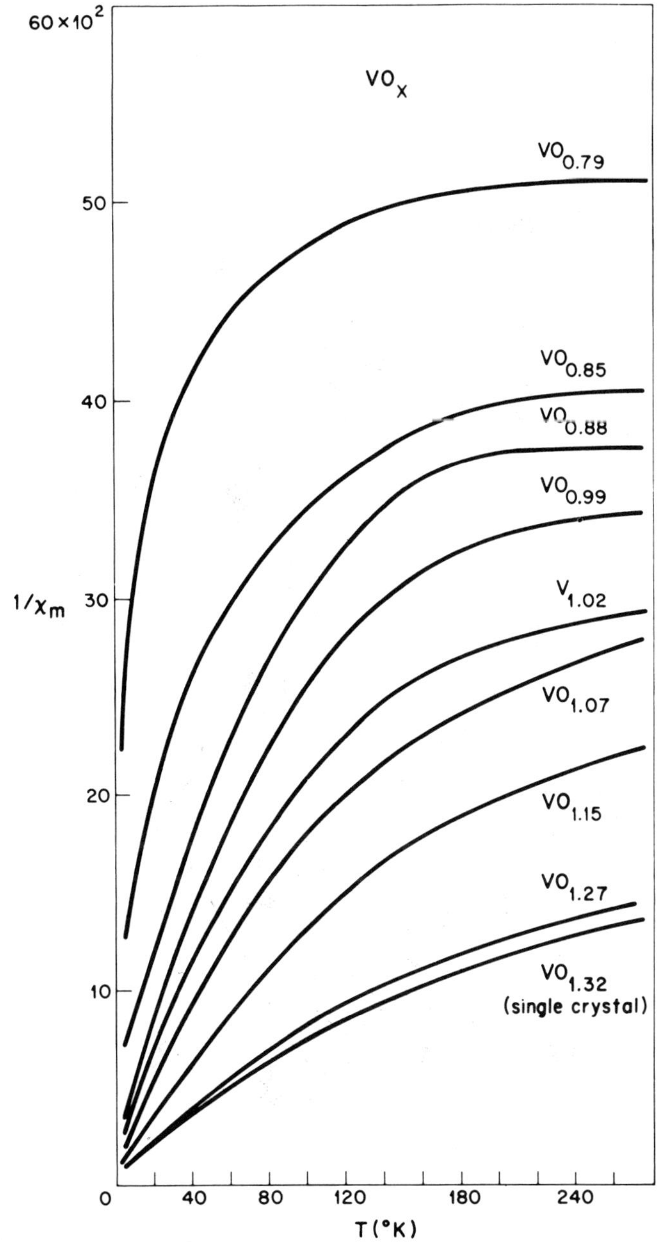

FIG. 18
Reciprocal magnetic susceptibility versus temperature for VO_x. (After M. D. Banus and T. B. Reed, Ref. [8].

into localized and itinerant orbitals in basal planes) and has the magnetic order shown in Fig. 19. The abrupt drop in both $\chi_{||c}$ and $\chi_{|c}$ below T_t is apparently due not only to the fact that T_N for the low-

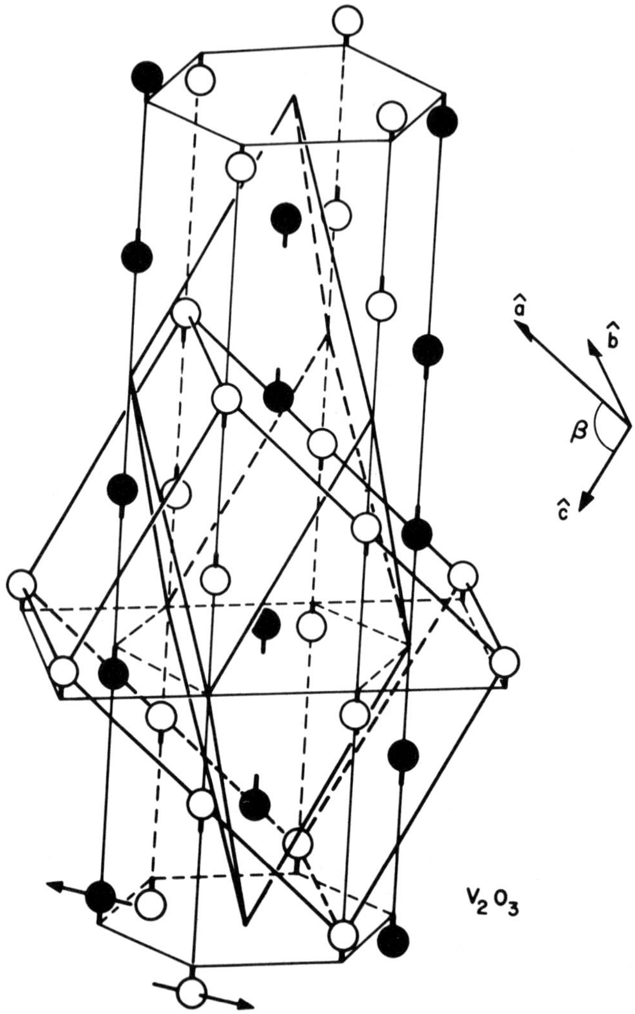

FIG. 19
The magnetic order of low-temperature (monoclinic) V_2O_3 as determined by neutron diffraction: R. M. Moon, J. Appl. Phys. 41, 883 (1970). The spin axis is in the ferromagnetic (010) planes about 75° from the pseudohexagonal c-axis. High-temperature rhombohedral cell (corundum structure of Fig. 6) also shown. Anions omitted. Full and open circles have opposite spins.

temperature phase extrapolates to a T_N(extrap) >> T_t, but also to the fact that the spin axis is neither parallel nor perpendicular to the crystallographic c-axis. Further, $V_{2-x}Cr_xO_3$ has localized d electrons for x > 0.08, which means that $R \simeq R_c$ in the basal plane. Application of hydrostatic pressure to $V_{1.92}Cr_{0.08}O_3$ produces a first-order semiconducting $\rightleftarrows$ metallic transition without changing the corundum structure[12]. Nevertheless this does not prove that a localized-electron $\rightleftarrows$ itinerant-electron transition is generally first-order. In this case there are overlapping d bands, and changes in the c/a ratio change the relative populations of these bands. It is this population change that makes the transition first-order in this case[13].

The perovskites $A^{2+}CrO_3$ have a $Cr^{4+} - O^{2-} - Cr^{4+}$ $b \simeq N_\pi \varepsilon_\pi \lambda_\pi^2$ that increases with increasing ionicity of the A^{2+} ion. This is because the $A^{2+} - O^{2-}$ covalent mixing competes with λ_π. In $LaVO_3$, where V^{3+} is isoelectric with Cr^{4+}, there is a Jahn-Teller distortion below T_N characteristic of spin-orbit coupling and collinear ordering of localized-electron spins. Although $PbCrO_3$ is antiferromagnetic below T_N = 240°K and has collinear spins, there is no associated crystallographic distortion[14]. This indicates band states in $PbCrO_3$ even though it is a semiconductor with $\mu_{Cr^{4+}} \simeq 2\mu_B$ and a temperature-dependent susceptibility that obeys a Curie-Weiss law. $CaCrO_3$ is also antiferromagnetic, but with a weak ferromagnetic component below $T_N \simeq 80$°K because it is orthorhombic. It has a $dT_N/dP < 0$, as would be expected from Fig. 17 for $b_c < b < b_m$. Finally, $SrCrO_3$ is metallic and Pauli paramagnetic, indicating a $b > b_m$. Although some contamination of the $PbCrO_3$ and $CaCrO_3$ samples by Cr^{3+} ions cannot be ruled out, these results seem to indicate that the transition region of

Fig. 17 is extremely narrow, since changes in the bandwidth on going from Pb^{2+} to Ca^{2+} to Sr^{2+} for the A cation cannot be too large.

The $A^{2+}Ru^{4+}O_3$ perovskites contain four outer d electrons per Ru^{4+} ion, and $SrRuO_3$ is a metallic ferromagnet with a lower moment than that predicted for two holes per cation in the three overlapping bands formed from the t_{2g} orbitals[15]. $CaRuO_3$, which should have the narrower d bands, is an antiferromagnetic metal. These results suggest that the band exchange splitting Δ_{ex}, which should be larger for $CaRuO_3$, is as shown in Fig. 20 for a band that is two-thirds filled. Splitting of the band states by the electrostatic energy U could place the Fermi energy E_F near or in the β-spin band gap in $SrRuO_3$, thus giving ferromagnetism with a reduced moment. $CaRuO_3$, on the other hand, would have partially filled α-spin and β-spin bands. This allows for the formation of a standing spin-density wave with wavelength commensurate with the Fermi surface, which would account for the antiferromagnetic order. This conjecture has not been confirmed experimentally, but it illustrates how different concepts will be required to account for the atomic moments and magnetic ordering that occur among itinerant electrons.

VII. FINAL REMARKS

This brief survey of the theory of antiferromagnetism and ferrimagnetism in transition-metal ceramics has necessarily omitted several interesting topics. The susceptibility of $LaCoO_3$, for example, shows a maximum as a function of temperature that is <u>not</u> due to antiferromagnetic ordering at a T_N, but to an energy separation $\Delta\varepsilon \simeq 0.08$ eV between the diamagnetic low-spin $^1A_{2g}(t_{2g}^6 e_g^0)$ ground state of trivalent cobalt and the high-spin $^5T_{2g}(t_{2g}^4 e_g^2)$ first-excited

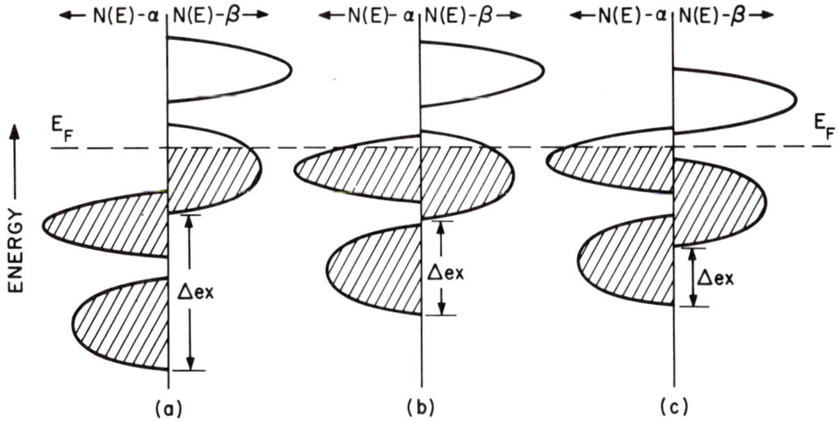

FIG. 20
Density-of-states versus energy for a band two-thirds filled having $b_c < b < b_g$ and (a) $\Delta_{ex} > fw_b$ (b) $fw_b > \Delta_{ex} > (f - \frac{1}{2})w_b$, and (c) $\Delta_{ex} \simeq (f - \frac{1}{2})w_b$.

state. At T = 0°K, the cobalt atoms are all diamagnetic. The population of the paramagnetic excited state increases with T. In fact, an anomaly in χ_m versus T above room temperature indicates that the population approaches 50-50 and that long-range ordering of high-spin and low-spin cations takes place at the higher temperatures![16].

Although the optical spectra due to excitations at isolated ions are well described by localized-electron theory, which takes account of the local point-group symmetry, an adequate description for magnetically coupled cations must introduce the crystal space-group symmetry. To do this the excited states are treated not as localized excitations, but as band states described by a wave vector **k** and having energies $\varepsilon(\mathbf{k})$ that occupy a narrow band of energies. These excited states are called excitons, and these exciton bands may be split by the translational symmetry of the crystal structure. These are called Davydov splittings, and such splittings have been observed and identified in the antiferromagnet Cr_2O_3[17], which has four Cr^{3+} ions per unit cell of the corundum structure (see Fig. 6). Although the excited-state single-ion energy levels are thus broadened into narrow bands by the magnetic coupling, this represents only a small perturbation of the basic localized-electron theory.

REFERENCES

[1]. L. Néel, Ann. Phys. [12], 3, 137 (1948).

[2]. I. E. Dzialoshinskii, J. Exptl. Theoret. Phys. (USSR) 33, 807 (1957); transl. Soviet Phys. JETP, 6, 621 (1958).

[3]. D. H. Lyons and T. A. Kaplan, Phys. Rev., 120, 1580 (1960); T. A. Kaplan, K. Dwight, D. H. Lyons, N. Menyuk, J. Appl. Phys. Suppl., 32, 13 S (1961).

[4]. U. Enz, J. Appl. Phys. Suppl., 32, 22 S (1961).

[5]. J. B. Goodenough, Bull. Soc. Chim France, 1200 (1965).

[6]. P. M. Levy, Phys. Rev. Letters, 20, 1366 (1968).

[7]. J. B. Goodenough, Progress in Solid State Chemistry, Vol. 5, H. Reiss, ed. (Pergamon Press, 1971). Chapter 4 gives a detailed discussion of the various problems raised here.

[8]. M. D. Banus and T. B. Reed, The Chemistry of Extended Defects in Non-metallic Solids, L. Eyring and M. O'Keeffe, eds. (North Holland, Amsterdam, 1970), p. 488.

[9]. J. B. Goodenough, Czech. J. Phys., B17, 304 (1967).

[10]. W. W. Warren, A. C. Gossard, and M. D. Banus, J. Appl. Phys., 41, 881 (1970).

[11]. J. B. Goodenough, Phys. Rev., 117, 1442 (1960); Colloque Internationaux du C. N. R. S. No. 157, Orsay, 28 Sept. - 1 Oct. 1965 (Edition CNRS, Paris, 1967); D. Adler and H. Brooks, Phys. Rev. 155, 826 (1967).

[12]. D. B. McWhan, T. M. Rice, and J. P. Remeika, Phys. Rev. Letters 23, 1384(1969).

[13]. J. B. Goodenough, Proc. 10th International Conf. Phys. Semiconductors, Cambridge, Mass., Aug. 17-21, 1970, S. P. Keller, J. C. Hensel, and F. Stern, eds. (Oak Ridge: Dir. of Tech. Inf., USAEC CONF-700801) p. 304.

[14]. W. L. Roth and R. C. DeVries, J. Appl. Phys., 38, 951 (1967).

[15]. A. Callaghan, C. W. Moeller, and R. Ward, Inorg. Chem., 5, 1572 (1966); J. M. Longo, P. M. Raccah, and J. B. Goodenough, J. Appl. Phys., 39, 1327 (1968).

[16]. P. M. Raccah and J. B. Goodenough, Phys. Rev., 155, 932 (1967).

[17]. J. W. Allen, R. M. MacFarlane, and R. L. White, Phys. Rev., 179, 523 (1969).

Chapter 25

MICROSTRUCTURE AND PROCESSING OF FERRITES

Frank J. Schnettler

Bell Telephone Laboratories
Murray Hill, New Jersey

I. INTRODUCTION

In our present society there are few of us who do not have an every day use of magnetic ceramics. At a recent symposium on Magnetism and Technology, Jacobs[1] estimated that the 1968 market for nonmetallic magnetic materials would exceed 3×10^8 dollars. This is for the material alone, not the devices incorporating it. His breakdown of the various categories of magnetic ceramics is listed in Table I. From this listing it is apparent that the use of ferrites is

TABLE I

CATEGORIES OF MAGNETIC CERAMICS

Television receivers	\$ 20 x 10^6
Communication and radio components, recording heads, transducers	20 x 10^6
Telephone communications	12 x 10^6
Microwave ferrite components	3 x 10^6
Computer memories	55 x 10^6
Permanent magnets	15 x 10^6
Recording tape	180 x 10^6
	\$305 x 10^6

broad. A great number of different materials have been developed to serve these many applications and in each of them the most desired properties have been optimized for their intended use. This discussion will attempt to cover the more interesting aspects of the chemistry, microstructure and properties of a particular material that was developed for the Touch-Tone telephone (Fig. 1).

Of the some ten million telephones manufactured by Western Electric in 1970, about three million will have the Touch-Tone dial. This dial is a dual-frequency signal generator designed to transmit two frequencies simultaneously. Using a 4 × 4 code, 16 combinations are available of which the telephone shown uses 10. Each pushbutton activates a single combination which is then transmitted to the central office for processing. At the heart of the dial are two ferrite core transformers which are shown in Fig. 2. This is a rear view of

FIG. 1
Telephone handset with touch-tone dial.

the dial. Surrounding it are an assembled transformer and the various piece parts of the assembly. Since this transformer is part of a precisely tuned circuit its requirements[2] are rigid. They are listed in Table II. The first three requirements, high Q, good stability with time and temperature, and small size are all concerned with the magnetic core. The next three requirements are concerned with the design of the transformer. Low cost, of course, affects both the core material and the transformer design. At the present time MnZn ferrite is the only material that meets these requirements.

FIG. 2
Rear view of touch-tone dial including transformer piece parts and assembly.

TABLE II

TRANSFORMER REQUIREMENTS

1. High Q
2. Good stability with time and temperature
3. Small size
4. High coupling coefficient
5. Adjustable inductance
6. Negligible external field
7. Low cost

II. MANGANESE ZINC FERRITE

Manganese zinc ferrite is a naturally occurring mineral (Fig. 3). Called franklinite by Berthier in 1819, it has been found in only two places, each as a large deposit and each within a few miles of Franklin, New Jersey. Because of contamination and oxidation, however, the mineral is only slightly magnetic. It was not until its synthesis by Snoek[3] that its potential as a high frequency core material was recognized. Manganese zinc ferrite is an isomorph of magnetite and as such has the cubic spinel structure. It exists over a wide range of compositions as shown in Fig. 4. For simplicity the

FIG. 3
Franklinite crystals in a bed of calcite (courtesy of Franklin Mineral Museum).

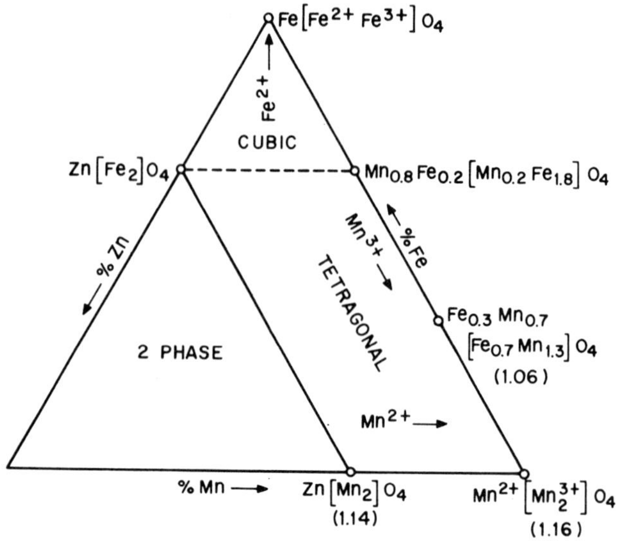

FIG. 4
Cation distribution in the (Fe, Mn, Zn) O_4 system; numbers in parenthesis are c/a ratio.

oxygen content has been held constant at a value of four atoms per three of metal. Ternary coordinates can thus be used to express the composition with reference only to metal components. Structure analyses have been reported[4],[5],[6],[7],[8] at each of the circled positions. Following the accepted convention, the metal ions occupying tetrahedral sites are given outside of the brackets and those in the octahedral sites are given within the brackets. A portion of the diagram has been labeled cubic and within it the components form a continuous series of solid solutions. When the manganese content is increased beyond that of the zinc ferrite-manganese ferrite join, trivalent manganese is introduced into the structure. The Mn^{3+} has a d^4 electron configuration[9],[10] and a preference for octahedral sites, which it distorts as a result of the Jahn-Teller effect. As long as the structure is held above some critical temperature, the distortions will be randomly oriented and the symmetry will remain cubic. Below that temperature, however, the distortions become cooperative[11] and a first order transition takes place. The result in this system is a structure with tetragonal symmetry. Some of the c/a ratios are shown in the diagram. In effect, we can consider the entire enclosed area a continuous series of solid solutions.

This system will also tolerate an extensive solubility of the fourth component, oxygen. The solution of oxygen occurs by a subtractional mechanism[12] which creates a vacant cation site for every three divalent iron or manganese ions oxidized to the trivalent state. Lattice vacancies have a pronounced effect on the cubic-tetragonal transition temperature as well as on such intrinsic magnetic properties as the magnetocrystalline anisotropy and magnetostriction.

An enlarged view of the cubic area is shown in Fig. 5. Again the mol fraction of oxygen is shown constant at 4/7. The scale at

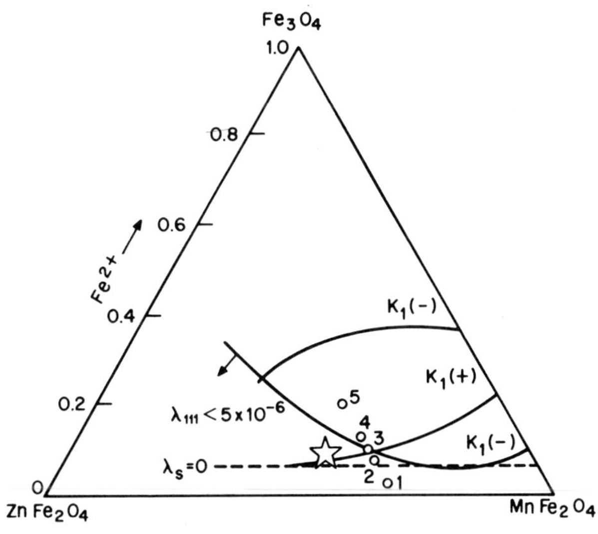

FIG. 5

Crystalline anisotropy and magnetostriction constants in the (Fe, Mn, Zn) O_4 system (after Ohta, Kobayashi and Shichijo; Ref. [13][14][15]).

the left leg of the triangle indicates the divalent iron content expressed as atoms per formula unit. We have included in this diagram some room temperature values of magnetocrystalline anisotropy[13] and of magnetostriction[14]. Two lines, drawn through the compositions for which $K_1 = 0$, enclose the region of positive anisotropy. The easy direction of magnetization within this region is [100]. It is [111] in materials with negative anisotropy. A line has also been included, below which the compositions have a magnetostriction in the [111] direction that is less than five parts per million. These measurements were all made on single crystals grown by the Bridgeman technique. Only an average value of magnetostriction (λ_s) can be measured on polycrystalline material. The dotted line indicates the compositions for which $\lambda_s = 0$[15].

Magnetic materials in general and ferrites in particular are burdened with magnetic imperfections. Voids in bodies that are not completely densified, nonmagnetic inclusions, precipitates, and the like are all a source of free magnetic poles and give rise to an internal magnetostatic energy. The stable domain configuration in a material is always such as to minimize this energy. The interaction of a moving domain wall and an imperfection is shown in Fig. 6. This is from some work by Williams[16] following the predictions of Neel[17] that such would be the case. The material is a single crystal of Fe — 3.8% Si. In the frame at the left the wall is shown intersecting an imperfection. That is, it would intersect but for the closure domains that form to minimize the magnetostatic energy. This constitutes a stable configuration. In the second frame an applied field has moved the wall away from the imperfection. The closure domains, however, have clung to the wall and would, if permitted,

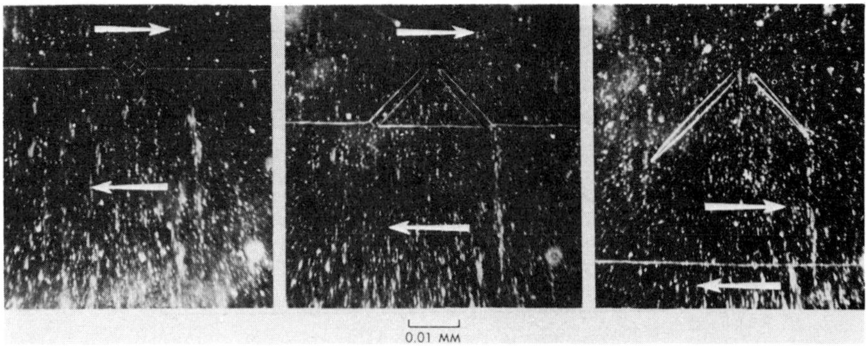

FIG. 6
Interaction of a domain wall with an imperfection in a single crystal of Fe-3.8% Si (after Williams, et al.; Ref. [16]).

draw it back. These spikelike domains continue to stretch as the wall is forced away until finally they snap off as shown in the third frame. This behavior imparts a stiffness to the motion of the domain walls and it is the primary source of hysteresis in magnetic materials. Hysteresis is directly related to microstructure and the loss that accompanies it is proportional to the imperfections of the material as well as its anisotropy, both crystalline and strain. Returning again to Fig. 5, a high permeability should be expected in polycrystalline materials when they are homogeneous and dense and have compositions that have very low values of K_1 and λ_s. The composition chosen for the Touch-Tone transformer core is indicated by the star. It does not lie at the exact point where K_1 and λ_s are zero at room temperature but has been modified, as will be discussed later, to meet the stability requirement.

Anisotropy is a temperature dependent property. The effect of this can be illustrated with two materials; one, with a composition such as point 2 where K_1 is negative at room temperature; and a second, such as that of point 4 where K_1 is positive. Turning to Fig. 7, the anisotropy of both materials is zero at the Curie tem-

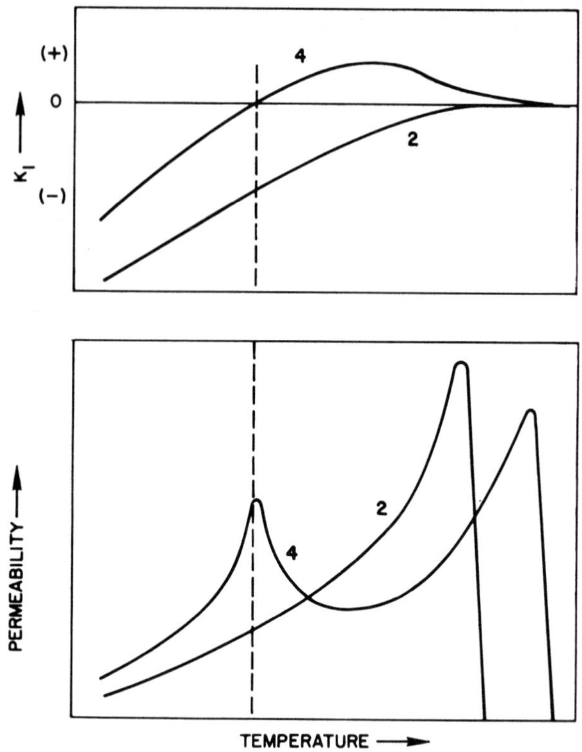

FIG. 7
Temperature dependence of the anisotropy constant K_1 and permeability of; (2) $Mn_{.61}Zn_{.31}Fe_{2.08}O_4$; and 4, $Mn_{0.56}Zn_{0.31}Fe_{2.13}O_4$ (after Ohta, Ref. [13]).

perature. It decreases monotonically with decreasing temperature in material "2." Material "4" which is higher in Fe^{2+} than "2," has positive value just below the Curie temperature. It rises to a maximum as the temperature drops, then drops through zero to negative values. Permeability varies inversely with the absolute value of K_1. This is apparent in the lower set of curves. Both materials have permeabilities near unity above the Curie temperature. Both show a sharp rise to a maximum just below the Curie temperature where K_1 is near zero. The permeabilities of both then decrease with decreasing temperature. Composition "4," however, shows a saddle corresponding to the temperature where K_1 has a maximum positive value, and a second maximum corresponding to $K_1 = 0$. This effect which can be controlled by the Fe^{2+} content will be exploited later to achieve the desired temperature stability.

Permeability, of course, is the ratio of the flux density induced in a material to the inducing magnetic field. In an alternating magnetic field, however, the induced flux will not be in phase with the field because of loss. It is then covenient to define a complex permeability[18],

$$\mu^* = \frac{B_0 e^{i(\omega t - \delta)}}{H_0 e^{i\omega t}} = \mu' - i\mu'' \tag{1}$$

where at any value of the magnetic field as given by $H_0 e^{i\omega t}$, the flux density will be delayed by some phase angle δ, and its value will be given by $B_0 e^{i(\omega t - \delta)}$. The μ^* can then be reduced to a real term, μ', which contains the component of B in phase with H and is therefore the normal permeability; and an imaginary term, μ'', which contains the component of B that is 90° out of phase with H and, therefore, is a measure of the loss per cycle. The loss factor of a material is usually expressed by

$$\tan \delta = \frac{\mu''}{\mu'} \tag{2}$$

The reciprocal of tan δ, Q, is another common term which is equal to the ratio of the energy stored per cycle to the energy dissipated.

$$\frac{1}{\tan \delta} = Q = \frac{\omega L}{\Delta R} \tag{3}$$

The quality of a core material is often expressed by the product μQ. It is a material constant that has the property of being independent of such air gaps as might be included in a magnetic circuit.

The frequency dependence of μ^* is shown in Fig. 8. Relaxation behavior of this type is typical of MnZn ferrite. Conductivity is

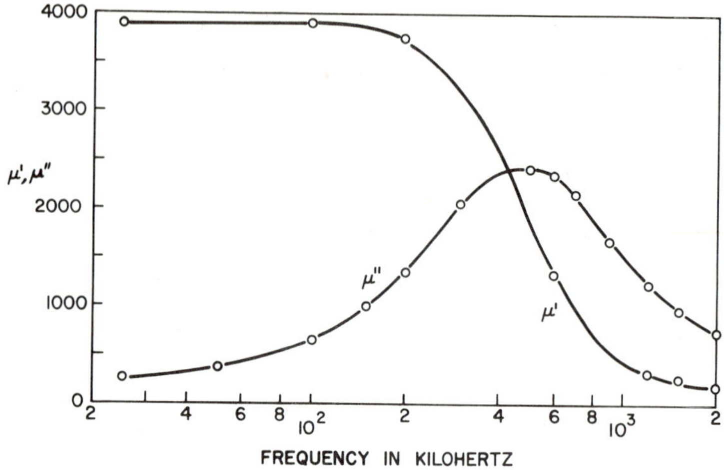

FIG. 8
Complex permeability versus frequency of $Mn_{0.5}Zn_{0.5}Fe_2O_4$.

probably the largest contributor to the viscous damping responsible for the relaxation. When divalent iron is present in a ferrite it occupies lattice sites crystallographically equivalent to those of trivalent iron. This permits a low energy mechanism for electron transfer and, hence, an electrical conductivity proportional to the divalent iron content. Since the relaxation frequency is inversely proportional to the viscous damping coefficient, it is also inversely proportional to the divalent iron content. The relaxation frequency is also directly proportional to the stiffness of the domain walls and, therefore, to the number of magnetic imperfections present. We find then that the course of μ' and μ'' are both dependent on the chemistry of the material as well as its microstructure. Let us examine now the methods used in manufacturing ferrite cores and what controls one can exercise to achieve the desired properties.

III. MANUFACTURING PROCESS

Cores for the Touch-Tone dial are manufactured in the solid state[19]. A 1500 pound charge of iron oxide, zinc oxide, and manganese carbonate is weighed out and then blended with a small amount of water in a mix-muller. After drying, the blended powder is calcined in a rotary furnace from which it is conveyed into a holding bin. Five hundred pound batches of this powder are ground in ball mills to a submicron size. At the same time a plasticizer is added. The resulting slurry is pumped into an agitated holding tank from where it is fed in a continuous manner through a spray dryer. At this stage, the product is in the form of spherical agglomerates suitable for dry moulding. The parts are formed in automatic presses and then fired in a continuous furnace under a protective atmosphere. After grinding to provide the proper air gap, the ferrite cores are ready for assembly into transformers.

The chemistry of this seemingly simple process can be rather complex. I have attempted a pictorial description in Fig. 9. Hematite, zincite, and manganite each with a different crystal structure, are mixed and the mixture converted to a single structure—MnZn ferrite—by heating. To do this efficiently the mixture is made as uniform as possible and then compacted to form point to point contacts between the various particles. The components interdiffuse on heating to form intermediate solid solutions, subcompounds and then regions of ferrite. It can be expected that the composition of the ferrite first formed is as varied as permitted by the phase boundaries of the system. With continued interdiffusion of the components, the ferrite regions grow in size and uniformity; at the same time the compact densifies as the pores diffuse outward. All of this proceeds irreversibly toward a dense homogeneous body. Homogeneous, that is, if the reaction is carried out under the correct partial pressure of oxygen. Actually the oxygen content of the ferrite must be controlled

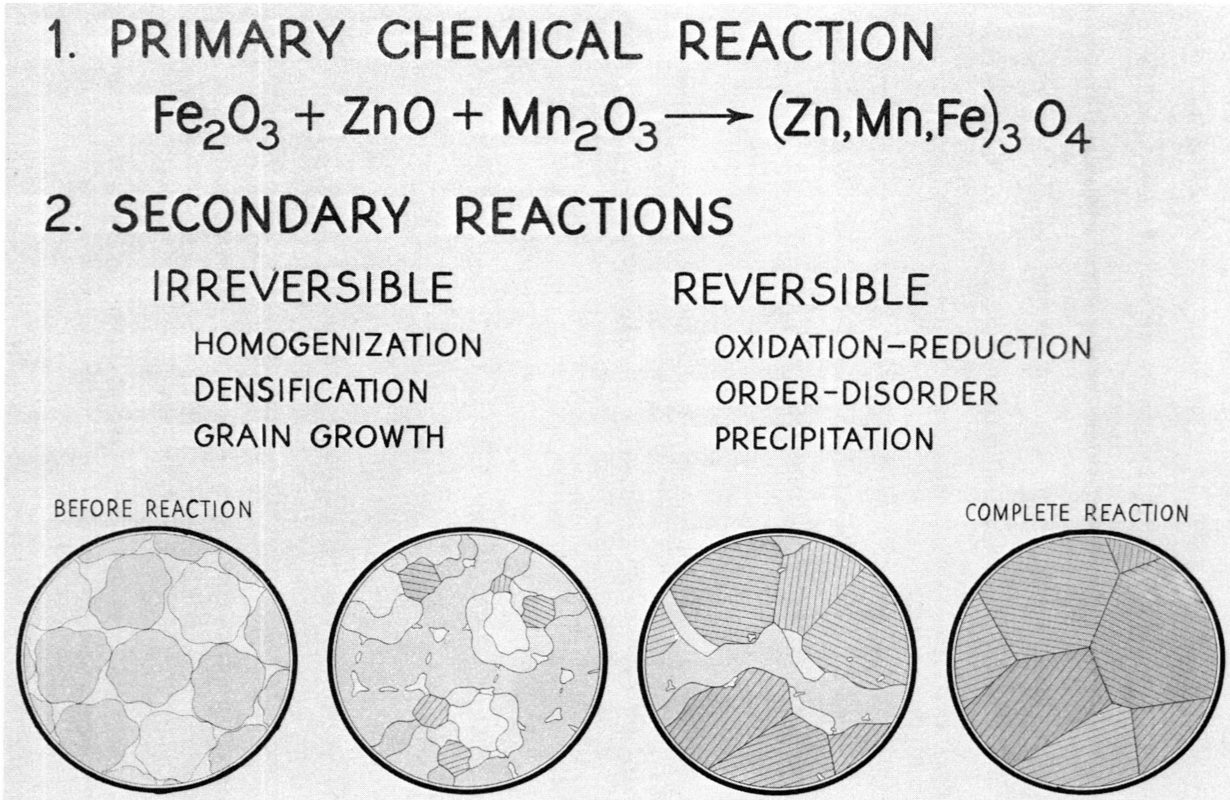

FIG. 9
Solid state reactions in the processing of manganese zinc ferrite.

to a much finer degree than just that necessary to keep the ferrite a single phase. Since it is in a trilateral equilibrium with the divalent iron and cation vacancies in the structure and since they both have a pronounced effect on the anisotropy of the ferrite, its control through the furnace atmosphere, is an important aspect of the manufacturing process.

If the oxygen content established during firing exceeds the solubility limit at some lower temperature, a second phase can be precipitated during cooling. This of course, would have a pronounced effect on the structure sensitive properties of the ferrite. Even before the actual separation of a precipitate, however, the stress inhomogeneites that arise induce local anisotropies and a corresponding reaction with the domain structure. Let us now consider the various factors of process control in terms of the properties required in the core of the Touch-Tone transformer.

IV. PROPERTY CONTROL

A. The Q and the μQ Product

It has already been shown that permeability increases as magnetic imperfections decrease, that is, it is directly dependent on the progress of densification, homogenization, and grain growth. The Q, on the other hand, has an opposite behavior as shown in Fig. 10. The length of grain boundary in a unit of surface was used as the structure parameter. This data was obtained from a MnZn ferrite fired at 1100°C under 5 mm of oxygen for various lengths of time. Some representative micrographs showing grain size and the onset of discontinuous grain growth have been included. The diameter of the field is 0.1 mm.

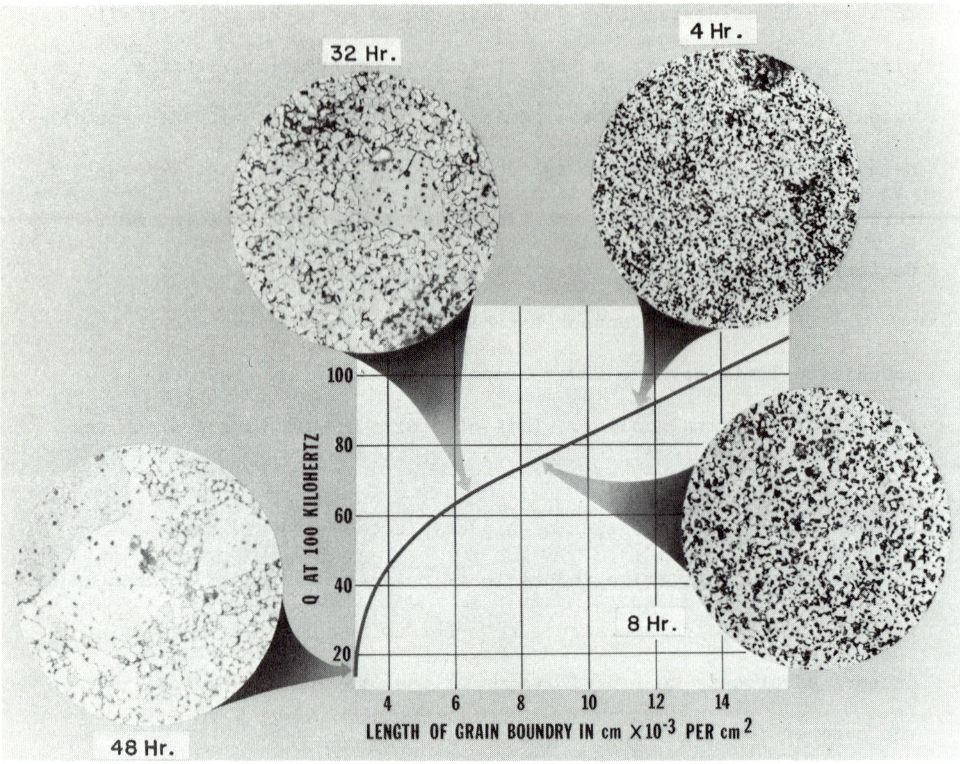

FIG. 10
Q versus length of grain boundary in MnZn ferrite fired at 1100°C (micrographs at 1000X).

The decrease in μQ as the grain boundaries are consumed by grain growth can be largely attributed to an increase in eddy current loss. Guilland and co-workers[20] found that certain additives, particularly calcium oxide, segregated to the grain boundaries in the ferrite and formed a high resistivity barrier. The loss of this barrier by homogenization and grain growth causes the observed effect. The quality of a ferrite then, as gauged by the μQ product, is at a maximum value at some intermediate stage of firing. Akashi[21] extended this work by looking at the combined effect of SiO_2 and CaO. His data showing the optimum values of μQ that can be obtained with various levels of Ca and Si are given in Fig. 11. The importance

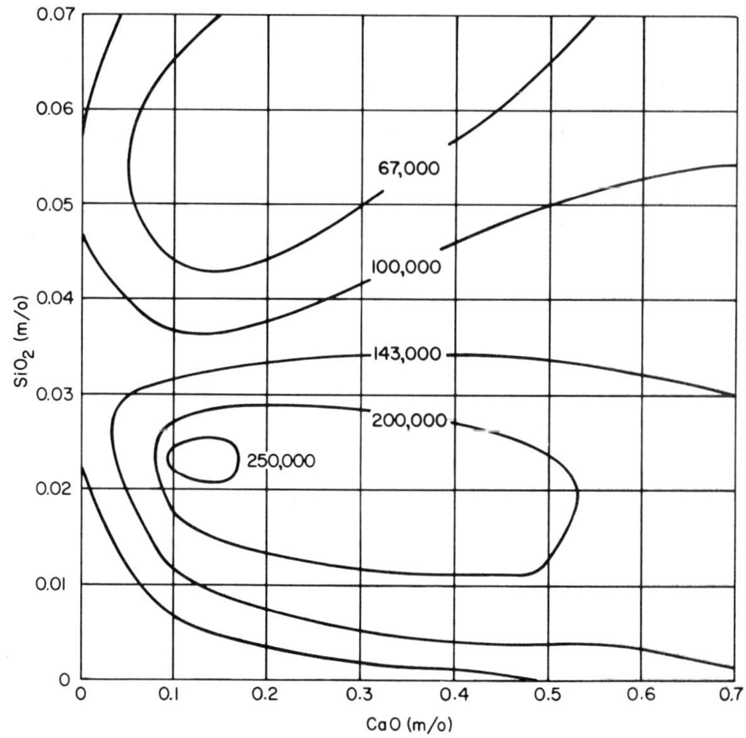

FIG. 11
Effect of SiO_2 and CaO additions on the μQ product of MnZn ferrite (after Akashi, Ref. [21]).

of controlling impurities in the manufacturing process is quite evident considering the extent to which such small additions can effect the magnetic properties.

B. Disaccommodation

There are a number of time dependent mechanisms by which a ferrite seeks to establish equilibrium between its crystal and magnetic structures. These range from electron transfer processes with activation energies less than 0.1 eV[22], through Jahn-Teller distortions, to ionic diffusion where the activation energy may exceed 1.0 eV[23]. Considering the range of time constants involved, these mechanisms contribute to loss over the entire frequency spectrum.

They also account for magnetic annealing effects[24] and for the time decrease in permeability that is observed in many ferrites. This latter effect is referred to as disaccommodation, a named coined by Snoek[25] who reported observing it in MnZn ferrite. His data is reproduced in Fig. 12. The decrease in permeability is the result of a time induced anisotropy conforming everywhere to the direction of the local spin system. This, in effect, freezes the domain walls to their position of rest by minimizing the energy at that position. More than one mechanism must be involved as the data cannot be described with just a single time constant. Iida and Inove[26] have shown the dependence of disaccommodation in MnZn ferrite on the degree of oxidation as shown in Fig. 13. The change in permeability with time of a commercial ferrite is shown in curve 1; oxidation increases disaccommodation as shown in curve 2. By removing just the right amount of oxygen from the sample they were able to reduce the disaccommodation to less than 0.02% per decade as shown in curve 3. Since more than half of the total change in permeability will occur in the first six weeks in the life of a core, a disaccommodation factor of 0.1% will guarantee an aging of less than 1/2% in twenty years.

C. Temperature Stability

The temperature stability of a ferrite is sensitive to both chemistry and microstructure. In the manufacturing process it is controlled by: first, selecting a suitable composition; then, obtaining the optimum degree of homogeneity; and, finally, adjusting both the composition and homogeneity by controlling the degree of oxidation during firing and cooling. Pelochek and Perduijn[27] show

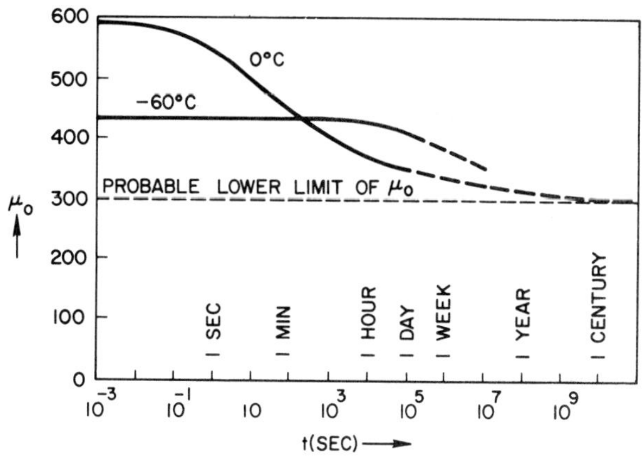

FIG. 12
Disaccommodation of MnZn ferrite (after Snoek, Ref. [25]).

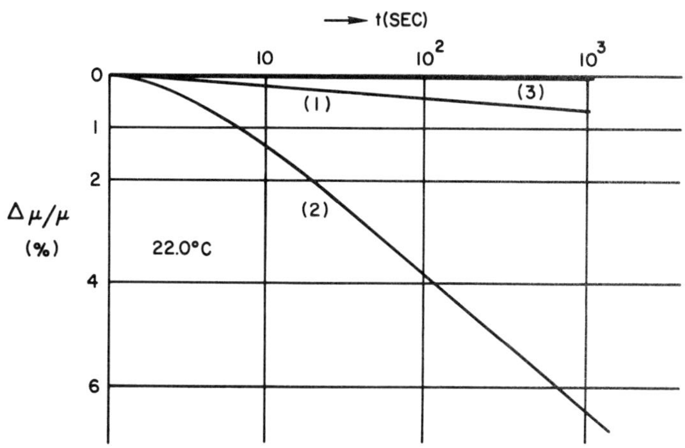

FIG. 13
Effect of oxidation on the disaccommodation of MnAn ferrite: (1) Commercial ferrite; (2) after oxidation; (3) after optimum reduction (after Iida and Inove, Ref. [26]).

the use of all three steps in Fig. 14 where they are used in obtaining a flat temperature coefficient in a high permeability MnZn ferrite. We have already discussed the temperature dependence of crystalline anisotropy. Its effect on the shape of the permeability—temperature curve is again illustrated in the upper left hand set of curves. The

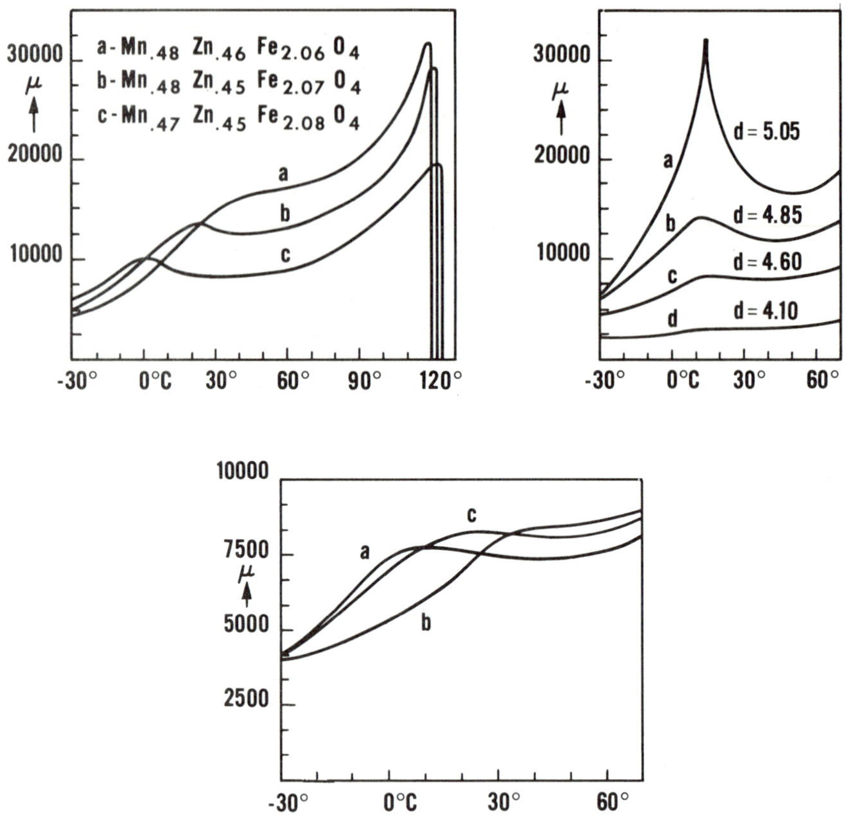

FIG. 14

Effect of composition, homogenity, and oxidation on the permeability-temperature relationship (after Peluschek and Perduijn, Ref. [27]).

temperature of the secondary peak in depressed as the divalent iron content increases from composition a to c. This temperature as well as the Curie temperature is an intrinsic property and as such is constant for any given composition. In a composite material the intrinsic properties at any temperature are the average of the composite. This has the effect of broadening the response with temperature of any dependent property such as permeability. The curves in the upper right corner were developed from composition b by firing for different periods of time. The structure parameter shown is density which follows the same course with time as homogeneity. One can then select a firing schedule which, consistent with the other magnetic properties, will give a minimum temperature coefficient. The set of curves at the bottom of the figure illustrate the effect of oxidation. The temperature of the secondary peak is again controlled by the divalent iron present. Here, however, the adjustment is made through the oxygen content, the material being the same for all three curves.

V. SUMMARY

It was probably evident in this discussion that the methods used to meet the requirements of one property could degrade another. This usually establishes a compromise in the end requirements which means even closer process control. Although our discussion has been limited to a single material the methods and the approach are in general the same for all materials. The control that can be achieved is shown in Fig. 15 where the actual stability performance of the Touch-Tone transformer is superimposed on the area of permissable deviation. This

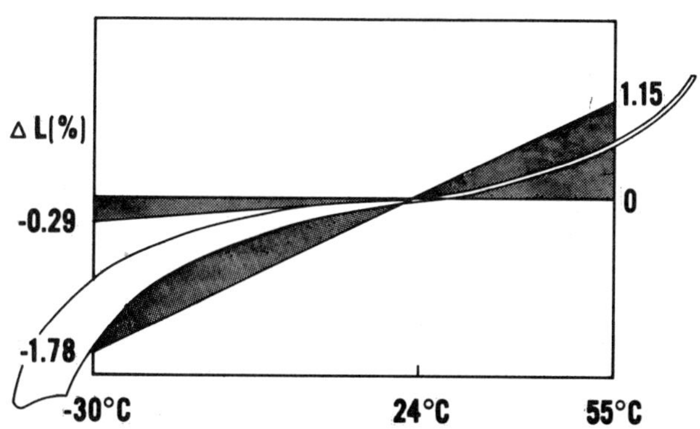

FIG. 15
Variation of inductance with temperature of commercial Touch-Tone transformers superposed on the permissable limits within the operating range.

is given as the percent change in inductance of a tuned transformer from the room temperature value. The positive slope was specified to compensate for the negative behavior of the capacitors and varistors in the circuit.

ACKNOWLEDGMENT

The author wishes to express his gratitude to his colleagues at Bell Telephone Laboratories and Western Electric Company for their many contributions to this presentation. In particular, the help of J. H. Ham, G. J. Maxson, E. J. Motyl, C. D. Owens, and P. I. Slick is acknowledged.

REFERENCES

[1]. I. S. Jacobs, J. of App. Phy., 40, 917 (1969).

[2]. J. H. Ham and F. West, IEEE Transaction on Communication and Electronics. Vol. 82, Mar. 1963, pp. 17-24.

[3]. J. L. Snoek, Philips Technical Review, 8, 353-384 (1946).

[4]. E. J. W. Verwey and E. L. Heilmann, Jr., Chem. Phys., 15, 174-180 (1947).

[5]. F. C. Romeijm, Philips Res. Repts., 8, 304-320 and 321-342 (1953).

[6]. S. C. Abrahams and B. A. Calhoun, Acta Cryst., 6, 105-106 (1953).

[7]. E. W. Gorter, Philips Res. Repts., 9, 403-443 (1954).

[8]. A. P. B. Sinha, N. R. Sinjana, and A. B. Biswas, Acta Cryst., 10, 439-440 (1957).

[9]. J. D. Dunitz and L. E. Orgel, Phys. Chem. Solids, 3, 318-324 (1957).

[10]. D. S. McClure, Phys. Chem. Solids, 3, 311-318 (1957).

[11]. P. J. Wojtowica, Phys. Rev., 116, 32-45 (1959).

[12]. G. Hagg and I. Sucksdorff, Z. Phys. Chem., 22B, 444-452 (1933).

[13]. K. Ohta, J. Phys. Soc. Jap., 18, 685-690 (1963).

[14]. K. Ohta and N. Kobayashi, Jap. J. App. Phys., 3, 576-580 (1964).

[15]. Y. Shichijo, Trans. Japan Inst. Metals, 2, 204 (1961).

[16]. H. J. Williams, R. M. Bozorth, and W. Shockley, Phys. Rev., 75, 155 (1949).

[17]. L. Neel, Cahiers Phys., 25, 21-44 (1944).

[18]. S. Chikazumi, Physics of Magnetism, 327, Wiley, New York (1964).

[19]. E. J. Motyl, The W. E. Eng, 7, #3, 3-11, 1963.

[20]. C. Guilland, R. Paulus, and R. Vaotier, C. R. Acad. Sci. Fr., 23 (242): 2712-2715 (1956).

[21]. T. Akashi, Trans. Japan Inst. Metals, 2, 171-176 (1961).

[22]. H. P. J. Wijn and H. van der Heide, Rev. of Mod. Phys., 25, 98-99 (1953).

[23]. R. M. Bozorth, E. T. Tilden, and A. J. Williams, Phys. Rev., 99, 1788-1798 (1955).

[24]. J. C. Slonczewski, Magnetism, Vol. 1, 205-241, Academic, New York.

[25]. J. L. Snoek, New Dev. in Ferromag. Mat., 54, Elsvier Pub. Co.

[26]. S. Iida and T. Inove, J. Phys. Soc. Japan, 17, Supp. B-1, 281-287 (1962).

[27]. H. P. Peluschek and D. J. Perduijn, IEEE Trans. on Mag., 4, 453-455 (1968).

Chapter 26

MICROWAVE GARNET COMPOUNDS

Gordon R. Harrison

and

L. Reese Hodges, Jr.

Sperry Microwave Electronics Division
Sperry Rand Corporation
Clearwater, Florida

I. INTRODUCTION

Magnetic ceramics of the ferrimagnetic structure are widely used in the microwave field. Many types are used so little attempt will be made here to cover all aspects of microwave ferrimagnetic ceramic materials. Three basic types or classes of magnetic crystal structures have found wide application in microwave components. Each of these types contains families of completely different compositions all of which possess their unique properties and characteristics applicable to microwave uses. These three classes of ferrimagnetic oxides are ferrites (spinels), garnets, and hexagonal materials. The ferrite materials (both spinels and hexagonals) have been treated in regard to structure and microwave properties in fairly complete detail in excellent reference books[1]-[8] and the readily available literature[9]-[13]. Emphasis here, therefore, is not placed on these materials; however, this should not be taken to negate the importance of these materials to the microwave industry. The garnet compounds are the most up-to-date ferrimagnetic materials being employed in microwave applications and prime treatment will be devoted to this class of microwave ferrites.

A. Yttrium Iron Garnet as a Microwave Material

Ferrimagnetic garnet materials have aroused much interest since their discovery[14][15], especially for their application in microwave devices. This interest has led to the substitution of over one half[16] of the metal ions in the periodic table into the garnet crystal structure. Of these ferrimagnetic garnet materials derived, yttrium iron garnet is by far the most useful for microwave applications. Many useful hybrid compounds have also been formulated in which yttrium iron garnet has been used as the basic composition to investigate the microwave properties of a number of polycrystalline mixed garnets. Some of the compositions which have been investigated include solid solutions of yttrium iron garnet with gadolinium iron[17]-[19][25], ytterbium iron, erbium iron, samarium iron[17][19][20][25], dysprosium iron, holmium iron[19], terbium iron, neodymium iron[16], yttrium aluminum[21]-[22][25]-[26], yttrium gallium[21][23]-[26], yttrium chromium[21][25]-[26], and gadolinium aluminum[26]. These materials are characterized by the following formula:

$$3[(1-x)Y_2O_3 \cdot xRe_2O_3] \cdot 5[(1-w)Fe_2O_3 \cdot wTr_2O_3]$$

where Re represents a rare earth substitution such as Gd, Ho, etc., and Tr represents a substitutional metal ion such as Al, Ga, etc. The microwave properties (magnetization, linewidth, Curie temperature, and g-factor) have been studied primarily from an engineering standpoint. This necessarily implies an investigation of the temperature and high power characteristics of these materials.

B. Hybrid Vanadium Iron Garnet as a Microwave Material

Another very useful family of materials has resulted from the studies of metal ion substitutions in the garnet structure[36]. These ferrimagnetic garnet compounds are characterized by the presence of pentavalent vanadium substitutions[37]-[42]. Many hybrid vanadium compounds have been formulated. Some of the more useful ones which have been prepared and evaluated are characterized by the following formulas:

YCaVIG	$\{Y_{3-2x}Ca_{2x}\}[Fe_2](Fe_{3-x}V_x)O_{12}$
BiCaVIG	$\{Bi_{3-2y}Ca_{2y}\}[Fe_2](Fe_{3-y}V_y)O_{12}$
YGdCaVIG	$\{(Y_{0.5}Gd_{0.5})_{3-2x}Ca_{2x}\}[Fe_2](Fe_{3-x}V_x)O_{12}$
BiCaV(Al)IG	$\{Bi_{0.4}Ca_{2.6}\}[Fe_2](V_{1.3}Fe_{1.7-w}Al_w)O_{12}$

The brackets indicate the garnet lattice site preference as follows: {24c} = dodecahedral sites; [16a] = octahedral sites; and (24d) = tetrahedral sites.

C. Structural and Ionic Properties of Substituted Materials

1. Structural Properties of Substituted Ions (YIG)

The structural configuration of the metal ions substituted into the garnet structure is very important in predicting the microwave properties of the metal ions substituted. The three lattice sites available for metal ions in the structure are known as 16a, 24c, and 24d sites. The lattice site location is in part determined by the size of the ion. In yttrium iron garnet, $3Y_2O_3 \cdot 5Fe_2O_3$,* all of the

*This formula can be written as $\{Y_3\}[Fe_2](Fe_3)O_{12}$.

yttrium ions are located on the 24c sites and the iron ions (ten per formula unit) are split between the other two sites (six on the 24d site and four on the 16a site). Gadolinium, ytterbium, erbium, samarium, dysprosium, holmium, terbium, and neodymium ions prefer the same site as yttrium. Gallium and aluminum ions, both smaller in size than iron, locate primarily on the 24d site, which is the smaller of the two iron sublattices. Even though chromium is smaller than iron, it prefers the 16a site, presumably because of its electronic configuration.

2. Ionic Location in Crystalline Structure (YIG)

Essentially all the microwave properties of ferrimagnetic garnet materials can be interpreted from the standpoint of ionic location in the crystalline structure. A strong antiparallel (antiferromagnetic) interaction exists between the ions on the 16a and 24d sites and a weaker, also antiferromagnetic, coupling between the 24c and 24d sites as indicated by the arrows in Table I. These interactions between the magnetic moments of the metallic ions concerned can be used to explain the magnetization of the material. According to Neel[27], a simple algebraic sum of these magnetic moments should predict the magnetization. Curie temperatures are primarily determined by the number of iron-iron interactions[21]. According to the theory of Kittel[28], linewidths are controlled predominantly by the ion located in the 24c site; the iron and rare earth ions both contribute to the g-factor of the material.

3. Structural Properties of Substituted Ions (BiCaVIG)

The metal ions normally used in such microwave garnets as yttrium iron garnet are all trivalent ions (refer to Table I).

TABLE I

STRUCTURAL PROPERTIES OF THE SUBSTITUTED IONS

Ion	Lattice site preference in garnets			No. of unpaired electrons[a]	Ionic radii[b] (Å)
	↓ 16a	↓ 24c	↑ 24d		
Y^{+++}		1.0		0	0.95
Gd^{+++}		1.0		7	0.97
Yb^{+++}		1.0		1	0.93
Er^{+++}		1.0		3	0.97
Sm^{+++}		1.0		5	1.00
Dy^{+++}		1.0		5	0.92
Ho^{+++}		1.0		4	0.91
Tb^{+++}		1.0		6	0.93
Nd^{+++}		1.0		3	1.04
Fe^{+++}	0.4		0.6	5	0.66
Al^{+++}			1.0	0	0.51
Ga^{+++}			1.0	0	0.62
Cr^{+++}	1.0			3	0.64
V^{+++++}			1.0	0	0.59
Bi^{+++}		1.0		0	0.93
Ca^{++}		1.0		0	0.99

[a]These unpaired electrons determine the magnetic moments per ion.

[b]Average values after Goldschmidt, Pauling, Zachariasen, and Ahrens, Table of Ionic Radii, TEM-PRES, Inc., State College, Pa.

Each metal ion generally exhibits a preference to one of the three lattice sites in the garnet structure. This preference is usually predictable. In the compositions of interest here, the pentavalent vanadium ion is substituted in the garnet structure. In so doing, and in order to maintain electrical neutrality, the divalent calcium ion is also substituted in appropriate proportions. In all compounds, therefore, calcium and vanadium are simultaneously substituted into the structure. One vanadium and two calcium ions replace three of the trivalent ions normally found in the structure.

Calcium and vanadium are both nonmagnetic. Vanadium prefers the 24d site. In pure YIG, three yttrium ions locate on the 24c site and there are two iron ions on the 16a site and three iron ions on the 24d site per formula unit, $\{Y_3\}[Fe_2](Fe_3)O_{12}$. Vanadium, therefore, is substituted for the iron on the dominant site. The calcium ion prefers the 24c site (generally referred to as the rare earth or yttrium site in garnet materials). In garnets, the magnetic coupling is antiferromagnetic; that is, the magnetic moments of the 16a and 24d sites are antiparallel. The magnetic moments, where present, of the 24c sites are antiparallel to the 24d sites. Since vanadium locates on the site normally occupied by the magnetic iron ions, the resulting compositions are expected to exhibit magnetizations that decrease with increasing vanadium substitution until the 16a site becomes dominant. It is also noted that similar situations should occur when gallium and aluminum are substituted in the garnet materials, because here again, nonmagnetic ions are being substituted for the highly magnetic iron ions.

4. Ionic Location in Crystalline Structure (BiCaVIG)

In the substitution of the calcium and vanadium into yttrium iron garnets, the magnetization of these compounds has been noted to decrease in a near linear fashion, going to 0 for a vanadium content of 1, and increasing again as the vanadium content is increased further. In a normal formula unit $\{Y_3\}[Fe_2](Fe_3)O_{12}$, the five iron ions split between the 16a and 24d sites, two iron ions being located on a 16a site and three on the 24d site. Vanadium has a strong preference for the 24d site and is nonmagnetic. When one vanadium ion per formula unit is substituted into the material, an equal number of magnetic iron ions locate on the 24d and the 16a sites. Since yttrium, calcium, and vanadium have no magnetic moment, this produces a magnetization equal exactly to 0. As more vanadium is substituted into the structure, the 16a site becomes the dominant magnetic site and the magnetization of the material is a result of the predominance of the iron ions located on this site. Of particular interest in the calcium-vanadium compounds is the fact that vanadium ions, even when substituted in large quantities, prefer the 24d site. This is a unique situation wherein the 16a site can be made the dominant magnetic site.

II. YTTRIUM IRON GARNETS

A. Materials Preparation Procedure

The preparation process begins with the formulation and weighing of the required oxide ingredients. (High purity and reagent grade oxides or carbonates are used throughout.) These oxides are mixed with water in stainless steel, ball mill jars containing stainless steel balls. After mixing and vacuum-filtering, the material is dried, screened, placed into platinum dishes and presintered at approximately 1200°C. After suitable lubricant (water) and binder are added, the presintered powder is pelletized and dry-pressed into the desired shape by hydraulically or isostatically pressing. These pressed samples are fired between 1400°C and 1500°C in muffle-type silicon-carbide element furnaces containing an oxidizing atmosphere to produce greater than 97% (of theoretical) dense magnetic ceramics.

In addition to ionic substitutions, the microwave properties of garnets may be controlled somewhat by the ceramic treatment given the material. Linewidth is a very strong function of density. In fact, all of the properties vary somewhat with density.

B. Important Properties and Measurement Techniques

State-of-the-art quality in microwave materials requires the accurate measurement and control of a minimum of eight magnetic, microwave and physical properties. These properties are shown in Table II.

Column 2 of Table II lists the equipment used to perform these measurements. The ASTM references are shown where standard methods exist. Various sample sizes are required for these measurements. A sample of any regular dimensions (rod or bar) may be used to measure density. Dielectric constant and loss tangent measurements at X band require a 0.040-inch-diameter rod, 0.5-inch long. Linewidth, g-factor, and Curie temperature measurements are made on sphere samples 20 ± 2 mils in diameter using a reflection type, X band cavity[29]. All samples should have 4/0 surface finishes. Saturation magnetization data are collected on sphere samples approximately 80

TABLE II

MATERIALS TESTS TOLERANCES

Material property	Equipment used	Measurement tolerance	Expected measured property variations for obtaining>90% yield
1. Density (gm/cm^3)	Calipers and micrometers for volume. Chain balance for weight. Accuracy: 4 mg/div Max load: 200 gm Weight: NBS Class P	±0.02 gm/cm^3	±0.05 gm/cm^3
2. a. Dielectric constant b. Dielectric loss tengent	X band transmission cavity, ASTM C525-63T	a. ±0.3 b. ±0.0002 for values <0.001	a. ±0.4 b. ±0.0008 max
3. Magnetization $4\pi M_s$ (gauss)	Vibrating sample magnetometer, ASTM C527-63T	3 to 5% (accuracy improves as value of $4\pi M_s$ increases)	±50 gauss for $4\pi M_s$ = 800 to 2,000 gauss
4. Linewidth ΔH (oe)	X band reflection cavity, ASTM C524-63T	±3%	±20 oe for ΔH = 40 to 150 oe
5. g_{eff}-factor	X band reflection cavity, ASTM C524-63T	±0.01	±0.01
6. Average grain size (μ)	Reichert metallograph	±2μ	±5μ
7. Curie temp.	Temp. chamber, and $4\pi M_s$ versus temp. characteristics	±5°C	±10°C
8. a. Remanent magnetization $4\pi M_r$ (gauss) b. Coercive field H_c (oe)	Square loop tester	a. ±5% b. ±0.1 oe	a. ±50 gauss for $4\pi M_r$ = 500 to 1,200 gauss b. ±0.2 oe for H_c = 0.5 to 2.0 oe

mils in diameter using a pendulum[30] or vibrating sample magnetometer[31] calibrated with pure nickel and pure iron spheres. High power measurements are made on sphere samples 30 ± 1 mils in diameter in an X band transmission cavity[32]. Average grain size is measured from a polished etched surface of the sample. Square loop and remanent properties are generally measured on toroidal-shaped samples.

C. Rare Earth Substitutions

Partial substitutions of rare earth ions for yttrium in yttrium iron garnet are quite interesting, particularly from the standpoint of temperature and high power characteristics. All of the trivalent rare earth ions investigated in this study possess unpaired electrons in their outermost shells and as a result have magnetic moments (refer to Table I). Rare earth substitutions, therefore, imply substituting a magnetic ion in the 24c site of the structure for the nonmagnetic trivalent yttrium ion. This site is not strongly coupled to the two iron sublattices; as a result, the magnetic moment of this site relaxes much faster as a function of temperature than do the iron ions. Typically, this fast relaxation of the magnetic moment of the rare earth ions, along with the antiferromagnetic coupling between the various lattice sites in the structure, produces compensation points in the magnetization versus temperature characteristics of the materials. The positions of these compensation points can be controlled by ionic substitutions and utilized to produce temperature-stable magnetizations in some materials.

D. Partial Substitutions of Rare Earth Ions for Yttrium in Yttrium Iron Garnet

1. Gadolinium Substitutions

One of the most inviting substitutions is gadolinium. Gadolinium has seven unpaired electrons and, thus, from the Neel[27] theory, gadolinium iron garnet has a magnetic moment of 32 Bohr magnetons per formula unit at 0°K. However, the magnetic moment of the gadolinium ions relaxes much faster as a function of temperature than do the iron ions, producing a compensation point at approximately 15°C. Variation of magnetization, linewidth, and g-factor with gadolinium content for mixed yttrium gadolinium iron garnet is shown in Fig. 1. Variations in linewidth and magnetization are observed since gadolinium locates in 24c site in the structure. At room temperature, linewidths remain essentially constant up to around 25% substitution of gadolinium while magnetization decreases steadily according to the percent of gadolinium substituted. The g-factor increases as higher gadolinium substitutions are made.

Figure 2 shows the variation of magnetization of these yttrium gadolinium garnets as a function of temperature. No change in the Curie temperature is observed, as expected, since the number of iron-iron interactions has not been altered. Variation of the magnetization with temperature in these materials is very interesting from an application standpoint. Particularly note the 60% gadolinium-substituted garnet. Here the magnetization of the material is constant over a fairly large temperature range. The compensation point for this composition is approximately -120°C. The magnetization can be varied at room temperature from the 1760 gauss of yttrium iron garnet to around 100 gauss of gadolinium iron garnet.

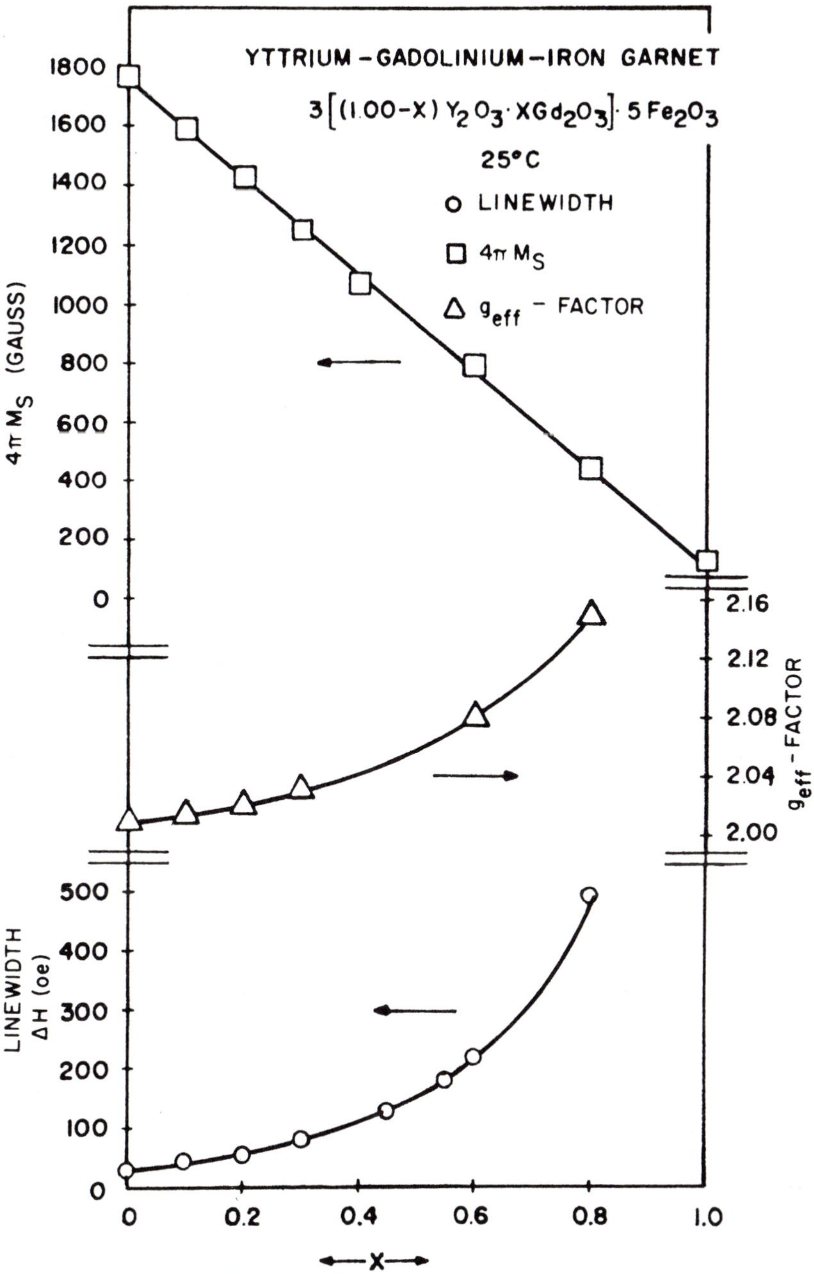

FIG. 1
Variation of linewidth, saturation magnetization, and effective g-factor with gadolinium content in yttrium gadolinium iron garnet.

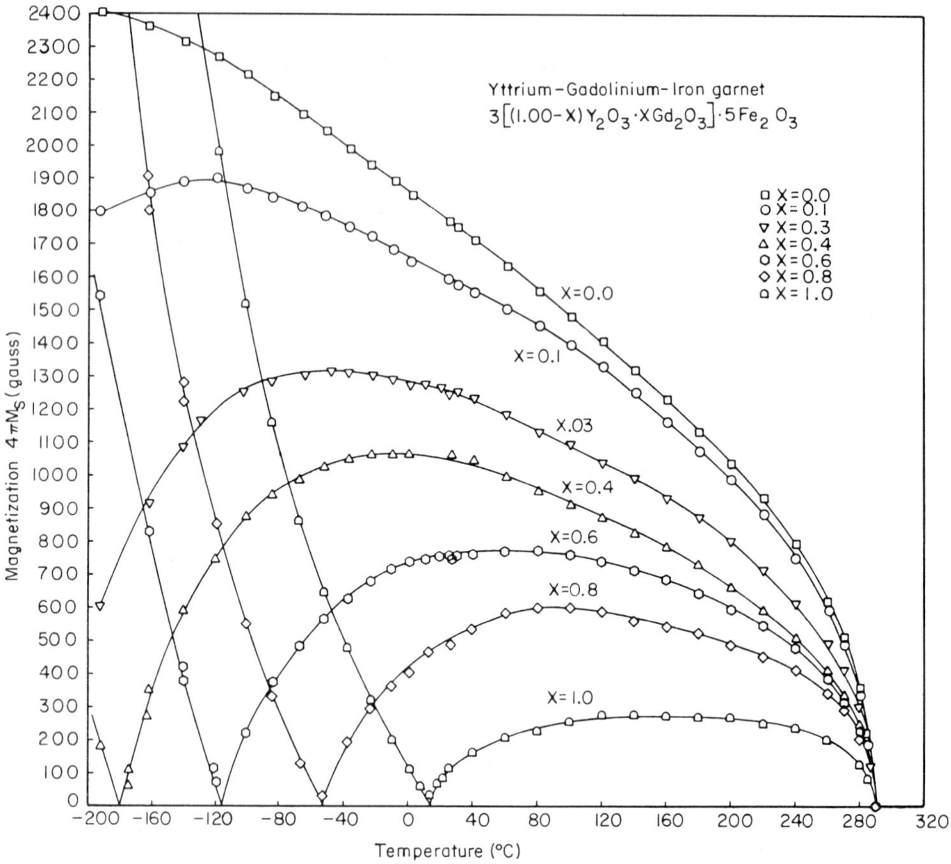

FIG. 2
Variation of saturation magnetization with temperature for mixed yttrium gadolinium iron garnet.

Figure 3 shows the linewidth variation as a function of temperature. Linewidths are observed to decrease in all cases as temperature is increased. At room temperature, gadolinium iron garnet has a very broad linewidth (<2000 oe) as compared to that of yttrium iron garnet (≃30 oe). The broad linewidth at room temperature for gadolinium iron garnet is due to the large anisotropy fields that exist at and near the proximity of the compensation point. Typically, this is true in regions near the compensation points for any of these materials.

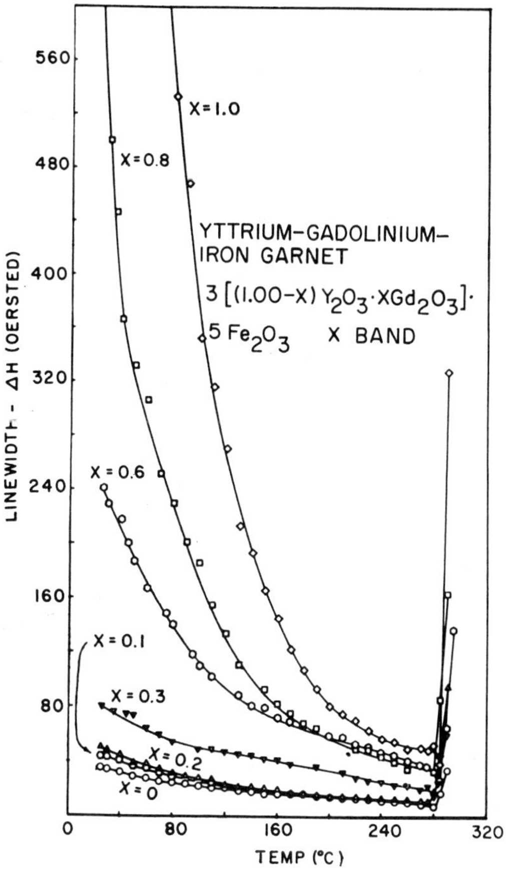

FIG. 3
Variation of linewidth with temperature for mixed yttrium gadolinium iron garnet.

Yttrium gadolinium iron garnets are quite useful in applications where medium magetizations are desired with low to moderate linewidths, and in devices where some temperature compensation of the magnetization is desired. Some difficulty is encountered, however, in designing a device which operates over a fairly large temperature range when the operation of the device is quite dependent on linewidth.

2. Ytterbium Substitutions

Figure 4 depicts the microwave properties of yttrium ytterbium iron garnet as a function of ytterbium content (substitutions up to 50% are shown). Ytterbium possesses properties very close to those of yttrium; the main difference is the one unpaired electron possessed by ytterbium. Moderate changes in linewidth and g-factor are observed.

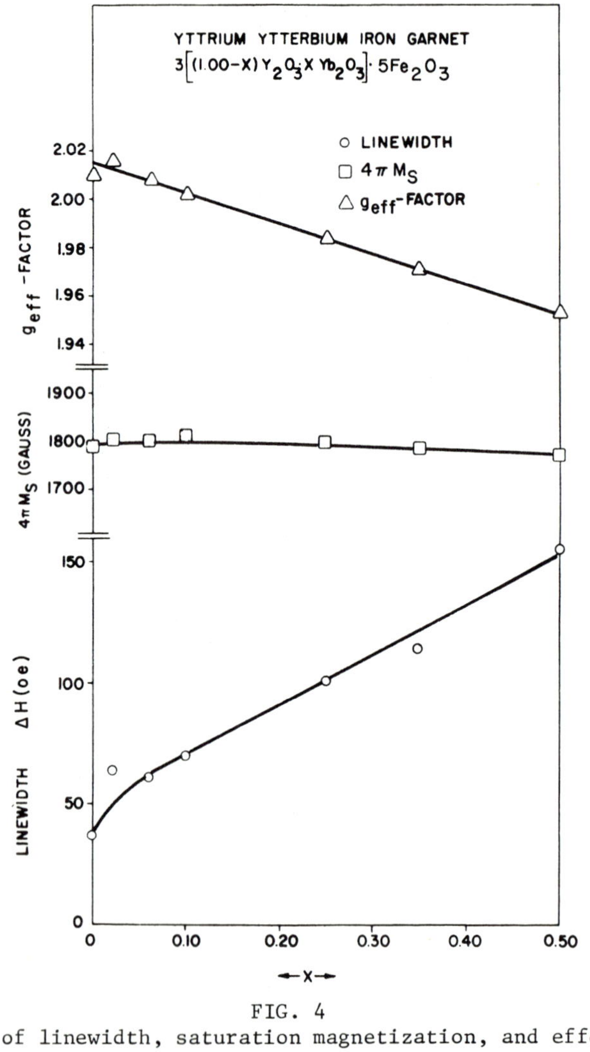

FIG. 4
Variation of linewidth, saturation magnetization, and effective g-factor with ytterbium content in yttrium ytterbium iron garnet.

(Note the degree of variation in these parameters as various rare earths are substituted in the structure.) This substitution produces no change in Curie temperature. The magnetization is also observed to remain essentially constant with composition.

3. Erbium Substitutions

The properties of yttrium erbium iron garnet are shown in Fig. 5. Measurements were made on substitutions up to 25%. Sharper

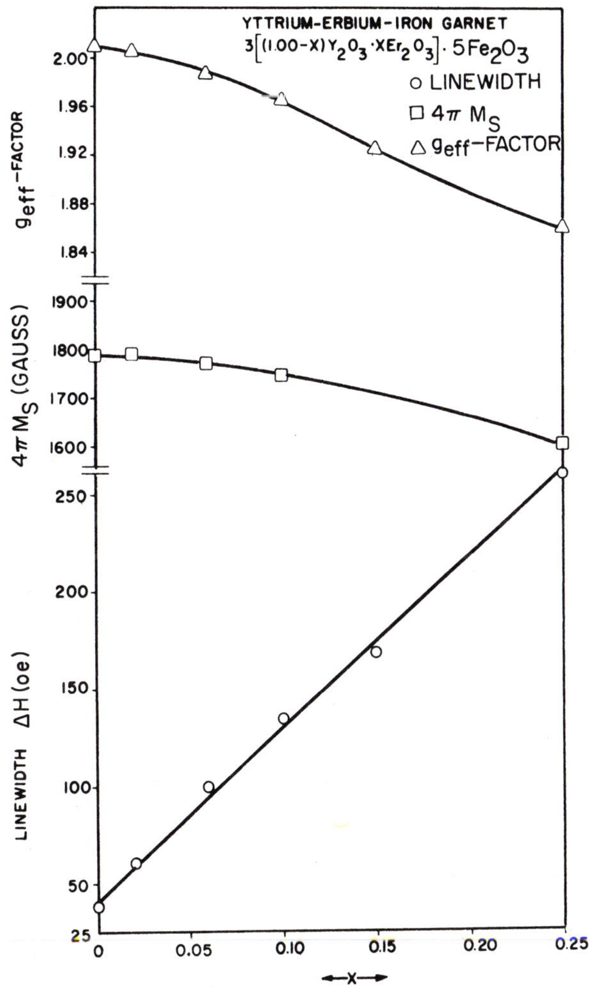

FIG. 5
Variation of linewidth, saturation magnetization, and effective g-factor with erbium content in yttrium erbium iron garnet.

variations in linewidth and g-factor are observed. The g-factor decreases and the linewidth increases with erbium content. Some moderate changes in magnetization are observed; the Curie temperature of 285°C is unaffected by erbium substitution.

4. Samarium Substitutions

Samarium surprisingly behaves very much like yttrium in the garnet structure. No compensation point has been observed even though samarium has five unpaired electrons. The magnetization and Curie temperature of samarium iron garnet are essentially the same as for yttrium iron garnet. The odd behavior of the magnetization here evidently arises from the electronic configuration of the samarium ion. The properties of yttrium samarium iron garnets have been studied for substitution up to 10% samarium as shown in Fig. 6. Linewidth is observed to increase in a linear fashion with increasing samarium additions. In the small substitutions studies, little change in the g-factor is observed. Little change in the magnetization is observed for all substitutions studied. These samarium-substituted garnets exhibit the following properties: constant magnetization for all substitutions; linear control of linewidth over very broad limits. No temperature compensation in the material, as such, can be obtained.

5. Dysprosium Substitutions

Dysprosium iron garnet possesses a compensation point at approximately -50°C. For temperature compensation, dysprosium substitutions in yttrium iron garnet are most promising next to that of gadolinium. The microwave properties of yttrium dysprosium iron garnet are shown in Fig. 7. The linewidth variation with dysprosium content is greater than that with gadolinium. In fact, before good

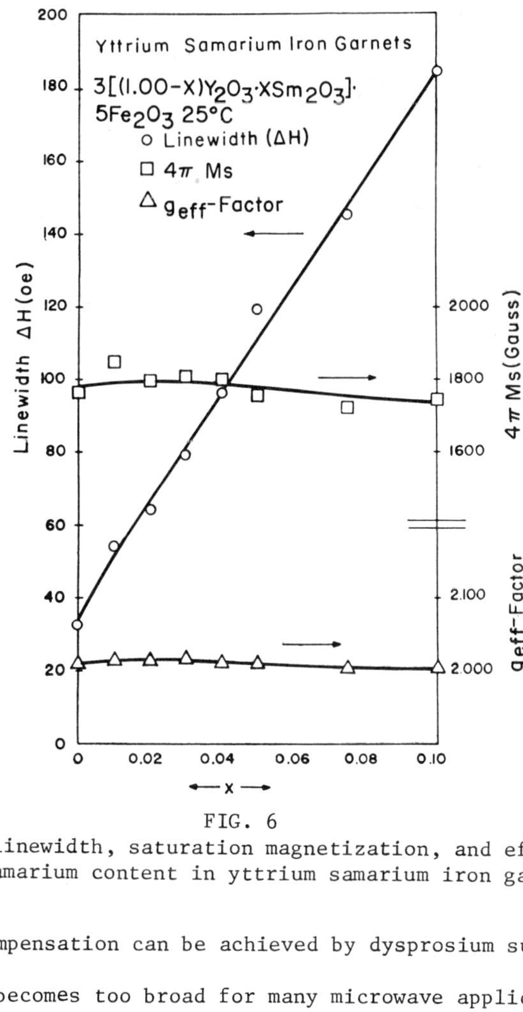

FIG. 6
Variation of linewidth, saturation magnetization, and effective g-factor with samarium content in yttrium samarium iron garnet.

temperature compensation can be achieved by dysprosium substitutions, the linewidth becomes too broad for many microwave applications. The g-factor decreases with dysprosium content and the Curie temperature for all substitutions is approximately 285°C.

6. Holmium Substitutions

Holmiun substitutions in yttrium iron garnet produce much the same results as dysprosium. The linewidth increases with holmium content faster than with previously mentioned substitutions. The g-factor decreases and magnetization and Curie temperatures are essentially constant for substitutions up to 15% as shown in Fig. 8.

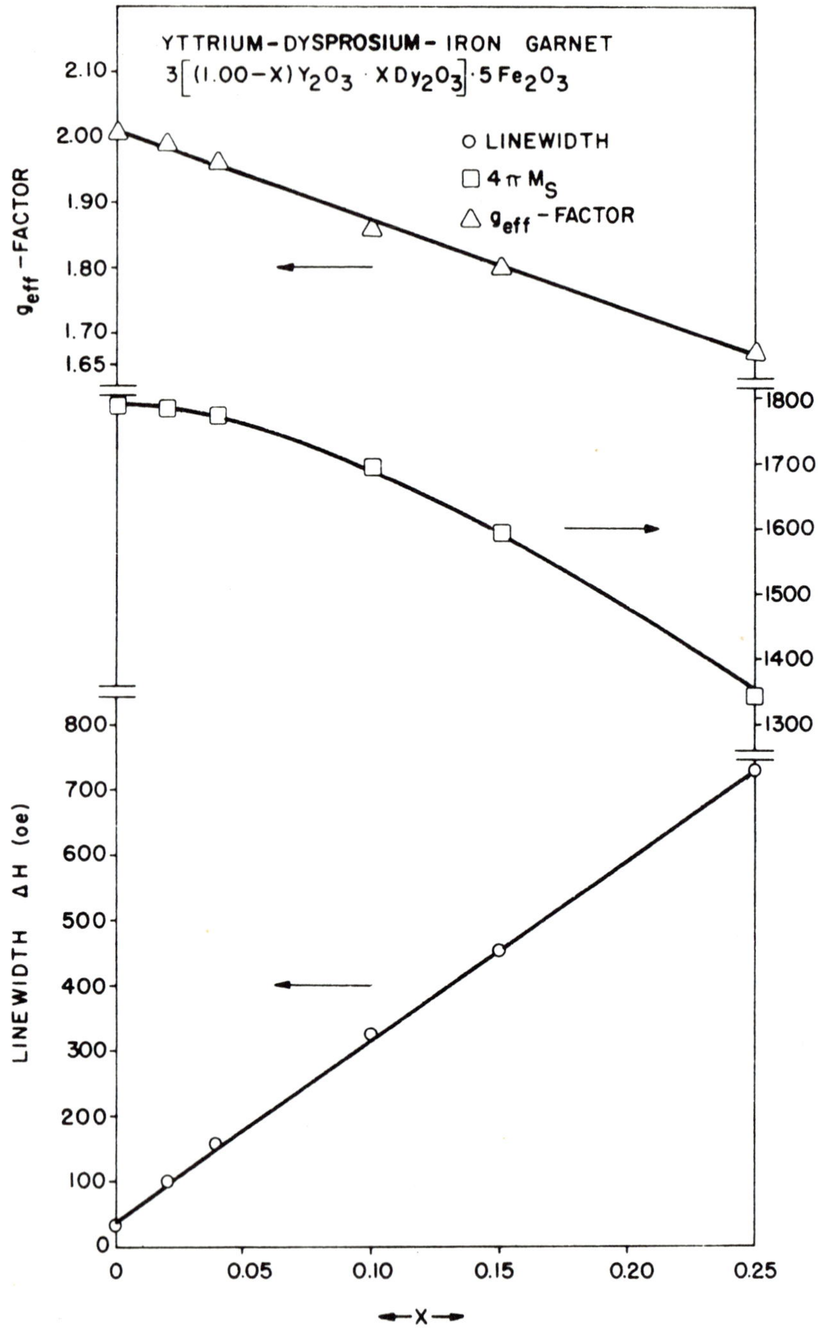

FIG. 7
Variation of linewidth, saturation magnetization, and effective g-factor with dysprosium content in yttrium dysprosium iron garnet.

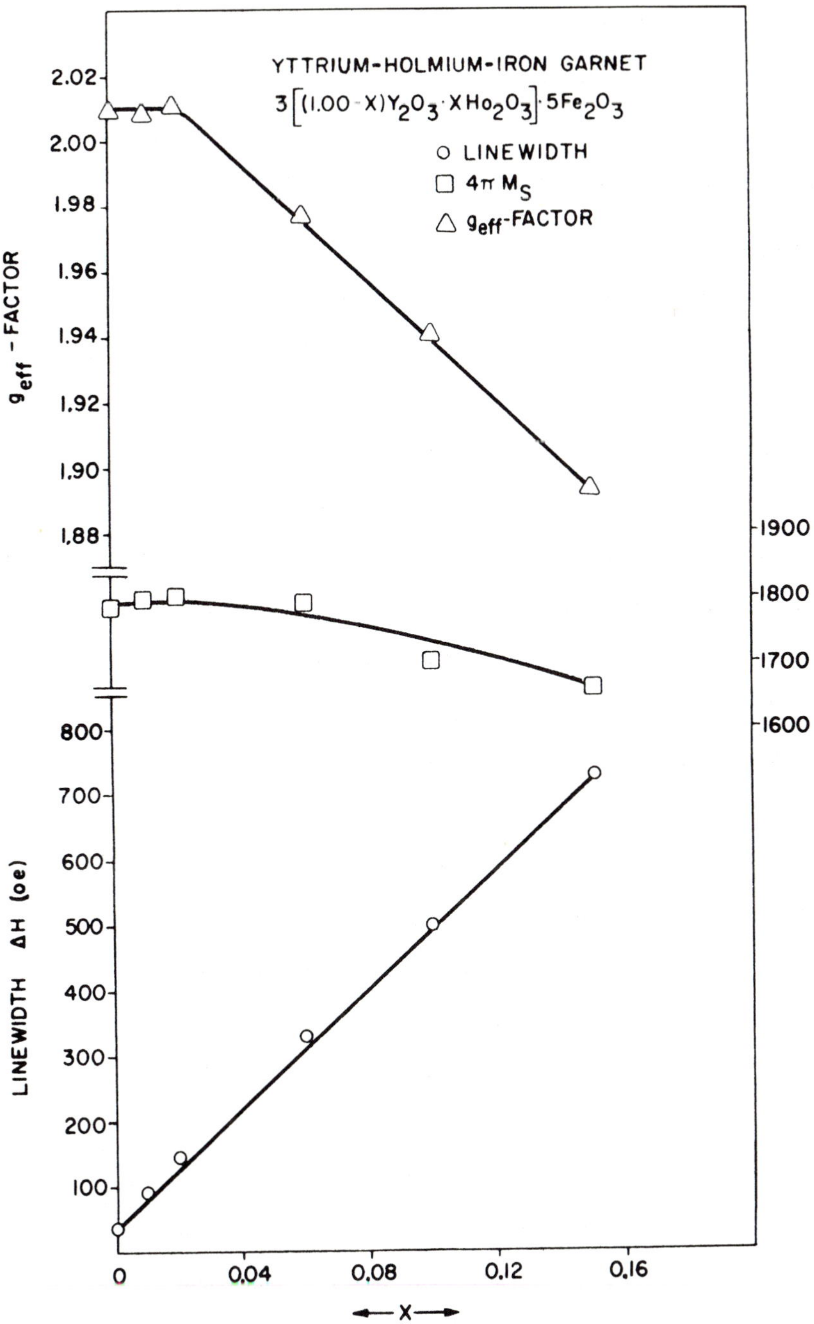

FIG. 8
Variation of linewidth, saturation magnetization, and effective g-factor with holmium in yttrium holmium iron garnet.

7. Terbium Substitutions

This same variation is noted in terbium substitutions (up to 10%) as depicted in Fig. 9. Linewidth variation with percent terbium is greater than any substitutions previously discussed.

8. Neodymium Substitutions

Figure 10 depicts the microwave properties of yttrium neodymium iron garnet. X-ray data reveal a second phase in these materials for substitutions greater than 10%. Neodymium is a large ion as noted from Table I. As a result, only small additions may be made in the structure without putting smaller ions elsewhere in the structure[16]. Nevertheless, the properties of the multiphase material are promising for microwave purposes. The reproducibility of the properties is somewhat more difficult, however, in a multiphase material. No data were collected on the reproducibility of the properties reported here.

E. Effects of Substitutions on Material Characteristics

1. Linewidth Variations as a Function of Temperature

Linewidth versus temperature data were collected on all samples from room temperature to the Curie temperature. Linewidth in all cases decreases with increasing temperature. Typical data (in this case, ytterbium and dysprosium substitutions) are shown in Figs. 11 and 12 for completeness. The variation in linewidth with temperature is not of the same order for all substitutions as can be noted by comparing these two figures with Fig. 3. The magnitude of the variation depends upon the ion substituted and the position of the compensation points.

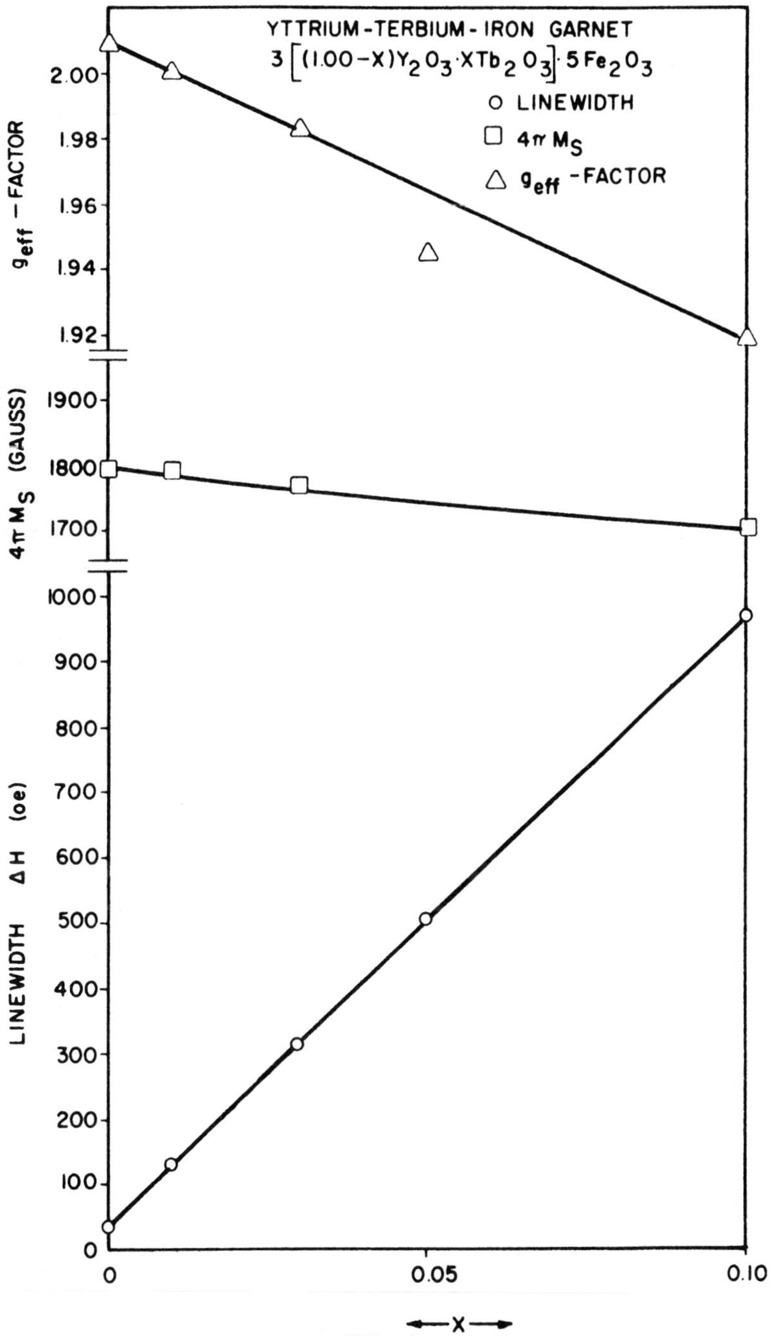

FIG. 9
Variation of linewidth, saturation magnetization, and effective g-factor with terbium content in yttrium terbium iron garnet.

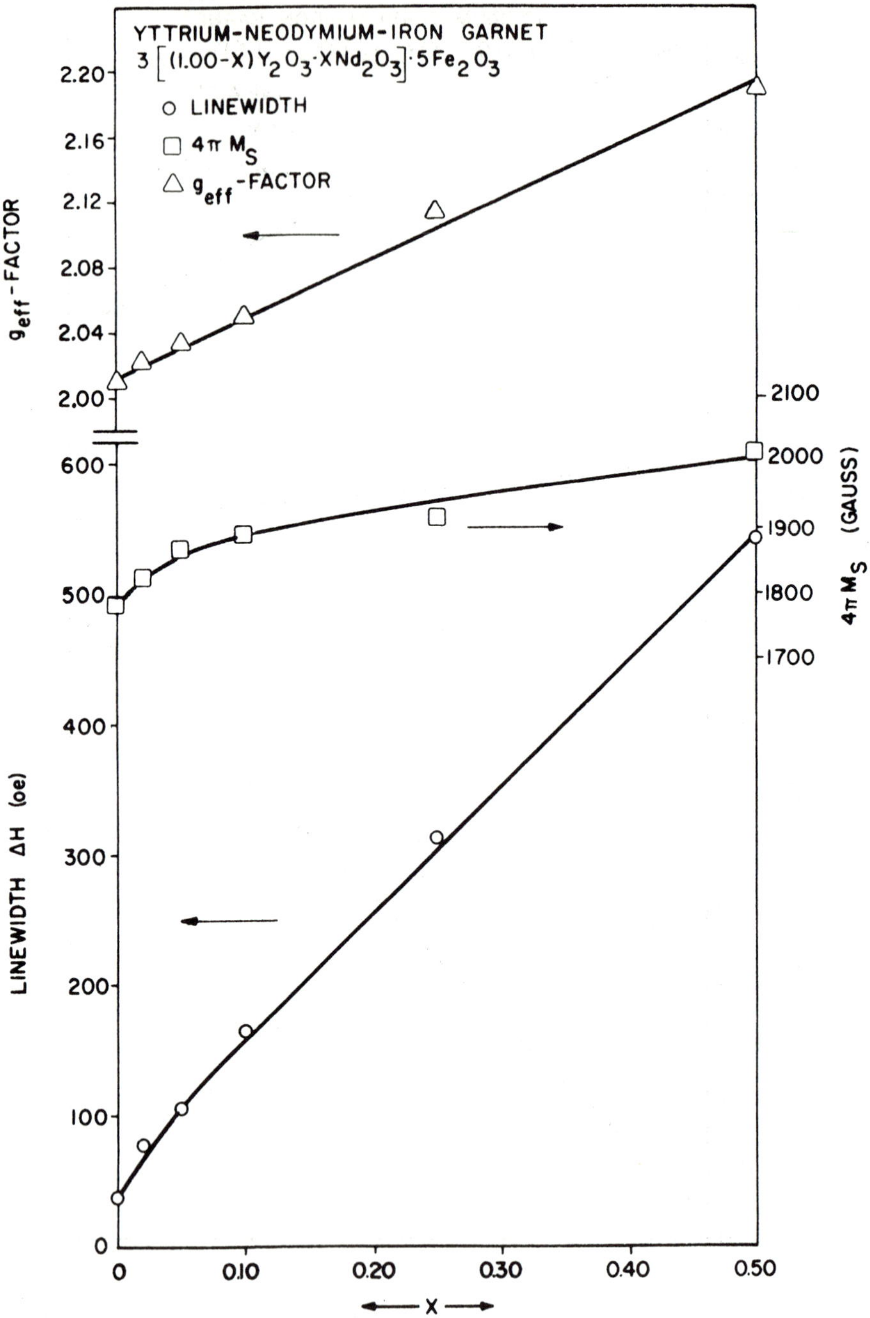

FIG. 10
Variation of linewidth, saturation magnetization, and effective g-factor with neodymium content in yttrium neodymium iron garnet.

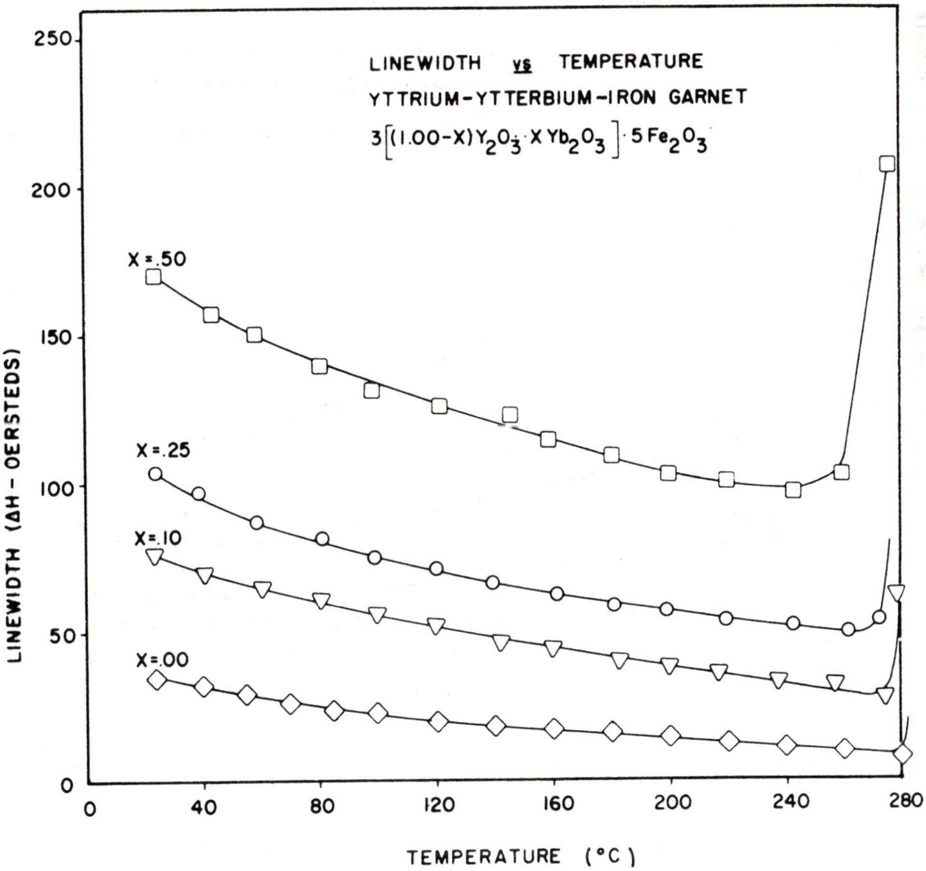

FIG. 11
Temperature variation of linewidth for yttrium ytterbium iron garnet.

2. Magnetization Variations as a Function of Temperature

Figures 13, 14, and 15 present typical data on the variation of magnetization ($4\pi M_s$) with temperature. Figure 13 depicts ytterbium and holmium substitutions; very little change is noted in the shape of the curves. Figure 14 presents the dysprosium data. Some temperature compensation is observed for 25% substitutions; higher substitutions would produce better temperature characteristics but the resulting broad linewidths would limit the usefulness of the material.

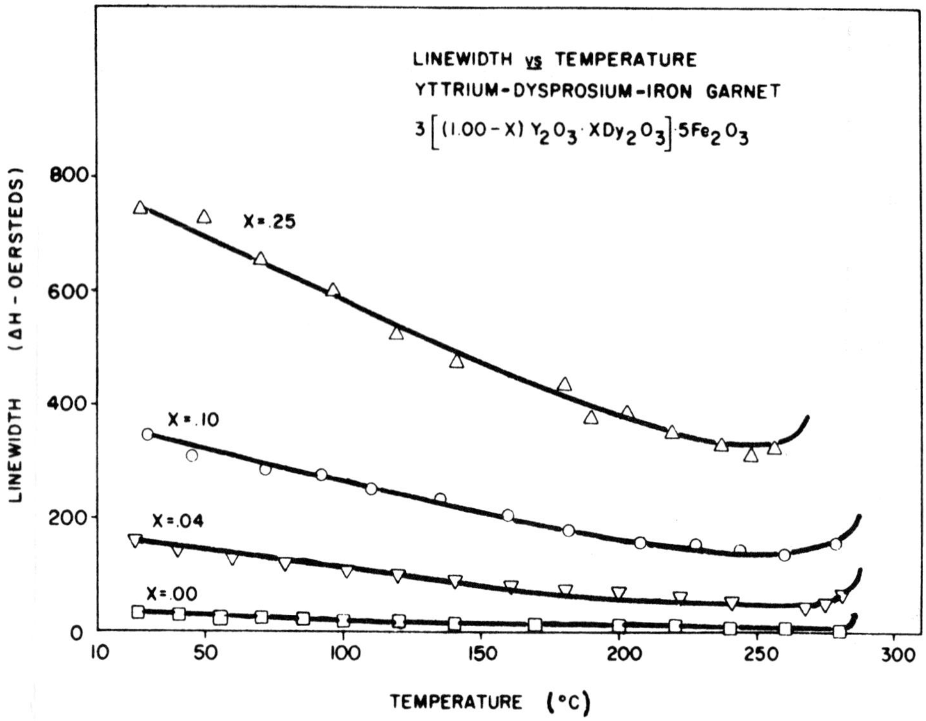

FIG. 12
Temperature variation of linewidth for yttrium dysprosium iron garnet.

Figure 15 presents the data on terbium- and neodymium-substitued garnets. These data are similar to that shown in Fig. 13 and are typical for the other rare earth substitutions.

3. Linewidth Variations as a Function of Composition

For comparison purposes, the linewidth as a function of percent rare earth substitution is presented in Fig. 16 for all the rare earth substitutions studied. All rare earth substitutions in yttrium iron garnet increase the linewidth to some degree. This is in agreement with arguments presented previously that the ions located in the 24c site control almost entirely the linewidth of the material[28]. To add to this argument, linewidth versus aluminum

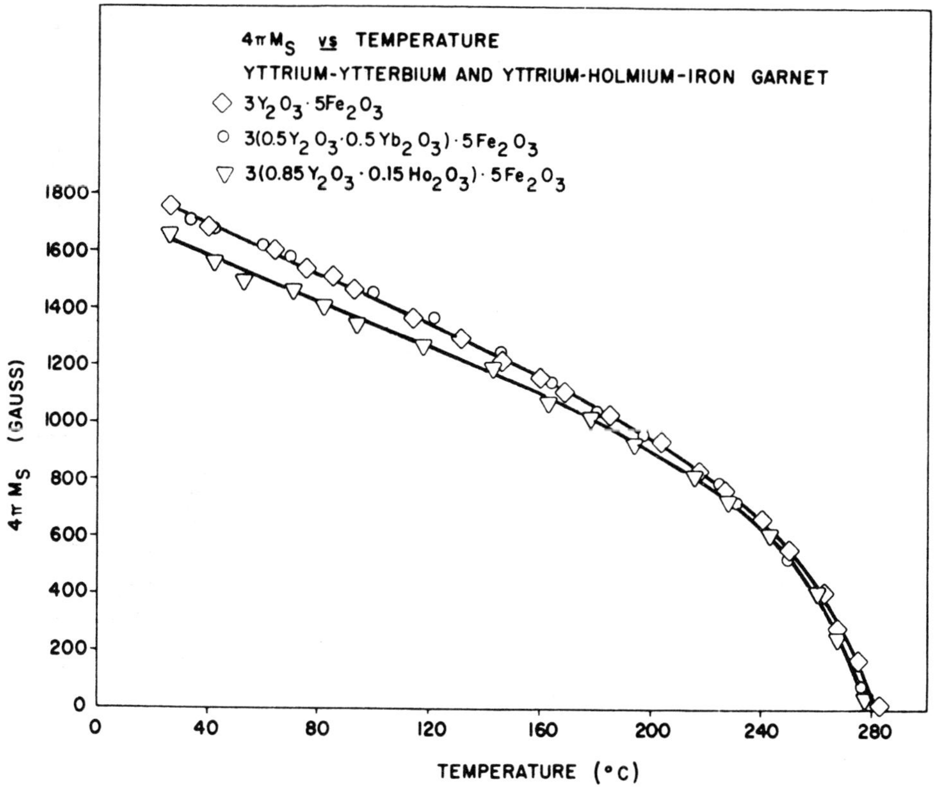

FIG. 13
Variation of saturation magnetization with temperature for yttrium ytterbium and yttrium holmium iron garnet.

content (to be discussed later) is presented also in Fig. 16. Aluminum locates primarily in the 24d site in the structure. This substitution does not affect linewidth to any appreciable extent. Since aluminum replaces iron in the structure, Curie temperature and magnetization are decreased as will be shown later.

4. High Power Thresholds as a Function of Composition and Linewidth

High power studies conducted at X band on some of these materials are shown in Fig. 17. Here the measured h_{rf} required for the onset of high-power nonlinear instability (H_c) is plotted as a

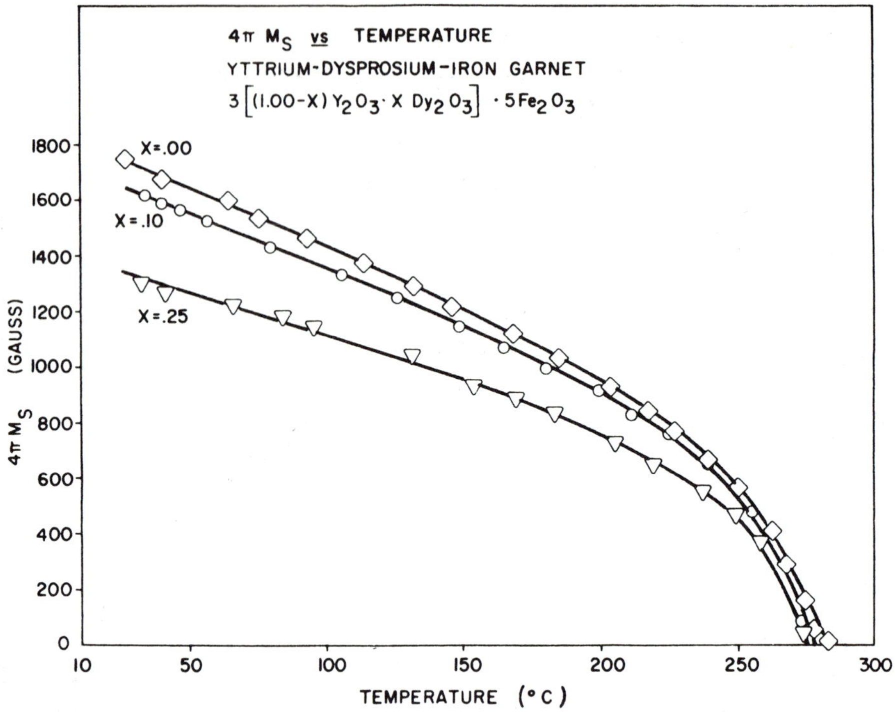

FIG. 14
Temperature variation of saturation magnetization for yttrium dysprosium iron garnet.

function of composition. Spheres placed in a waveguide cavity were used as samples; the conditions were such that the main and subsidiary resonance were noncoincident[33][34]. Figure 17 also shows the power required in X band waveguide to produce the threshold rf field. These data show that the thresholds for high power instability increase with increasing rare earth content. The high power thresholds increase in the same manner as do linewidths with rare earth substitution (compare Figs. 17 and 16). The threshold for high power instability can be increased to conveniently high values if suitable rare earth impurities are substituted in the yttrium iron garnet structure[32][34][35]. Alu-

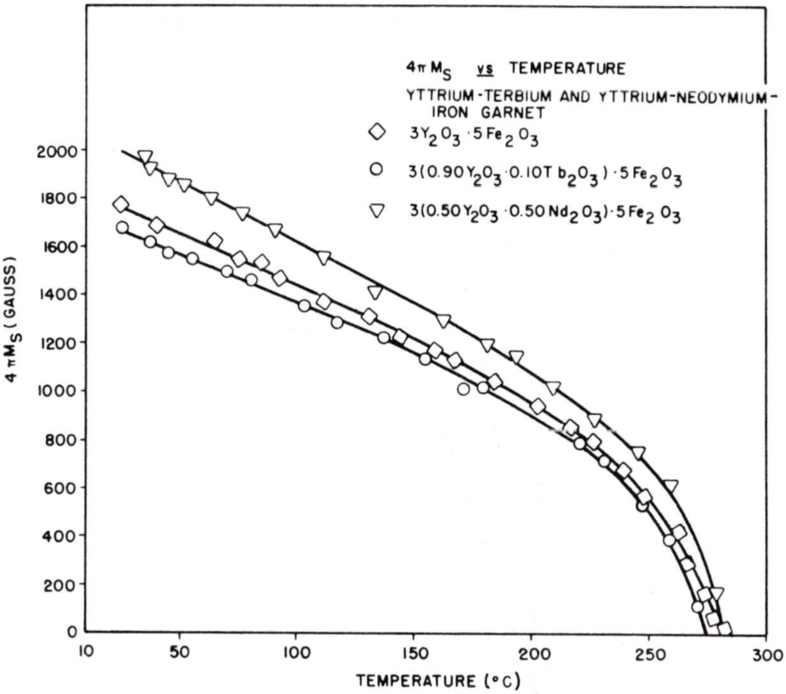

FIG. 15
Temperature variation of saturation magnetization for yttrium terbium and yttrium neodymium iron garnet.

minum substitutions have little effect on the high power threshold as noted in Fig. 17.

The data in Figs. 15 and 17 can be combined to derive a linear relationship between the rf magnetic field (H_c) required for the onset of nonlinear instability and the linewidth of the material. This empirical relationship is as follows: $H_c = 0.135\ \Delta H - 3.6$ for $\Delta H \geq 30$ oe, where H_c and ΔH are in oersteds. This empirical equation is accurate to ±10% for all rare earth substitutions previously discussed, except gadolinium. The linewidth and thus H_c of the gadolinium-substituted garnets are not linear with gadolinium content due to

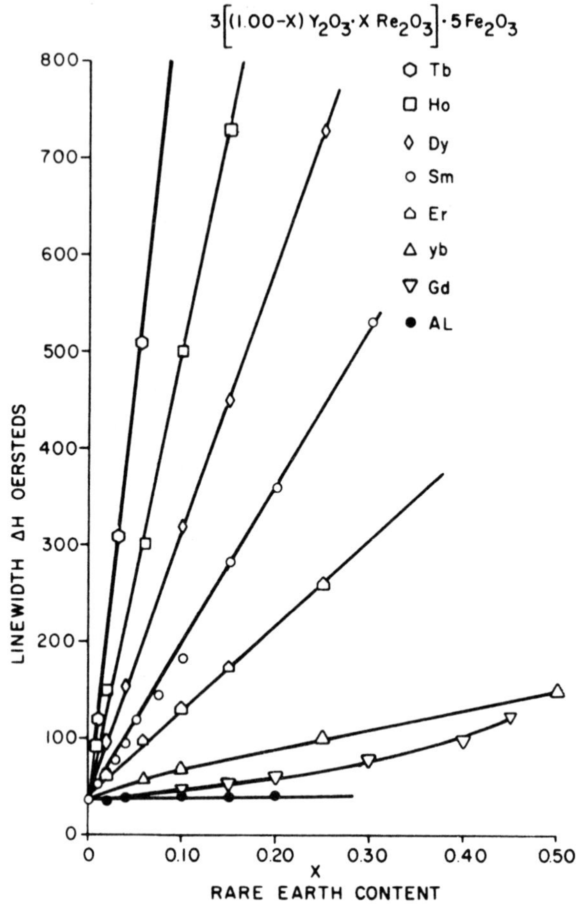

FIG. 16

Linewidth variation as a function of composition for terbium, holmium, dysprosium, samarium, erbium, ytterbium, gadolinium, and aluminum-substituted yttrium iron garnet.

the close proximity of the compensation points to room temperature and the linear decrease of $4\pi M_s$ with composition. The above equation is applicable for determining the microwave rf magnetic field or power required to observe decline of main resonance (material magnetically biased to resonance) under the conditions that main and subsidiary resonance are noncoincident.

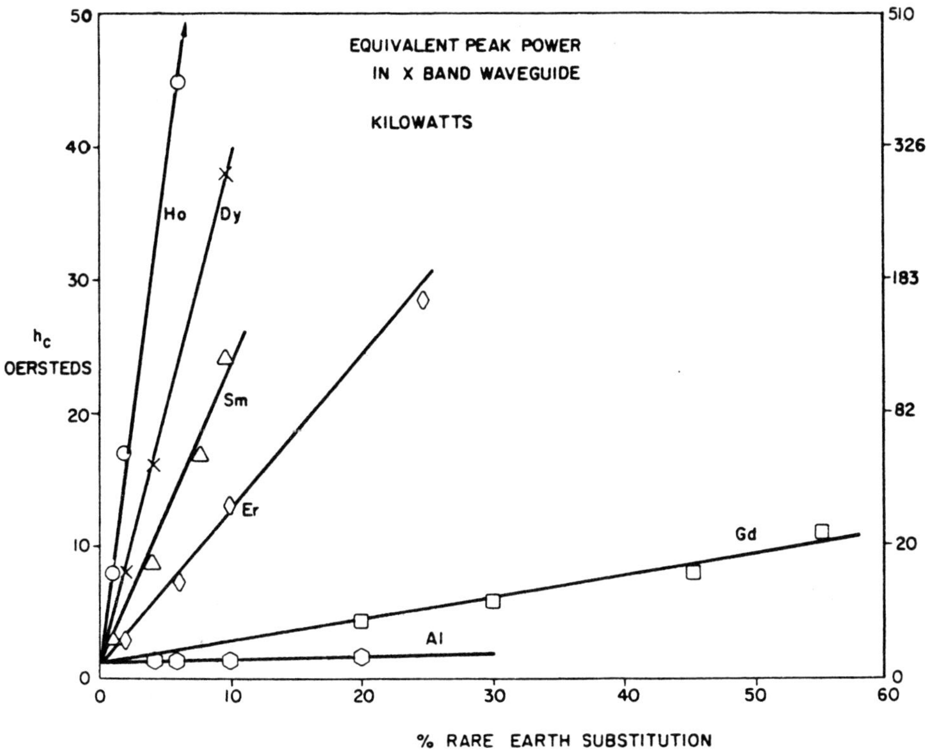

FIG. 17
High power thresholds as a function of composition in yttrium rare earth iron garnets.

F. Transition Metal Ion Substitutions

Some very interesting properties of substituted garnets can be obtained when trivalent transition metal ions are substituted for iron in the yttrium iron garnet structure.

1. Yttrium Aluminum Iron Garnet

The composition $3Y_2O_3 \cdot 5\,[(1.00 - x)\ Fe_2O_3 \cdot xAl_2O_3]$ has been investigated for x = 0 to x = 0.25. The variations of saturation magnetization, linewidth, effective g-factor, and Curie temperature are shown in Fig. 18. Trivalent aluminum goes into the structure primarily on the 24d site and has a zero magnetic moment. Therefore,

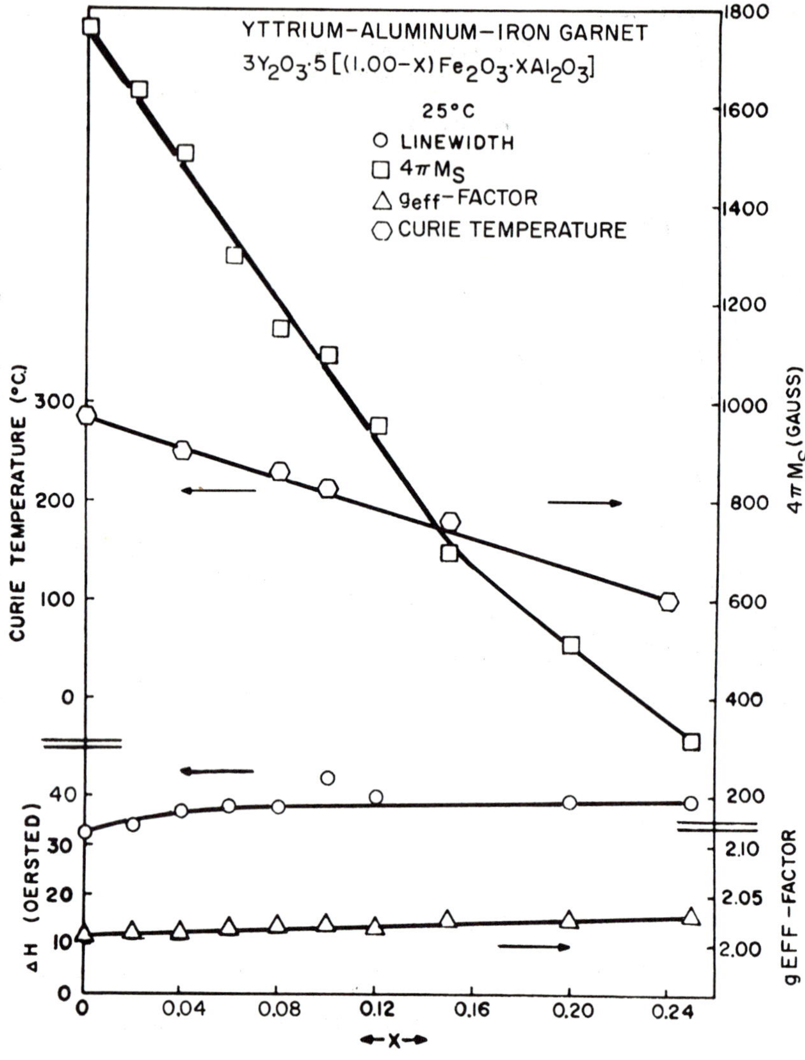

FIG. 18
Variation of linewidth, saturation magnetization, effective g-factor, and Curie temperature with aluminum content in yttrium iron garnet.

partial substitution of aluminum for iron on the 24d site should reduce the magnetization as observed. Since the number of iron-iron interactions has been reduced, the Curie temperature is also expected to decrease as shown. Since no change has been made in the ions located on the 24c site, very little variation in linewidth is observed. No change in the g-factor is observed for the substitutions made. Linewidth is observed to decrease with increasing temperature. There is little change in the g-factor with temperature. These materials are very useful in applications where a low linewidth and low magnetization are desirable.

2. Yttrium Gallium Iron Garnet

The composition $3Y_2O_3 \cdot 5[(1.00 - x)\ Fe_2O_3 \cdot xGa_2O_3]$ has been studied for $x = 0$ to $x = 0.20$. The variation in the microwave properties is shown in Fig. 19.

Gallium is very similar in structure to aluminum and predominately locates also on the 24d site in the garnet structure so this substitution produces approximately the same effect on the microwave properties. Linewidths are observed to increase slightly as a function of gallium substitutions; $4\pi M_s$ decreases and g-factors remain almost constant. Curie temperatures are noted again to decrease. Linewidths decrease as a function of increasing temperature and g-factors are observed to be essentially invariant with temperature. For a particular temperature, these materials appear to be a little more active than the yttrium aluminum iron garnets. However, this advantage is overridden by the much greater cost of gallium as compared to that of aluminum.

3. Yttrium Chromium Iron Garnet

The composition $3Y_2O_3 \cdot 5[(1.00 - x)\ Fe_2O_3 \cdot xCr_2O_3]$ has been studied for very small additions of chromium, as shown in Fig. 20. Observations suggest that the electronic configuration of chromium is favorable for forming octahedral bonds so, therefore, the magnetization

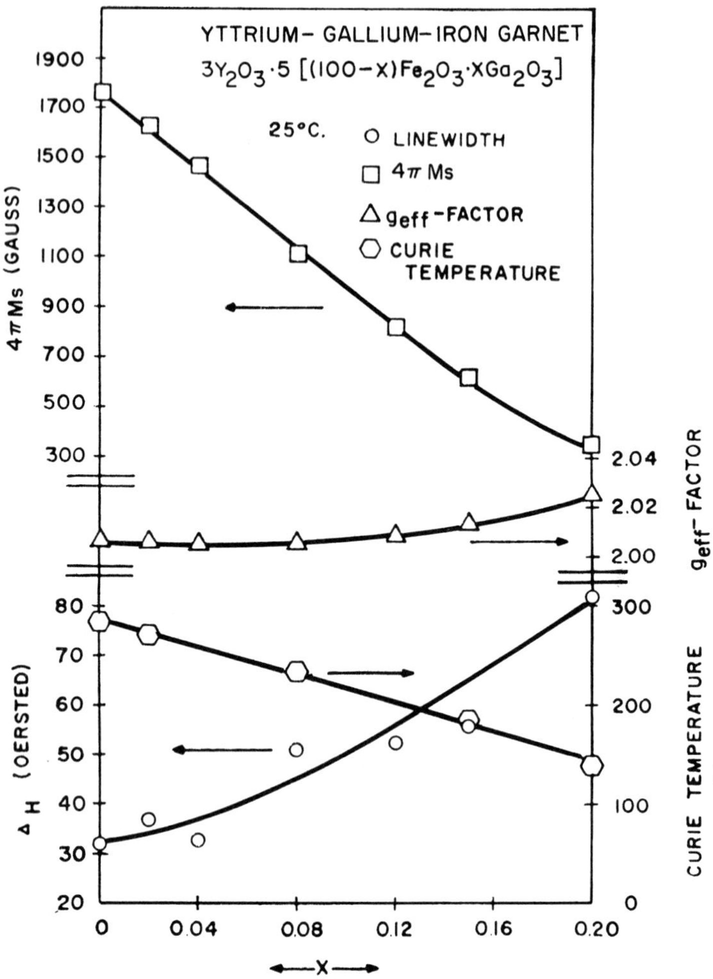

FIG. 19
Variation of linewidth, saturation magnetization, effective g-factor, and Curie temperature with gallium content in yttrium gallium iron garnet.

of yttrium chromium iron garnets is expected to increase. Since the number of iron-iron interactions is reduced, the Curie temperature is again decreased as shown in Fig. 20. Due to this decrease in Curie temperature, the magnetization, as measured at room temperature, appears to increase only slightly over the range of chromium added. It

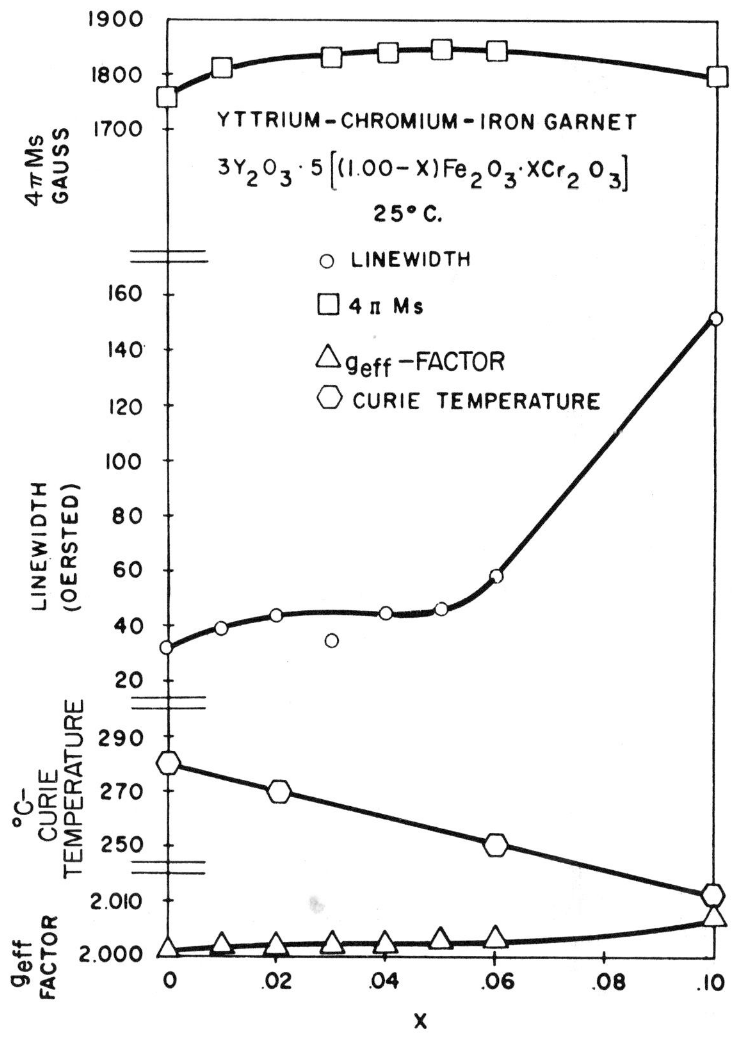

FIG. 20
Variation of linewidth, saturation magnetization, effective g-factor, and Curie temperature with chromium content in yttrium chromium iron garnet.

has been noted that only small substitutions of chromium can be made to the yttrium garnet structure[21]; the sharp change in linewidth for the highest chromium substitution attempted is attributed to this fact. Very little change in linewidth and g-factor is expected for chromium substitution. Linewidths are observed again to decrease with increasing temperature and g-factors remain constant with temperature. At present these materials appear to have almost no advantage over the yttrium aluminum or yttrium gallium iron garnets. It was thought in the beginning that these materials might be particularly useful in limiter applications. However, the data collected to date do not verify this assumption.

G. Five Ion Systems

The compositions discussed previously are classified as four ion systems, e.g., yttrium gadolinium iron garnet, the fourth ion is oxygen. For microwave applications it is desirable to control the properties of the materials as completely as possible. This desire invites the study of five and six ion systems.

1. Temperature Characteristics of Four Ion Systems

One of the most interesting four ion systems, from an application standpoint, is the yttrium gadolinium iron garnet series. The temperature characteristics of these compositions, as shown in Fig. 2, are particularly encouraging. Gadolinium iron garnet possesses a compensation point at approximately 15°C. The ionic distribution per formula unit ($3Gd_2O_3 \cdot 5Fe_2O_3$) is as follows:

$\downarrow$24c	$\uparrow$24d	$\downarrow$16a
$6Gd^{+++}$	$6Fe^{+++}$	$4Fe^{+++}$

Gd^{+++} has 7 Bohr magnetons per ion; Fe^{+++} has 5 Bohr magnetons per ion. The two iron sublattices are strongly antiferromagnetically coupled together and the resultant magnetic moment of these two sites of 10 Bohr magnetons is in the direction given by the 24d site. The magnetic moment of the 24c site relaxes much faster as a function of temperature than does the resultant of the 16a and 24d sites. At some temperature, these magnetic moments are equal and a compensation point results.

When yttrium (Y^{+++} has zero Bohr magnetons per ion) is substituted in gadolinium iron garnet, the resultant magnetic moment of the 24c site is reduced and the compensation point is lowered in temperature. The 40% yttrium-60% gadolinium composition positions this compensation point (-120°C) such that the temperature characteristics of the $4\pi M_s$ around room temperature is relatively invariant. This substitution allows essentially only one value of the $4\pi M_s$ to be invariant.

2. Yttrium Gadolinium Aluminum Iron Garnet

To obtain temperature-stabilized magnetizations at various values, the following series of compositions was investigated: $3[0.5Y_2O_3 \cdot 0.5Gd_2O_3] \cdot 5[(1 - w)\ Fe_2O_3 \cdot wAl_2O_3]$, for w = 0 to 0.1. The magnetization versus temperature characteristics are shown in Figs. 21 and 22. For small substitutions, aluminum locates predominately on the 24d site in the garnet structure. This reduces the resultant magnetization of the iron sublattices and thus the magnetization at 25°C, reduces the Curie temperature[25], and positions the compensation point to higher temperatures as W increases. The result is a series of materials with temperature-stable magnetizations from 900 gauss downward. Other microwave properties of these compositions are as follows: linewidth ΔH (25°C, X band) = 165 to 300 oe; g-factor (25°C, X band) = 2.04 to 2.10; dielectric loss tangent

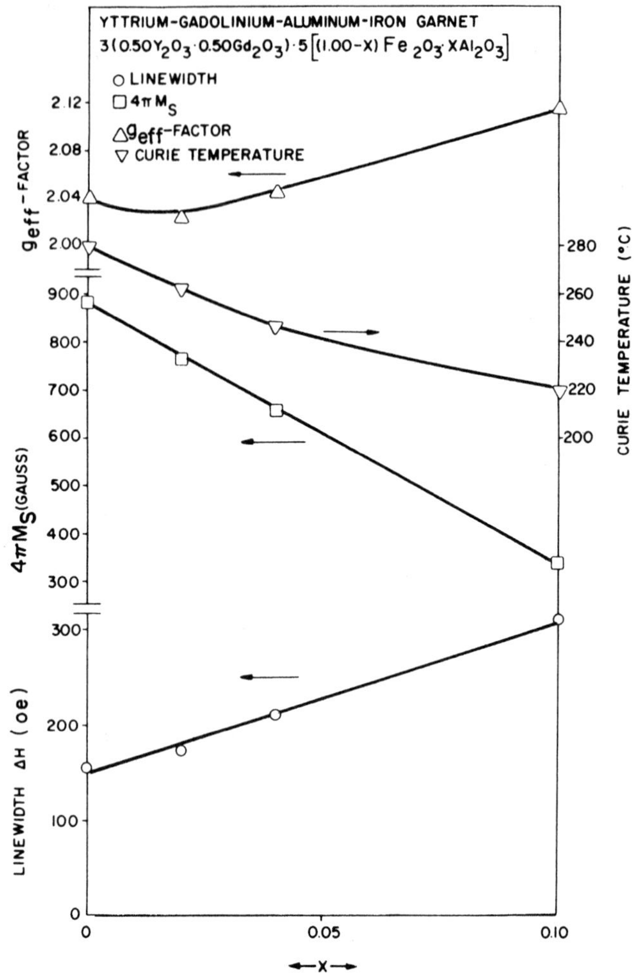

FIG. 21
Variation of linewidth, saturation magnetization, effective g-factor, and Curie temperature with aluminum content in yttrium gadolinium aluminum iron garnet.

(1 GHz) = 0.004. True compensation points were not observed for these compositions due presumably to a slight nonuniformity in composition from grain to grain. The compensation points did increase with increasing aluminum content as expected.

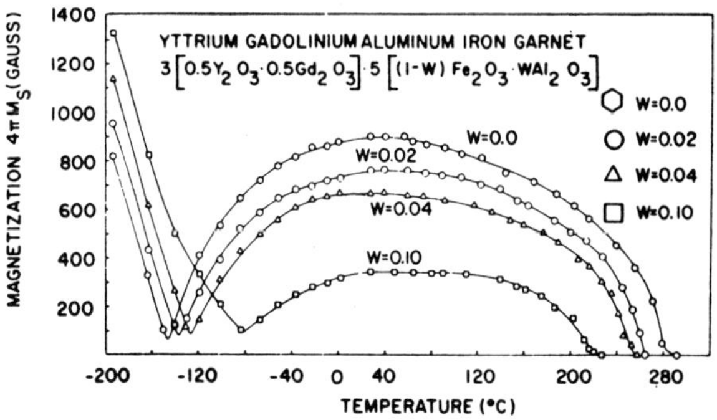

FIG. 22
Temperature variation of the saturation magnetization for yttrium gadolinium aluminum iron garnet.

3. Yttrium Gadolinium Dysprosium Iron Garnet

To obtain a series of materials with greater high power thresholds[34], solid solutions of the above compositions with dysprosium iron garnet were studied. The magnetic properties of the compositions $3[(0.9 - x)Y_2O_3 \cdot xGd_2O_3 \cdot 0.1Dy_2O_3] \cdot 5Fe_2O_3$ for x = 0 to 0.5 are shown in Figs. 23 and 24.

Temperature stability is again observed in compositions for X = 0.3 to 0.5; these materials also exhibit superior microwave high power handling capability. Other microwave properties of these compositions are: linewidth ΔH (25°C, X band) = 450 to 800 oe; g-factor (25°C, X band) = 2.00 to 1.80; dielectric loss tangent (1 GHz) = 0.004.

A combination of the above two systems would produce a still more versatile series of compositions. The result, yttrium gadolinium dysprosium aluminum iron garnet, would possess a range of magnetizations which could be temperature stabilized plus unusual high power handling capability.

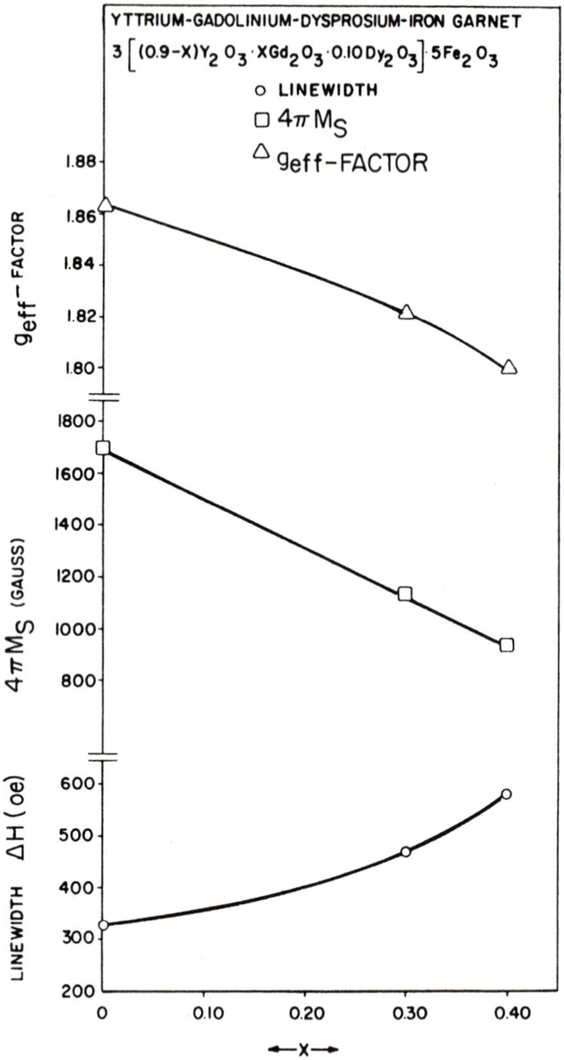

FIG. 23
Variation of linewidth, saturation magnetization, and effective g-factor with gadolinium content in yttrium gadolinium dysprosium iron garnet.

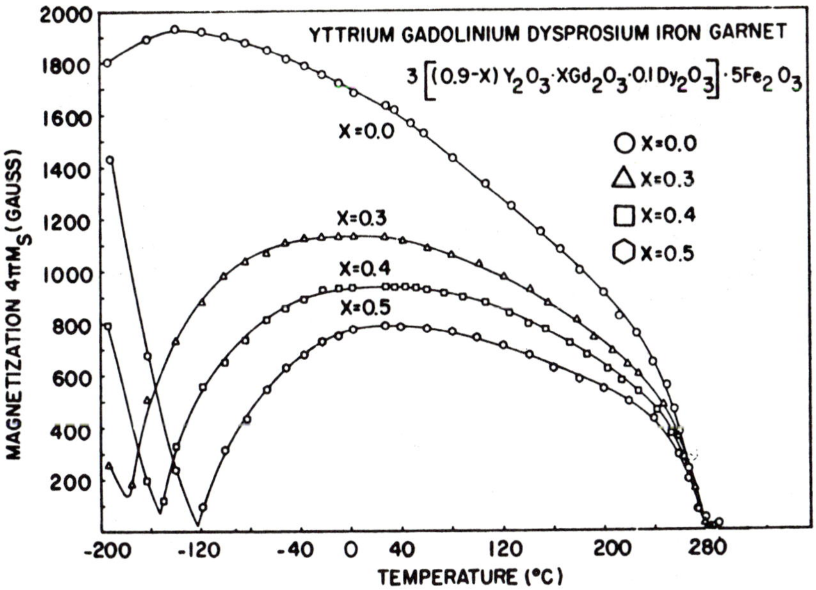

FIG. 24
Variation of saturation magnetization with temperature in yttrium gadolinium dysprosium iron garnet.

III. HYBRID VANADIUM IRON GARNETS

A. Compositions Studied: Calcium Vanadium-Substituted Garnet

Polycrystalline ferrimagnetic garnet compositions conforming to the formulas given previously have been prepared and evaluated for x values between 0 and 1.5 and y values from 0.75 to 1.5.

1. Yttrium Calcium Vanadium Iron Garnet

The data obtained on the yttrium calcium vanadium iron garnet compositions are depicted in Fig. 25. The magnetization of these compounds varies in a linear fashion, going to zero for a vanadium content of 1.0 and increasing as the vanadium content is increased to 1.5. This variation in magnetization is caused by a compensation of the spins of ions located at the various sites.

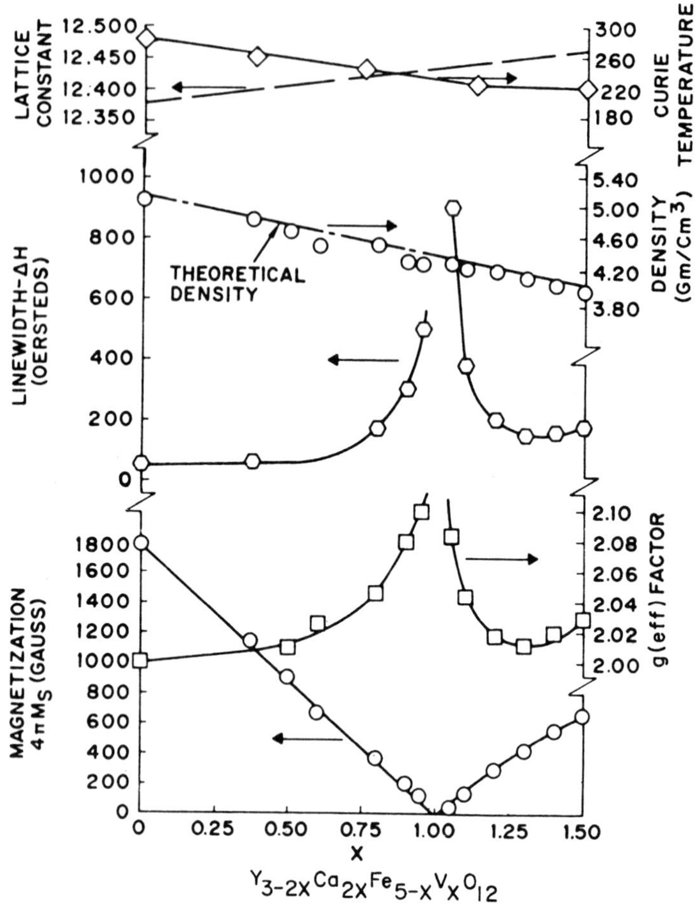

FIG. 25
The magnetic, microwave, and static properties of the YCaVIG compounds.

The linewidth for small calcium and vanadium substitutions was measured to be 50-oe, characteristic of pure YIG, and increased near the magnetic compensation point, varying approximately as M^{-1}. For x > 1, linewidths were measured to be in the 150- to 200-oe range.

Densities measured in nearly all cases were greater than 97% of theoretical density. The dielectric constant of these materials was observed to vary from a value of 16 for yttrium iron garnet to 13 for calcium vanadium iron garnet. Dielectric loss tangents measured at X band for single phase compositions were less than 0.0005, and some samples possessed loss tangents less than 0.0001. The measured Curie temperatures for these compounds are shown in Fig. 25. In all cases the Curie temperatures are higher than those of compositions derived from the yttrium aluminum or yttrium gallium iron garnet families with comparable magnetization.

2. Bismuth Calcium Vanadium Iron Garnet

Figure 26 contains data on the bismuth calcium vanadium iron garnets. Good ceramic samples of the BiCaVIG compounds are difficult to prepare, and the firing temperature required for these materials is more critical and generally lower (1000°C to 1100°C) than that required for the other garnets. Single phase compositions have not been prepared and may well not exist for y values less than 0.96. The microwave and magnetic properties of the bismuth calcium vanadium iron garnet family are particularly interesting, since these compounds contain no rare earth or yttrium ions. The zero magnetization point is again observed for a vanadium content of $y = 1$. Linewidth values on these compounds are, in general, higher than those obtained for the YCaVIG. However, comparable linewidths are measured for $1.2 < y < 1.4$. Relative dielectric constants up to 25 have been measured on single phase BiCaVIG compositions. Measured dielectric loss tangents are less than 0.0005 at X band.

3. Temperature Studies

Figure 27 compares the magnetization curves of members of both new garnet families with comparable YIG(Gd,AR) and YIG(AR) compounds. The vanadium compounds show promise of improved temperature stability,

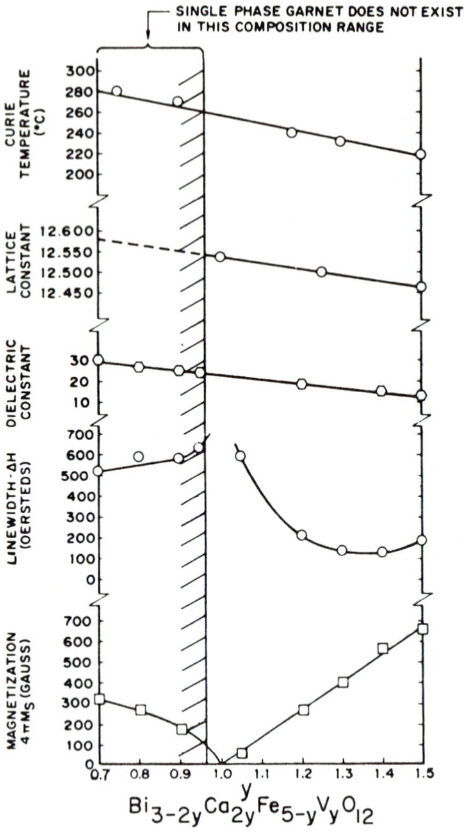

FIG. 26
The magnetic, microwave, and static properties of the BiCaVIG compounds.

particularly at low $4\pi M_s$ values. These materials have higher Curie temperatures than are obtained in garnets containing aluminum. The temperature stability of the yttrium gadolinium compounds is somewhat better, since these materials have compensation points in their temperature characteristics. While the linewidth of the vanadium compounds decreases with the increasing temperature, the slope of the linewidth versus temperature curve for the vanadium compounds is about half (0.4 oe/°C) the slope observed on the aluminum-doped garnets (1.0 oe/°C).

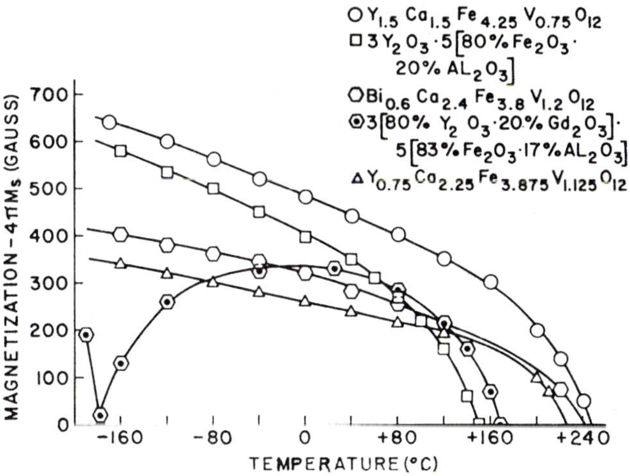

FIG. 27

The magnetization versus temperature characteristics of selected BiCaVIG and YCaVIG compounds compared to similar data of yttrium aluminum iron and yttrium gadolinium aluminum iron garnet.

B. Compositions Studied: Aluminum and Gadolinium-Substituted Calcium Vanadium Garnet

Hybrid compounds conforming to the formulas given previously have been prepared and evaluated for x values between 0.6 and 1.2 and w values from 0 to 0.5.

1. Aluminum-Substituted Calcium Vanadium Garnet

Figure 28 shows measured values of saturation magnetization, linewidth, and Curie temperature on a series of polycrystalline $Bi_{0.4}Ca_{2.6}V_{1.3}Al_wFe_{3.7-w}O_{12}$ materials with w values from 0 to 0.5. Aluminum locates preferentially on the 24d sites (at least for small substitutions); for a vanadium content less than 1.0, the 24d sites are normally dominant. Therefore, the presence of aluminum causes a reduction in $4\pi M_s$ value in this region. For a vanadium content greater than 1.0, where the 16a sites are dominant in the BiCaVIG compounds, the aluminum substitution causes an increase in magnetization. Thus,

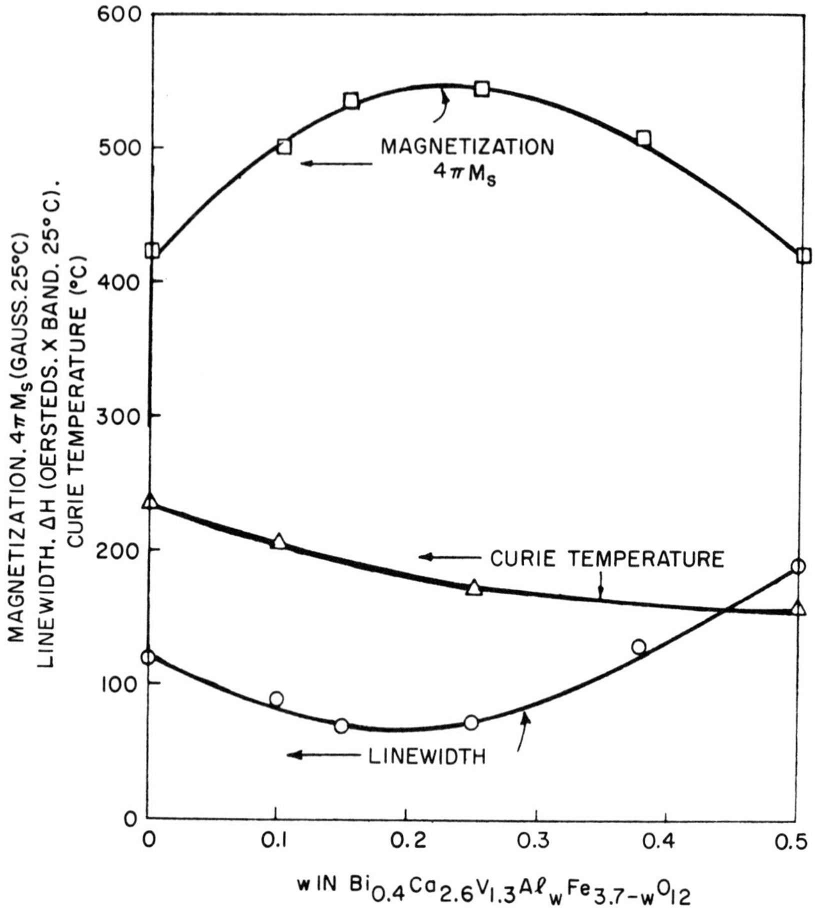

FIG. 28
Saturation magnetization, linewidth, and Curie temperatures as a function of aluminum content w in $Bi_{0.4}Ca_{2.6}V_{1.3}Al_w$-$Fe_{3.7-w}O_{12}$.

with increasing aluminum content, the magnetic compensation point ($4\pi M_s = 0$) shifts from a vanadium content of 1.0 toward lower values, and these materials with y = 1.3 initially show an increase in $4\pi M_s$ as the 24d sublattice magnetization is decreased.

For w > 0.25, however, the room temperature magnetization decreases. These peculiarities together with Curie temperature dependence may be associated with a breakdown in the lattice site preference of aluminum above a critical substitution. The linewidth is observed to go through a minimum of approximately 70 oe for a w value of 0.2.

2. Gadolinium-Substituted Calcium Vanadium Garnet

The substitution of gadolinium for yttrium follows much the same reasoning as the aluminum substitution. Since the magnetic Gd^{+++} ions locate at 24c sites, their magnetization will be aligned antiparallel to the 24d sites, and the addition of Gd^{+++} should cause the compensation point to shift from a vanadium content of 1.0 to lower values. However, since the magnetization of the Gd^{+++} ions on the 24c sublattice is a stronger function of temperature than the iron sublattices, the Gd^{+++}-substituted YCaVIG compounds exhibit a compensation point in the $4\pi M_s$ curves as a function of temperature. The aluminum-substituted or straight YCaVIG or BiCaVIG do not show compensation points in their magnetization curves because both iron sublattices become magnetized at essentially the same rate as a function of temperature.

Samples of $(Y_{0.5}Gd_{0.5})_{3-2x}Ca_{2x}V_xFe_{5-x}O_{12}$ with x values from 0 to 1.2 have Curie temperatures that are no different from those of YCaVIG. Magnetization and linewidth values are shown in Fig. 29 as a function of composition for $(Y_{0.5}Gd_{0.5})_{3-2x}Ca_{2x}V_xFe_{5-x}O_{12}$ and for $Y_{3-2x}Ca_{2x}V_xFe_{5-x}O_{12}$ (included here for comparison purposes)[41]. The gadolinium substitution shifts the compensation point from x = 1.0 for YCaVIG to approximately x = 0.9 for (Y-Gd)CaVIG.

3. Temperature Studies

Linewidth and magnetization measured as a function of temperature for the compositions $Bi_{0.4}Ca_{2.6}V_{1.3}Al_wFe_{3.7-w}O_{12}$ with w = 0, 0.25, and 0.5 and for $Y_3Al_{0.85}Fe_{4.15}O_{12}$(YAlIG) are shown in Fig. 30.

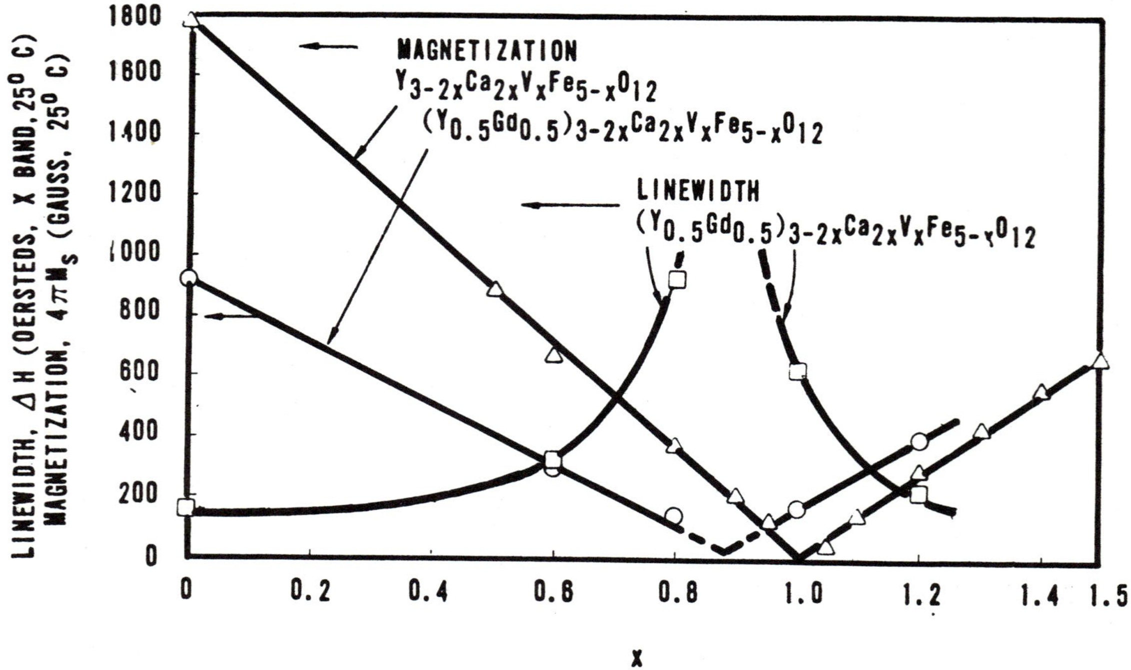

FIG. 29

Saturation magnetization $4\pi M_s$ and linewidth as a function of composition for $(Y_{0.5}Gd_{0.5})_{3-2x}Ca_{2x}V_xFe_{5-x}O_{12}$ ($4\pi M_s$ versus composition for $Y_{3-2x}Ca_{2x}V_xFe_{5-x}O_{12}$ included for reference).

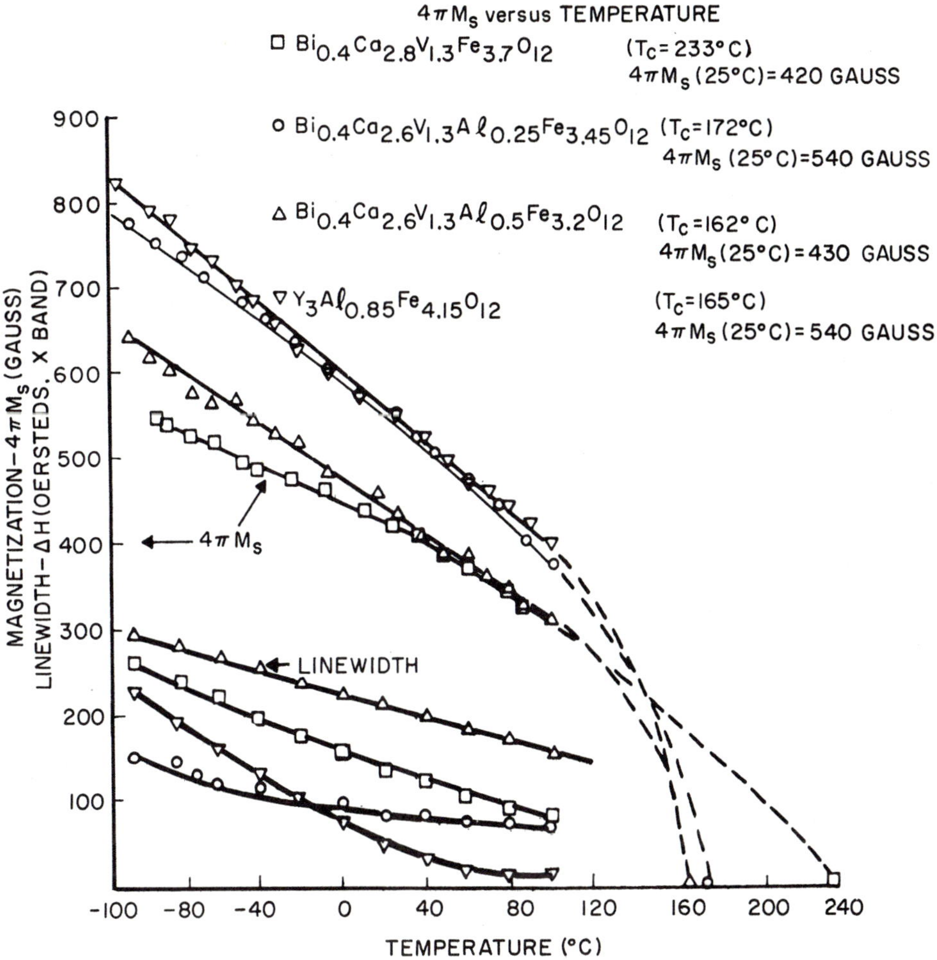

FIG. 30
Saturation magnetization and linewidth as a function of temperature for BiCaV(Al)IG and YIG(Al).

The 17% aluminum-substituted YIG was chosen for comparison with the BiCaVAlIG with w = 0.25 because these materials have equal room temperature $4\pi M_s$ values. YAlIG and BiCaVAlIG (w = 0.25) possess approximately the same $4\pi M_s$ values over the temperature range -100°C to Curie temperature (approximately 165°C). The slopes of the magnetization curves of BiCaVAlIG for w = 0.5 and for w = 0 are approximately equal (2 G/°C) in spite of their differing Curie temperatures.

The linewidths measured as a function of temperature for these same four materials also appear in Fig. 30. The linewidth of YAlIG, and the BiCaVAlIG for w = 0 and 0.5 vary approximately the same (1.1 oe/°C) from -100°C to room temperature, while BiCAVAlIG (w = 0.25) varies only 0.4 oe/°C from -100°C to +100°C. This composition, $Bi_{0.4}Ca_{2.6}V_{1.3}$-$Al_{0.25}Fe_{3.45}O_{12}$, appears very satisfactory for certain microwave applications such as circulators.

The linewidth, g-factor, and magnetization as a function of temperature for $(Y_{0.5}Gd_{0.5})_{1.8}Ca_{1.2}V_{0.6}Fe_{4.4}O_{12}$ are shown in Fig. 31. This composition has a compensation point at -123°C. Due to this compensation point, a temperature-stable magnetization is obtained from -20°C to +120°C varying only ±25 G from 340 G. The linewidth over this same temperature range drops at a rate of 2.5 oe/°C. The relatively large slope of the linewidth is typical of materials having magnetic compensation points.

IV. SUMMARY

The microwave properties (magnetization, linewidth, g-factor, and Curie temperature) have been studied for a number of polycrystalline hybrid garnets. The results of these studies are believed to constitute useful information on all significant facets of the materials, including the important topics of high power effects and temperature stability.

The properties reported herein have not included the square loop and remanent properties of these materials. These properties are very important to microwave digitally controlled components. Typical

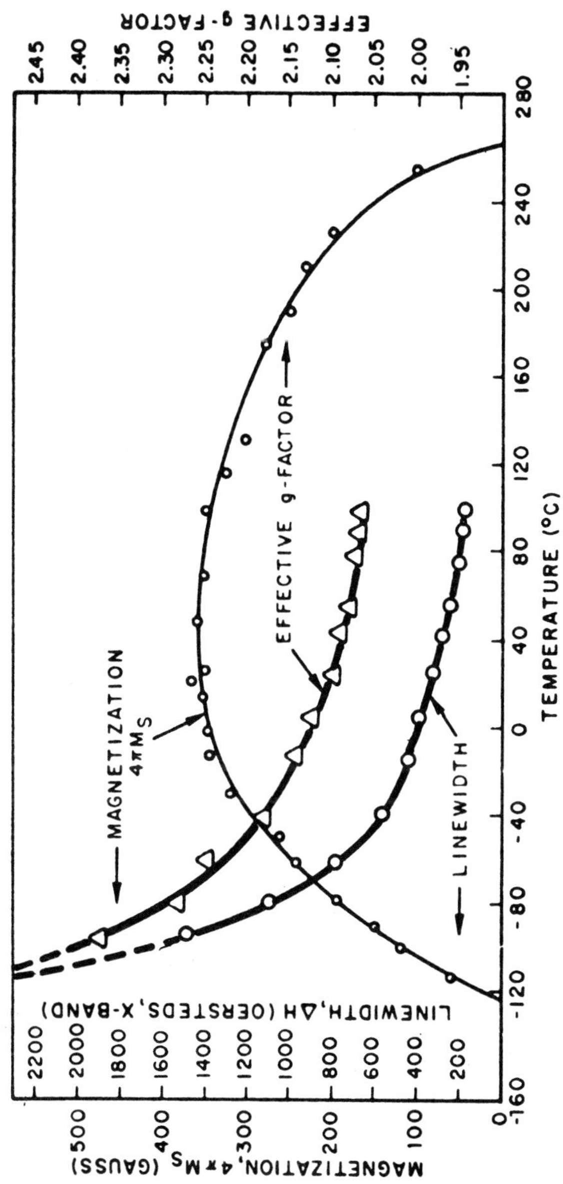

FIG. 31

Saturation magnetization, linewidth, and g-factor as a function of temperature for $(Y_{0.5}Gd_{0.5})_{1.8}Ca_{1.2}V_{0.6}Fe_{4.4}O_{12}$.

values of remanence ratio ($4\pi M_r/4\pi M_s$) and coercive fields are 0.6 and 1 oersted, respectively.

These studies illustrate how "molecular engineering" or "ionic substitutions" may be used to independently control the properties of garnet materials, to increase high power thresholds, and to temperature-stabilize materials for microwave applications. This wide control of the properties of ferrimagnetic garnets allows materials to be tailored for specific applications and leads to easier fabrication of fixed field high power microwave components whose operation is optimized over a large temperature range.

ACKNOWLEDGMENT

The authors would like to express their sincere appreciation to many colleagues of Sperry Microwave Electronics Div., Sperry Rand Corp., Clearwater, Florida, for their assistance in preparation of materials and the measurement of properties. The Air Force Systems Command (RADC) and the U. S. Army Electronics Command are acknowledged for the support under which many of the studies were conducted.

REFERENCES

[1]. Benjamin Lax and Kenneth J. Button, Microwave Ferrites and Ferrimagnetics, McGraw-Hill, New York, 1962.

[2]. Richard M. Bozorth, Ferromagnetism, Van Nostrand, Princeton, New Jersey, 1951, 5th printing, February 1959.

[3]. J. Smith and H. P. J. Wijn, Ferrites, Wiley, New York, 1959.

[4]. Ronald F. Soohoo, Theory and Applications of Ferrites, Prentice Hall, New York, 1960.

[5]. W. J. Polydoroff, High Frequency Magnetic Materials, Wiley, New York, 1960.

[6]. Wilhelm H. Von Aulock, Handbook of Microwave Ferrite Materials, Academic, New York, 1965.

[7]. A. G. Gurevich, Ferrites at Microwave Frequencies, Consultants Bureau, New York, 1960.

[8]. J. Roberts, High Frequency Applications of Ferrites, The English Universities Press, London, 1960.

[9]. D. R. Taft, G. R. Harrison, and L. R. Hodges, Jr., "Millimeter resonance isolators utilizing hexagonal ferrites," IEEE Trans. on Microwave Theory and Techniques, MTT-11, 346 (1963).

[10]. G. P. Rodrigue, "Magnetic materials for millimeter wave application," IEEE Trans. on Microwave Theory and Techniques, MTT-11, 351 (1963).

[11]. D. R. Taft, "Millimeter resonance isolators utilizing hexagonal ferrites," Electro-Technology (1964).

[12]. Gordon R. Harrison and L. Reese Hodges, Jr., "Anisotropic hexagonal ferrites for microwave uses," Sperry Engineering Review, 16, 45, (1963).

[13]. L. R. Hodges, Jr. and G. R. Harrison, "Oriented hexagonal ferrite compounds for application at millimeter wave frequencies," J. Am. Ceram. Soc., 47, 601 (1964).

[14]. F. Bertaut and F. Forrat, "The structure of ferrimagnetic ferrites of the rare earths," Compt. rend, 242, 382 (1956).

[15]. S. Geller and M. A. Gilleo, "The crystal structure and ferrimagnetism of yttrium iron garnet," Acta Cryst., 10, 239 (1957).

[16]. S. Geller, "Magnetic interactions and distribution of ions in the garnets," J. Appl. Phys. Suppl., 31, 30S (1960).

[17]. M. H. Sirvetz and J. E. Zneimer, "Microwave properties of polycrystalline rare earth garnets," J. Appl. Phys., 29, 431 (1958).

[18]. Elmer E. Anderson, J. Richard Cunningham, Jr., and G. E. McDuffie, "Magnetic properties of the mixed garnets $(3-x)Y_2O_3 \cdot xGd_2O_3 \cdot 5Fe_2O_3$," Phys. Rev., 116, 624 (1959).

[19]. G. P. Rodrigue, "Ferromagnetic resonance in yttrium and rare earth iron garnets," a thesis, Harvard University (1959).

[20]. J. Richard Cunningham, Jr. and Elmer E. Anderson, "Samarium substitutions in yttrium-iron garnet," J. Appl. Phys. Suppl., 31, 45S (1960).

[21]. M. A. Gilleo and S. Geller, "Magnetic and crystallographic properties of substituted yttrium-iron garnet, $3Y_2O_3 \cdot XM_2O_3 \cdot (5-X)Fe_2O_3$," Phys. Rev., 110, 73 (1958).

[22]. H. S. Yoder and M. L. Keith, "Complete substitution of aluminum for silicon: the system $3MnO \cdot Al_2O_3 \cdot 3SiO_2 \cdot 3Y_2O_3 \cdot 5Al_2O_3$," Am. Mineralogist, 36, 519 (1951).

[23]. M. L. Keith and R. Roy, "Structural relations among double oxides of trivalent elements," Am. Mineralogist, 39, 1 (1954).

[24]. Elmer E. Anderson, "Some electrical and magnetic properties of garnets," J. Appl. Phys. Suppl., 30, 299S (1959).

[25]. Gordon R. Harrison and L. R. Hodges, Jr., "Microwave properties of polycrystalline hybrid garnets," J. Am. Ceram. Soc., 44, 214 (1961); Microwave Journal, 4, 53 (1961).

[26]. G. Vielers, R. Pauthenet, and J. Loriers, "Magnetic properties of ferrites of the garnet type substituted with Al, Ga and Cr," J. Phys. Radium, 20, 382 (1959) in French.

[27]. L. Neel, "On the interpretation of the magnetic properties of rare earth ferrites," Compt. Rend., 239, 8 (1954).

[28]. C. Kittel, "Microwave resonance in rare earth iron garnets," J. Appl. Phys. Suppl., 31, 115 (1960).

[29]. J. O. Artman and P. E. Tannerwald, "Measurement of susceptibility tensor in ferrites," J. Appl. Phys., 26, 1124 (1955).

[30]. C. H. Nowlin, "Improvements in design of the five-string pendulum magnetometer," Scientific Report No. 14, Gordon McKay Laboratory, Harvard University (1958).

[31]. S. Foner, "Versatile and sensitive vibrating-sample magnetometer," Rev. Sci. Instr., 30, 548 (1959).

[32]. J. J. Green and E. Schlomann, "High power ferromagnetic resonance at X-band in polycrystalline garnets and ferrites," Inst. Radio Engrs. Trans. on Microwave Theory and Techniques, MTT-8, 100 (1960).

[33]. H. Suhl, "The non-linear behavior of ferrites at high microwave signal levels," Proc. Inst. Radio Engrs., 44, 1270 (1956).

[34]. E. Schlomann, J. J. Green, and U. Milano, "Recent developments in ferromagnetic resonance at high power levels," J. Appl. Phys. Suppl., 31, 386S (1960).

[35]. R. C. Fletcher, R. C. LeCraw, and E. G. Spencer, "Electron spin relaxation in ferromagnetic insulators," Phys. Rev., 117, 955 (1960).

[36]. S. Geller, G. P. Espinosa, H. J. Williams, R. C. Sherwood, and E. Z. Nesbitt, J. Appl. Phys., 35, 570 (1964).

[37]. S. Geller, G. P. Espinosa, H. J. Williams, R. C. Sherwood, and E. A. Nesbitt, Appl. Phys. Letters, 3, 60 (1963).

[38]. S. Geller, H. J. Williams, G. P. Espinosa, R. C. Sherwood, and M. A. Gilleo, Appl. Phys. Letters, 3, 21 (1963).

[39]. G. P. Espinosa and S. Geller, J. Appl. Phys., 35, 2551 (1964).

[40]. F. R. Montforte and W. W. Thodes, Bull. Am. Ceram. Soc., 43, 279 (1964).

[41]. L. R. Hodges, Jr., G. P. Rodrigue, G. R. Harrison, and A. D. Sanders. "Magnetic and microwave properties of calcium vanadium-substituted garnets," J. Appl. Phys., 37, 1085 (1966).

[42]. W. R. Wilson, L. R. Hodges, Jr., G. P. Rodrique, and G. R. Harrison, "Magnetic and microwave properties of aluminum- and gadolinium-substituted calcium vanadium garnets," J. Appl. Phys., 38, 1405 (1967).

IX. OPTICAL MATERIALS AND PHENOMENA

Chapter 27

THE OPTICAL ABSORPTION OF GLASSES

N. J. Kreidl

Department of Ceramic Engineering and
Materials Research Center
University of Missouri-Rolla
Rolla, Missouri

I. INTRODUCTION

The optical properties of glasses characterize some aspects of their interaction with electromagnetic radiation within and near the energy range to which the human eye is sensitive: near-infrared, visible, and ultraviolet "light." This narrow range is placed in the energy spectrum in Fig. 1. The selection of this narrow range in the

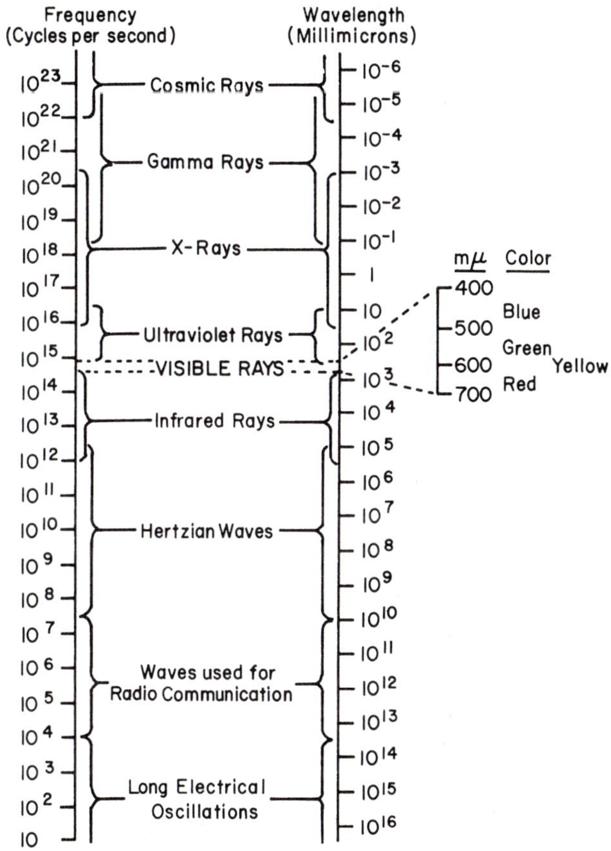

FIG. 1.
The electromagnetic spectrum. The optical range as indicated includes infrared, visible, and ultraviolet sections.

evolution of human orientation is chiefly associated with the abundance of electronic transitions in the environment and the raw materials of our physiology from which the eye and its sensitive receptors developed. The resulting system can be well described by a three coordinate array of signals which in turn can be sorted out by neural stations and brain interpretation into the qualitative sensations of light and color in millions of combinations. The system is not only technically remarkable, but also gratifying to the esthetic responses of our consciousness (Figs. 2,3).

The variety of manifestations of the interaction gives rise to descriptive property names such as refraction, reflection, absorption, emission, photochromism, etc. This survey is restricted to absorption, the loss aspect of the interaction; but one of the objectives of the survey is to stress the fundamentals of the interaction, unifying the manifestations of which absorption is only one. A survey of so vast a field must be restricted in some manner. Other points

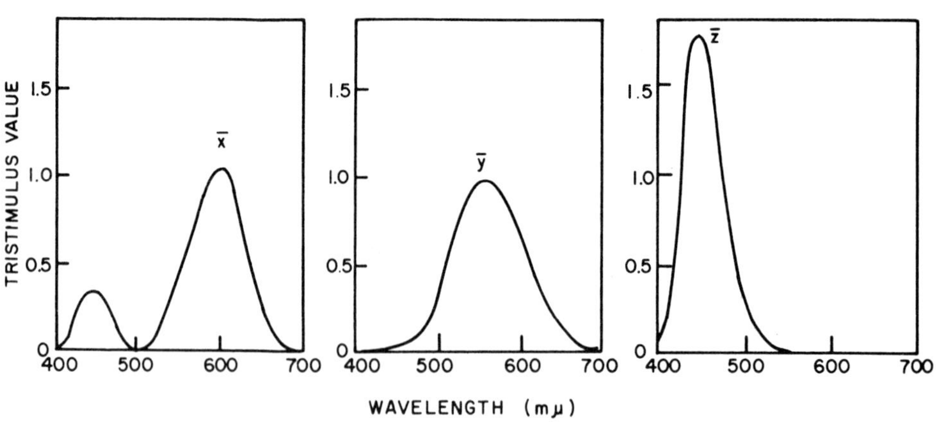

FIG. 2.
Every color can be combined from three "templates" simulating our eye reception system in regard to wavelength.

deserving emphasis are: use of absorption spectra in structural interpretation, the control of absorption in important technologies, and the most recent developments in some of the areas of interest.

The control of the optical absorption of glasses is important in many applications such as photographic filters, sun glasses, radiation shields, and, in the other direction, in highly transparent windows, lenses, and fibers. This includes the promotion or prevention of absorption in the ultraviolet and the infrared. In addition, the control of absorption affects that of stimulated emission in laser glasses.

Irreversible changes due to high energy radiation are utilized to measure integrated intensity in dosimeter glasses. Protected glass

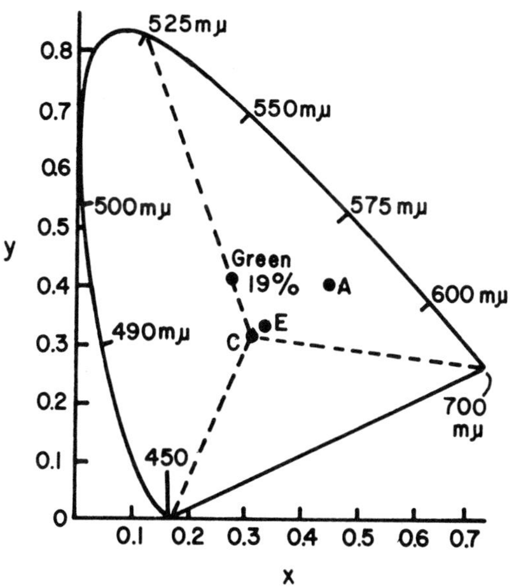

FIG. 3.
The relation of two of these data to the sum can be plotted in a two-dimensional diagram containing all colors. The spectral colors are on the "locus" indicated. Gray, white, or black is in the center intersection. Brightness would have to be labeled extra in this two-dimensional colorplot.

is the designation of glass unusually stable against induced absorption changes. Photosensitive glasses show various effects ranging from image production to nucleation. In photochromic glasses reversible absorption changes may be used to fabricate variable density windows or lenses, and, perhaps memory and display devices.

Conversely, the understanding of the absorption mechanism in the glass aids increasingly in the understanding of its constitution. The subtle variations of absorption with atomic environment are probes for this environment. High energy radiation induced changes in absorption can serve to detect structural detail present before irradiation or illuminate the mechanism of radiation effects. In this manner the nature of imperfections such as oxygen deficiences and metal or hydrogen impurities have been elucidated.

II. SOME RELATIONS DESCRIBING OSCILLATORS

The general relationships describing the interaction of the electromagnetic field with the electronic systems in the glass cannot be derived and detailed in this survey. For their study, the following texts are recommended: Benumof[1], Dekker[2], Feynmann[3], Longhurst[4]. It is useful, however, to recall that they are formally related to the general relations describing the repeated disturbance of any system held together by restoring forces, such as the plucking of a string. Most disturbances of this kind can be illustrated by the behavior of harmonic oscillators (or combinations of such) which can be treated easily because the derivatives of cosines are sines, and vice versa, so that the second derivatives appearing in the equations of motion become simple functions of the coordinate itself.

A. Free Harmonic Oscillation

In a disturbance against a restoring force k the acceleration decreases proportionally to the displacement and is inversely proportional to the mass:

$$\ddot{x} = -k/m \cdot x \tag{1}$$

Because of the described properties of trigonometric functions one correctly guesses the solution

$$x = \cos \sqrt{k/m} \cdot t \tag{2}$$

which does not change if 2π is added in the cosine. This means that, if for a time t_0,

$$2\pi = \sqrt{k/m} \cdot t_0 \tag{3}$$

x returns to the initial value and t_0 is the period,

$$t_0 = 2\pi/\sqrt{k/m} = 2\pi/\omega_0 \tag{4}$$

and ω_0 the characteristic frequency of the oscillation generated by the distrubance; and Eq. (2) can also be written:

$$x = \cos \omega_0 t \tag{5}$$

or, more generally

$$x = A \cos \omega_0 t + B \sin \omega_0 t \tag{6a}$$

which is identical with

$$x = a \cos \omega_0 (t - t_1) \tag{6b}$$

$$x = \cos (\omega_0 t + \Delta) \tag{6c}$$

where Δ is a phase difference related to the other constants by

$$A = a \cos \Delta, \; B = -a \sin \Delta, \; \Delta = -\omega_0 t_1 \tag{7}$$

$$A = x \text{ (for } t = o), \; B = -V_0/\omega_0 \tag{8}$$

A is the amplitude depending on the disturbance, not on the frequency of the system. The energy of the process is proportional to A^2.

B. Forced Oscillation

If the disturbance oscillates with a frequency ω, as in the case of a string in the field of another musical note, or, as is of interest here, a glass in the field of light, one speaks of forced oscillation and obtains:

$$m\ddot{x} = -kx + F_0 \cos \omega t \tag{9a}$$

or

$$\ddot{x} = -k/m \cdot x + F_0/m \cdot \cos \omega t \tag{9b}$$

which is

$$\ddot{x} = -\omega_0^2 x + F_0/m \cdot \cos \omega t \tag{9c}$$

It is easy to see that the equation is fulfilled if

$$x = C \cos \omega t = F_0/[m(\omega_0^2 - \omega^2)] \tag{10}$$

since $\ddot{x} = -\omega^2 C \cos \omega$, substituted in (9b), permits one to calculate that $C = F_0/[m(\omega_0^2 - \omega^2)]$. Inspection for this simple case of an external disturbing oscillation of frequency ω interacting with a characteristic oscillation ω_0 of the disturbed system shows that the displacement would increase towards infinity, if the disturbing frequency ω approaches the value of the characteristic frequency ω_0 of the disturbed system. This great increase of amplitude with tuning in is the phenomenon of resonance. In real systems the amplitude is damped by losses, for instance, friction in a mechanical vibration. The purpose of recalling this simple case is to point out that all more complicated classic descriptions of the interactions of electromagnetic fields (light) with materials (glass) show terms closely related to that appearing in Eq. (10), modified by a loss term.

Before introducing friction into the simple case of forced harmonic oscillation, it is useful to replace the trigonometric functions by imaginary exponentials with whom they are found equivalent in serial development:

$$\exp i\theta = \cos \theta + i \sin \theta \tag{11}$$

This is done because of the ease to calculate using exponentials. Forced harmonic oscillation then can be described by

$$x = \hat{F}/[m(\omega_0^2 - \omega^2)] \cdot e^{i\omega t} \tag{12}$$

If damping is assumed (as often is the case) proportional to velocity (and an electrical analogy exists), the equation of motion (9) is modified to

$$\ddot{x} = -g\dot{x} - \omega_0^2 x + \hat{F}/m \cdot e^{i\omega t} \tag{13}$$

and the solution (10) is modified to:

$$x = \hat{F}/m[\omega_0^2 - \omega^2 + ig\omega] \tag{14}$$

Resonance still occurs for ω approaching ω_0, but not to an infinite displacement because of the finite g.

Expressions resembling Eq. (14) are found in relations describing the interaction of light with matter, such as in those for refraction, dispersion and absorption of light in glasses.

III. REFRACTION, DISPERSION, AND ABSORPTION

A. Interaction of the Field of a Light Source with that of the Charges in a Material

The resonant charges in the glass are either electrons (UV or visible light) or ions (infrared). This discussion refers to the case of UV or visible light and electron transitions having 2—5eV energies. A classic relation which describes quite well the interaction of the field of a source of electromagnetic radiation with the charges in the glass can be readily compared with those derived for a forced, damped harmonic oscillation, and thus can be intuitively understood (Feynmann [3] p. 31-5, 1965).

If E_{gl} is the field produced by this interaction, q the charge per electron, z the coordinate perpendicular to the glass surface, Δz the thickness of the glass, E_0 the amplitude of the source, m the mass of the electron, ω the frequency of the light, ω_0 the resonant frequency of the electron bound in a glass atom, N the number of charges per unit volume in the glass, ε_0 an electrical conversion factor, c the velocity of light, and g the damping coefficient

$$E_{gl} = -\frac{N\cdot\Delta z\cdot q}{2\varepsilon_0 c}\left\{i\omega\frac{qE_0}{m(\omega_0^2 - \omega^2 + ig\omega)}e^{i\omega(t - z/c)}\right\} \tag{15}$$

B. Refractive Index

The refraction of glass can be "explained" by assuming a decrease of the velocity of propagation of the electromagnetic wave in the glass. If the decrease of velocity is described by the index of refraction

$$n = c/v \tag{16}$$

where the velocities in vacuum and glass are c and v respectively; the angle of refraction can be derived from Fig. 4. In vacuum, wave crests have distances:

$$\Lambda_0 = 2\pi c/\omega \tag{17}$$

in glass:

$$\Lambda = 2\pi v/\omega = 2\pi c/n\omega = \Lambda_0/n \tag{18}$$

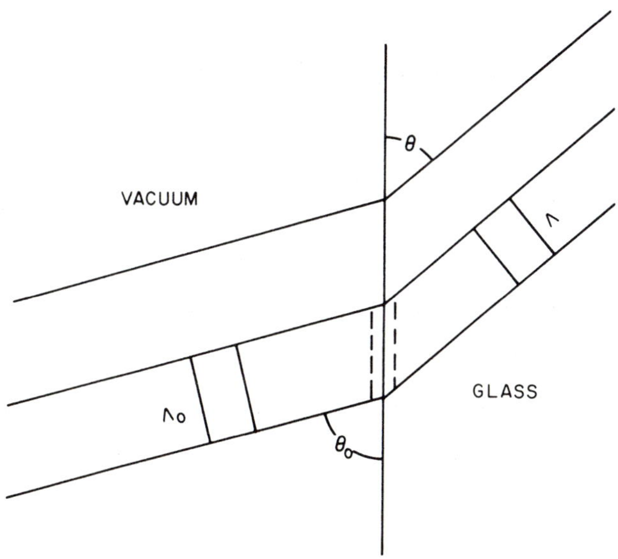

FIG. 4.
Refraction; $n = \Lambda_0/\Lambda = \sin\theta_0/\sin\theta$ from the equality of the dotted lines $\Lambda_0/\sin\theta_0 = \Lambda/\sin\theta$.

Since crests must be identical in the interface, then

$$a = \Lambda_0/\sin\theta_0 = \Lambda/\sin\theta \tag{19^a}$$

as indicated; and (from Eq. 18)

$$n = \sin\theta_0/\sin\theta \quad \text{(Snell's law)} \tag{19^b}$$

where θ_0 and θ are angles of incidence and refraction as indicated.

Under the assumption of the loss of velocity derived from this observation, using the measured index of refraction, it is possible with some labor[3] to assess the change of the electromagnetic wave after passing the glass as:

$$E_{\text{after glass}} = e^{-i\omega(n-1)\,\Delta z/c}\, E_0 e^{i\omega(t-z/c)} \tag{20^a}$$

or (since for small x, $e^x \simeq 1 + x$)

$$E_{\text{after glass}} = \underbrace{E_0 e^{i\omega(t-z/c)}}_{\text{E source}} - i\omega(n-1)\Delta z/c \cdot \underbrace{E_0 e^{i\omega(t-z/c)}}_{\text{E gl}} \tag{20^b}$$

(The symbols have meanings as in Eq. 15.) Since the first term describes the source field at the plate, the second term corresponds to the effect of the glass. Comparing this with the expression obtained in Eq. (15) the refractive index can be expressed or "explained" purely in terms of glass properties (resonant frequencies) and the frequency of light: (Eq. 21). At the same time in Eq. (22) k's have been inserted to account for more than one resonating electron.

$$(n-1)\Delta z = N\Delta z q^2/2E_0 m(\omega_0^2 - \omega^2 + ig\omega) \tag{21}$$

or

$$n = 1 + N_k q^2/2E_0 m(\omega_\kappa^2 - \omega^2 + ig_\kappa \omega) \tag{22}$$

Further refinements are necessary to account for the dipole moments in the mutual interaction of glass constituents. Also, the formulas usually listed are adapted to quantum mechanics introducing a sum of resonators with oscillator strength f_k. In these relations the index frequently appears in a form more directly related to the interaction field, called the polarizability (α).

$$\frac{n^2-1}{n^2+2} = \frac{1}{3}\sum_j \frac{N_j q_j}{m_j E_0}\sum_k \frac{f_k}{\omega_\kappa^2 - \omega^2 + ig_k\omega} = \sum \frac{N\alpha}{3} \tag{23}$$

But even in these more complex expressions the analogy to the response in a mechanical forced damped oscillation is recognized. This recognition is an objective of the introductory section.

C. Dispersion

The results of any evaluation of the interaction of light and glass, and, specifically, refraction, obviously depend upon the frequency of the interacting light, ω. This effect is called dispersion. The role of the resonating systems in the glass in the resulting change in refraction is of primary importance in the formulation of optical glasses but interests us here only in passing.

Figure 5 shows the typical relation of refractive index and frequency and its behavior at resonant frequencies in the glass. The refractive index normally increases with frequency in the manner shown except right at the resonant frequency. This descending branch

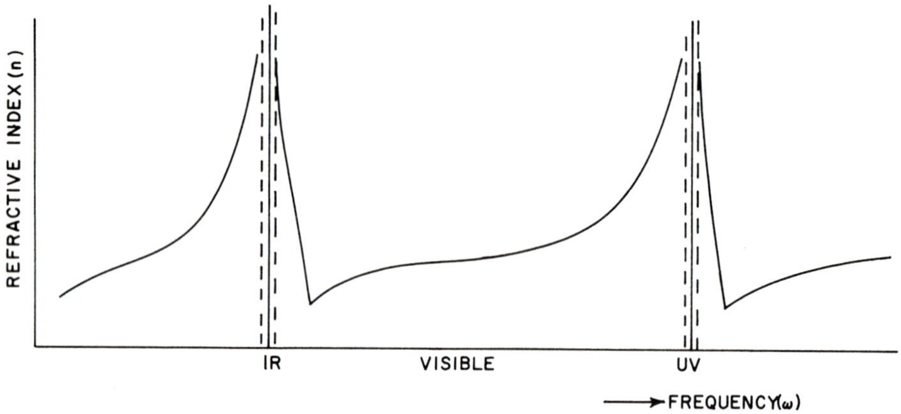

FIG. 5.
Typical dispersion curve.

is called "anomalous" dispersion, although it is normal in respect to the relations presented deriving from the change in sign of $\omega_0^2 - \omega^2$. Hensler at Bausch and Lomb[5], has successfully experimented with the insertion of a resonating center (that of cerium which is just barely affecting the visible color), to affect the dispersion curve in optical glasses as a matter of theoretical interest. In practice, dispersion of optical glasses is mostly determined by the fundamental transitions of oxygen electrons.

IV. ABSORPTION

A. Absorption, General

The introduction of damping or loss in the interaction relations has clearly caused the resulting quantities to be complex numbers in the selected notation for oscillation ($e^{-i\omega t}$).

To understand the role of damping for our specific evaluation of absorption we like to separate the loss part from the general relations. One relatively easy way to do this is to consider a complex index of refraction and introduce it in Eq. (20^a).

Defining real and imaginary parts in the complex index formulation of the type given in Eq. (22) or indicated in Eq. (23), one may write:

$$n = n' - in'' \tag{24}$$

$$E_{(\text{after glass})} = \exp\left[-i\omega(n' - in'' - 1)\ \Delta z/c\right] \cdot E_0 \exp i\omega(t-z/c) \tag{25}$$

$$\left[\text{since } -i\omega\ (n' - in'' -1) = -i\omega(n'-1) + ii\omega n'' = -i\omega(n'-1) - \omega n''\right] \tag{26}$$

can be written, separating the real exponent,

$$E_{\text{after glass}} = \exp\left[-\omega n''\Delta z/c\right] \exp\left[-i\omega(n'-1)\Delta z/c\right] \cdot E_0 \exp i\omega(t-z/c) \tag{27}$$

Now the first term no longer contains i and is a decaying, not a vibrating term. The n'' in the real, negative factor describes the decay or loss. It measures the decrease of magnitude with thickness expressed in terms of Δz. The n'' is called, at times, the absorption index, more frequently the extinction coefficient, written k. The amplitude is reduced by exp $- 2\pi k$ for a thickness change of one wavelength $\Lambda = 2\pi c/\omega$.

The determination of the extinction as a function of wavelength permits the detection of resonant frequencies in the glass. Their variation with structure can be used as a probe for the structure, or a control for the color of the glass.

B. Absorption of Glass in Practice

In practice, expressions more closely related to the measured data obtained are preferred. Using notations of Fig. 6 it follows (for 90° incidence):

I_0 = incident intensity

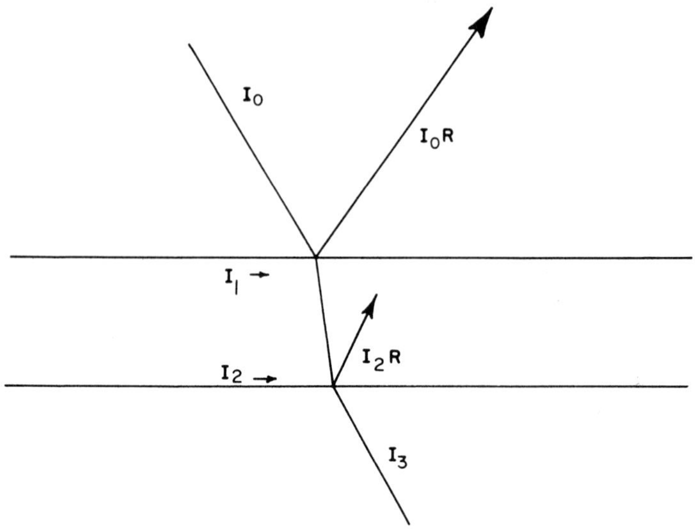

FIG. 6.
Absorption: I_0 = incident intensity; I_1 = reflected.

I_1 = intensity entering glass after reflection

I_2 = intensity at second face after absorption

I_3 = intensity leaving glass

d = thickness

$I_2 = I\ e^{-\alpha d}$

α = absorption coefficient = $\ln(I_2/I_1)d$ (28)

T = internal transmittance = $e^{-\alpha d}$ (29)

These relations to absorption are, however, not directly available from spectrophotometric data, because reflections take away intensity. Neglecting multiple reflections one further defines and calculates:

$$R = \text{reflection} \simeq \frac{(n-1)^2}{(n+1)^2}$$

Thus, T = (overall) transmittance = I_3/I_0 (30)

One obtains

$$I_3/I_0 \text{ (see Fig. 6)} = (I_2 - I_2R)/I_0 =$$

$$I\ e^{-\alpha d}(1-R)/I_0 = I_0(1-R)^2/I_0 \cdot e^{-\alpha d}$$

That is,

$$T = (1-R)^2 e^{-\alpha d} = \left[1 - \frac{(n-1)^2}{(n+2)^2}\right]^2 e^{-\alpha d} \tag{31}$$

Technologists also use optical density, which is based on the decadic logarithm:

$$D = \log^{10} 1/T = \alpha t \log_{10} e - \log_{10} (1-R)^2 \tag{32}$$

Density and transmittance compare as follows:

D	T	%T
2	0.01	1
1	0.1	10
0.5	0.316	31.6
0.1 (1↓)	0.794	79.4

In most glasses $g_k\omega << \omega_k^2 - \omega^2$ for visible light and the absorption is small. In some glasses $\omega_k^2 - \omega^2$ becomes small and the index becomes almost all imaginary (absorption dominates). This occurs in all glasses near the absorption frequencies in the UV.

The following problems are related to this subject

1. Transparency and dispersion (fiber optics).
2. Design of absorption patterns for photographic filters, signals, and sunglasses.

3. Analysis for composition.
4. Sensitivity to structure, fine structure, environment (temperature, fictive temperature, radiation field).
5. Color (art, etc.).

Much knowledge on absorption was, and still is, derived from the comparison of glass with solutions[6]. At present, progress is made in considering the quantitative influence of the field of surrounding ions (oxygen) on the electronic transitions of coloring ions (ligand field theory) (Section V).

Past successes are attributed to similarities of the oxygen environment in glass with the water environment in solutions, as well as to analogies derived from the study of crystals. But the new approach has given very promising detailed results since 1960.

V. LIGAND FIELD CONCEPTS APPLICABLE TO GLASS SCIENCE

The electronic transitions of transition metal ions are influenced by neighboring ions (ligands). Because of the rapid fall of the field with distance ($\simeq 1/r^5$), only immediate neighbors are considered.

A. Types of Transitions

The following types of transitions are observed and indicated in Fig. 7.

1. Strong transitions from
   a. Ligand excited state to central atom
   b. Ligand excited state to other ligand
   c. Central atom excited state to central atom ground state

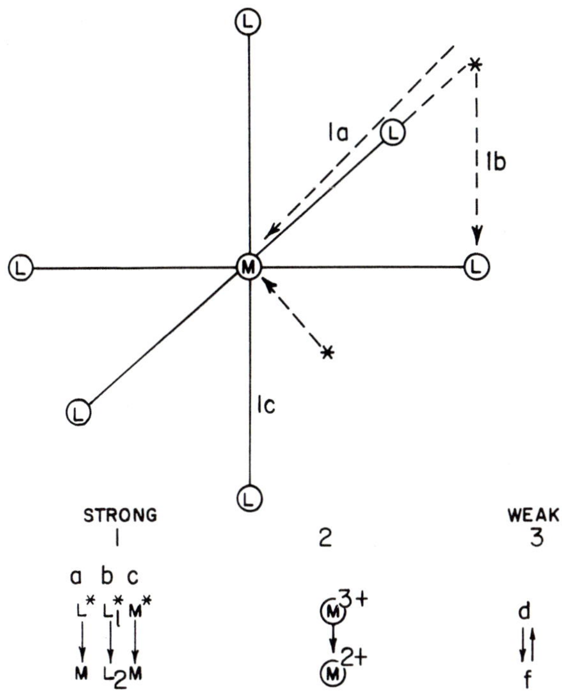

FIG. 7.
Types of transitions: Example $M^{2+} \rightarrow M^{3+}$; M = central atom, L = ligand, and * = excited state.

2. Strong transitions involving redox states, e.g., $Mn^{2+} \rightleftarrows Mn^{3+}$
3. Weak (forbidden) transitions between d or f levels of central atom.

B. Removal of Degeneracy in Octahedral and Tetrahedral Fields

A few examples will serve to illustrate the method and success of ligand field theory. A thorough text stressing glass science is offered by Bates[7].

In Fig. 8, five d orbitals are sketched, which are all degenerate in the free atom (i.e., exhibit only one level of energy in its spectroscopy). Two of these, $d(z^2)$ and $d(x^2-y^2)$ with lobes pointing to the corners of the octahedron are equally unfavorable in its

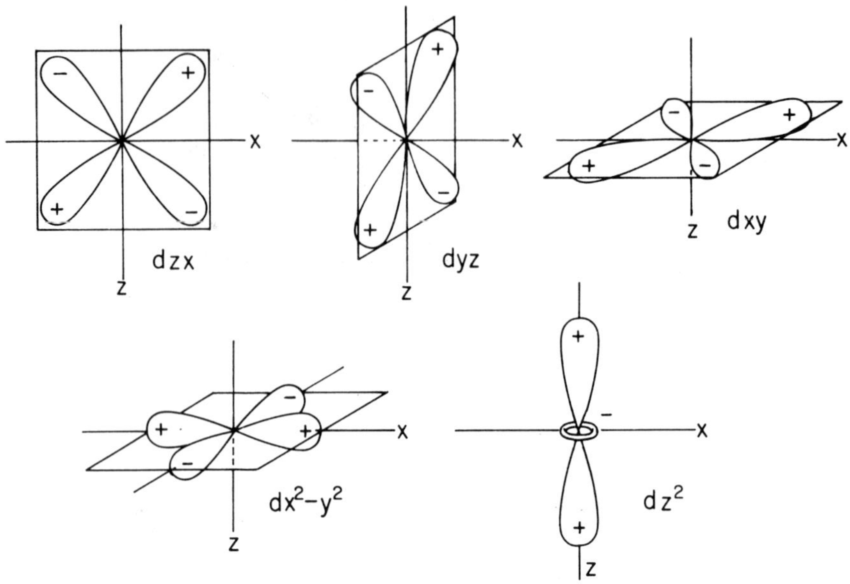

FIG. 8.
The d-orbitals in a free atom.

field, the three others are equal and favorable, having lobes between ligands. As a consequence, the d level is split in two in the octahedral field (Fig. 9).

In Fig. 9 the analogous splitting in a tetrahedral field is also illustrated. The split energies are labeled Δ_0 and Δ_T, and the

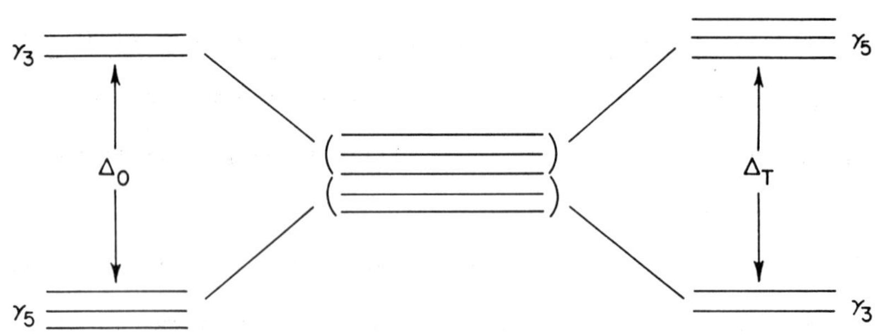

FIG. 9.
Schematic showing of five d orbitals of a free ion (center) in an octahedral (left) and in a tetrahedral (right) field.

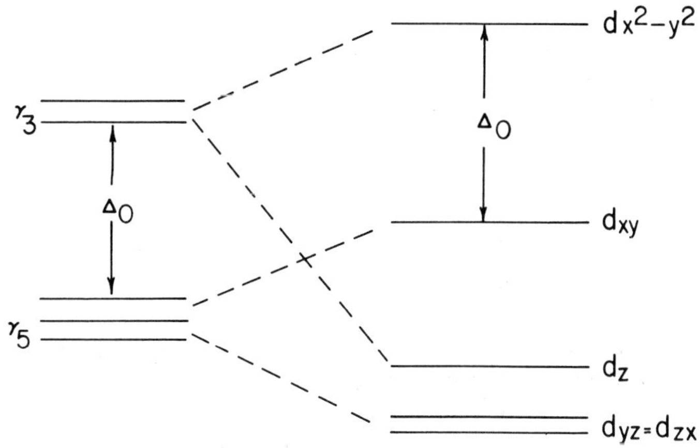

FIG. 10.
Distorted octahedral field (right). The degeneracy of d orbitals is further reduced.

split levels γ_3 and γ_5. The effect of a distorted octahedron exaggerated to the point that the top and bottom ligands are removed, is shown in Fig. 10.

C. Spin States in Octahedral Field

In populating the d shell with electrons, there are some choices between pairing electrons in γ_5 or extending the population to γ_3. The choice depends upon whether the pairing energy, P, or the energy difference of the higher state, ΔE, is greater.

Figure 11 illustrates the choices (high spin states, electrons in γ_3; low spin states, electrons paired in γ_5) for the d^5 configuration. Similar choices exist for d^4, d^6, and d^7 states. There are no choices for d^1 to d^3, and d^8 to d^{10}; in the first three because γ_5 levels are incomplete, in the last three because they are already occupied by two electrons each. Thus, buildup in the former is all in γ_5, in the latter all in γ_3. It has been possible to predict spin states correctly, because strong ligands favor the low spin

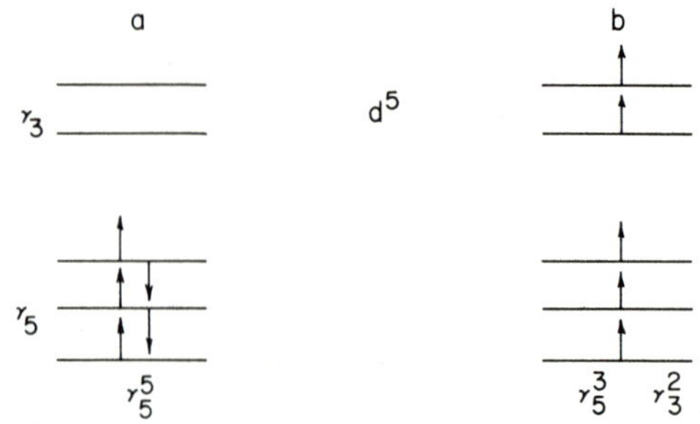

FIG. 11.
Possible distribution of spins in d^5 orbitals; (a) Distribution preferred in the vicinity of a negative ligand.

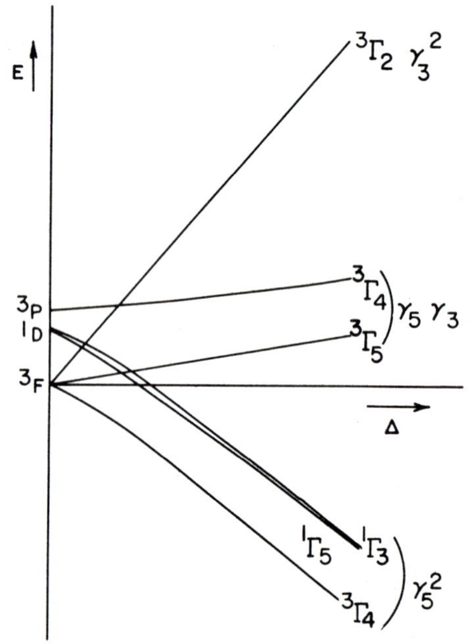

FIG. 12.
d^2 configuration in an octahedral field. Free ion levels ^3F ^1D ^3P are split into levels $^{2s+1}\Gamma_n$ (BETHE'S notation).

state, i.e., a pairing of electrons rather than electrons populating the γ_3 level of higher energy. Since pairing energy is known from spectroscopy some predictions can be made and the results are presented in Table I.

TABLE I

SPIN STATE PREDICTIONS

Ion	Type	Ligand	P	ΔE	State predicted	exp.
Mn^{3+}	d^4	H_2O	24000	21000	high	high
Fe^{2+}	d^6	H_2O	15000	10400	high	high
Fe^{2+}	d^6	CN^-	15000	33000	low	low

D. Mixed Orbitals

This picture is greatly simplified. But the more complicated situation involving interactions has been treated with success. The d states for total angular momentum L are:

0	1	2	3	4	5
^xS	^xP	^xD	^xF	^xG	^xH

where x = number of spin directions.

A d^2 ion has, for example, a configuration of: ^3F ^1D ^3P ^1G ^1S. In an octahedral field they split as shown in Fig. 12 for the three

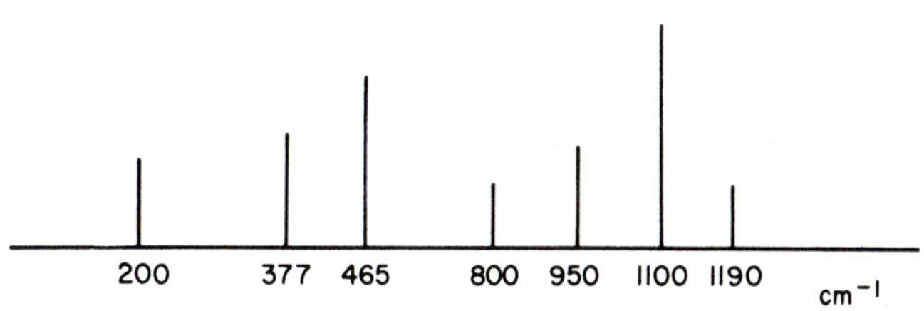

FIG. 13.

Maxima of ir absorption bands in vitreous SiO_2

lower states. The complexity is alleviated because transitions are generally limited to equal spin multiplicity states.

One interesting aspect is that some states are made more stable by ligands, with the energy of stabilization being often as high as 25-50 Kcal/mole. The d^2, d^7 states, for instance, have great energetic advantages in tetrahedral symmetry; most other states are favored in octahedral symmetry, some in neither.

Octahedral surroundings are found experimentally in accord with these conditions for Ti^{3+} (d^1), V^{3+} (d^2), Cr^{3+} (d^3), Fe^{2+}, Co^{3+} (d^6), Co^{2+} (d^7), Ni^{2+} (d^8), Mn^{3+} (d^4) (distorted), Cu^{2+} (d^9).

VI. SPECIFIC ABSORPTION: COLORED GLASSES

Glasses are colored mainly by: (a) absorption of ions in ligand fields, (b) absorption of nonoxide compounds formed or precipitated in oxide glasses, and (c) absorption of precipitated (colloidal) elements.

A. Ionic Absorption

Typical colorations are by V, Cr, Mn, Fe, Co, Ni, and Cu.

Ion	Chief bands around (nm)	Color of glass
Mn^{2+}	440 (weak)	Light yellow-brown
Cu	Broad to 650	Blue
Ni^{2+}	450, 580, 620	Gray, purple, brown
Co^{2+}	550, 610, 650	Blue
Mn^{3+}	480 (strong)	Purple
Cr^{3+}		Green

Cerium absorbs in the UV and violet, due to the transition Ce^{3+} -Ce^{4+} (or Ti^{4+}); rare earths in glasses exhibit sharp absorption bands due to inner shell transitions, causing beautiful colors.

The coloration by V^{3+} is similar to that by Cr^{3+}. Selections and blending provide most colored glasses. The chemistry of colorants in glass is amply described in a classic treatise by Weyl[6].

Examples from more recent work are the following data on Mn^{2+} and Fe^{2+} (see also Bamford[10]). Many other ions in glass are now studied with similar results.

Mn^{2+} has a d^5 configuration. Both octahedral and tetrahedral coordination are permitted. The weak (forbidden) bands are stable in glasses containing reducing agents (Sb_2O_3, As_2O_3, Ce_2O_3, FeO). The transitions usually observed seem to be ^4E (G) or ^4A (G) to ^6A (S). The ligand theory gives 10 levels of which only a few are involved in the observed transitions. This is why, in most cases, Linwood and Weyl's original assignment[11] based on comparison with the red and green fluorescence in crystals with known manganese configurations remains valid.

But new results yield some very interesting dttails[12]. The Mn^{2+} is found in distorted octahedral surroundings in calcium phosphate and sodium borate glasses that are high in network formers, and in tetrahedral surroundings in a lower network silicate glass. In a borophosphate glass both surroundings are found.

In most glasses an equilibrium of Mn^{2+} and Mn^{3+} exists. The Mn^{2+} is enhanced by Sb_2O_3, As_2O_3, Fe_2O_3, and CeO_2. At present, cerium is generally used because the reaction

$$Ce_2O_3 + Mn_2O_3 = 2CeO_2 + 2MnO$$

is not reversed by sunlight (solarization); Mn_2O_3 is enhanced by chromium oxide.

The Fe^{2+} and Fe^{3+} ions and their interactions give rise to very complex spectra and the familiar green coloration in silicate glasses. The Fe^{2+} ion causes an intense IR absorption reaching deep into the visible range. The Fe^{3+} produces an intense UV absorption, also reaching into the visible range. Both ions give rise to a complexity of transitions and interactions. In the case of Fe^{3+}, ligand interpretation is difficult because $3d^5$ ions have the same energy level diagram for octahedral and tetrahedral coordination[13]. Consequently little satisfaction has been derived[14] from the numerous optical and esr studies of iron in silicate glasses as far as the precise role of Fe^{3+} is concerned.

Mössbauer and optical data now support[14] the view that Fe^{3+} is tetrahedrally coordinated in silicate glasses; at low concentration epr gives a $g^1 = 4.3$ signal for distorted tetrahedral as well as (in phosphate glasses) octahedral coordination. At high concentration the $g' = 2$ resonance growing at the expense of $g' = 4.3$ is the result of spin-spin interaction, again for either coordination. Ligand field theory has not been able to satisfactorily detail these distortions in spite of many good theoretical analyses. For, many of them equally result in the observed 4.3 value. Those views are in contrast to those of Hirayama, Castle, and Kuriyama[15] who definitely interpret $g^1 = 2$ growth as indication of octahedral sites.

Mössbauer isomer shifts were evaluated on the basis of data for many compounds showing shifts of the order of 0.001 cm/second for tetrahedral, and 0.020 cm/second for octahedral sites. Theory suggests that the 4 s electron density is larger, amounting essentially to more covalency in the case of tetrahedral environment. Moreover, Kurkjian[4] studied crystalline systems with known coordination for further comparison.

A very interesting result of these comparisons is the fact that iron in octahedral sites in certain chain structures (pyroxene) is not so found in glasses of equal composition. This agrees with the view that silicate glasses with higher SiO_2 content tend towards greater difference from their crystals, and, finally, lead to immiscibility.

It also is interesting that manganese, too, prefers octahedral sites in phosphate glasses and tetrahedral sites in silicate glasses. Fe^{2+} has spin 2s, L = 2, so that no esr is to be expected. The transition $Fe^{2+} \rightarrow Fe^{3+}$ has been used to interpret the field of lunar surfaces[16]. Additional work on epr signals from Fe in glasses is by Tucker[17] and Volokhova[18].

Iron has long been known, or suspected to affect the uv absorption edge of the base glass, (e.g., at 230 nm in a $3Na_2O \cdot 7SiO_2$ glass[19]). Studies on uv absorption edges of different base glasses and of the same glass, with different oxidation states or high energy radiation histories, must always take into consideration the possible influence of trace ion impurities[20]. Bands due to ferrous and ferric iron combine in the visible to produce the familiar green color in glass. Sharper bands in some phosphate glasses leave most of the visible range transparent and can be used in colorless heat absorbing glasses.

B. Sulfide Colors

Nonoxide compound formation may be used in glasses to produce color (e.g., that of sulfides). This coloration is sensitive to heat treatment, indicating a precipitation of colloidal regions. Colorless Cd^{2+} forms a colored (yellow) sulfide. The absorption is characterized by a steep edge separating transmission in the yellow and red from absorption in the purple, blue, and green. Recently Vogel et al.[21] found that the CdS phase crystallizes in one of two immiscible glass phases. The control of this phase separation permits fine control of absorption band wavelength and steepness and the extension of the filter series to the ultraviolet using Zn in the phase. The addition of selenium allows the shift of this absorption edge to longer wavelengths as mixed cadmium sulfide-selenides form. In this manner a series of yellow, orange, and red filter glasses are obtained.

Numerous sulfide colors are known and used in glass. "Carbon amber," an important and long known bottle color, is based on the very complex equilibrium between sodium and iron polysulfides and sulfates in glasses containing more than 0.0001% of iron and carbon as a reducing agent. Recently a careful analysis of the equilibrium of four species (ferrous ion, ferric ion, sulfide sulfur, and sulfate sulfur) in the chromophore and two species (sulfide and ferric) outside were given by Douglas and Zaman[22].

C. Elemental Colors

Color may be obtained by the controlled precipitation of elemental gold, silver, or copper regions. The process is a complex one, involving solution of the metal ion in the glass, and subsequent reduction by agents such as Sb or Sn to the metallic state. Controlled reheating to effect nucleation of many regions and to keep the regions small is important as color varies with the size and shape of the regions. Examples are the deep rich colors of gold or copper ruby glasses.

VII. PHOTOSENSITIVE GLASSES

If a glass undergoes a primary, or secondary reaction starting with the absorption of a light quantum which is irreversible and, therefore, ends with a new product, it is called a photosensitive glass[23]. The reduction of silver and gold resulting, on proper heat treatment, in colored glasses can be sensitized by the photon-induced release of electrons from Ce^{3+}[24].

$$Ce^{3+} + h\nu \rightarrow Ce^{4+} + e^-$$

$$Au^+ + e^- \rightarrow Au^0$$

This phenomenon can be used as a photographic process or in the nucleation of crystallized glasses for other uses. Photosensitive glasses may typically contain 0.05 — 0.1% cerium oxides and 0.05% gold.

VIII. PHOTOCHROMIC GLASSES

A. General

The absorption of electromagnetic radiation may cause the more or less reversible change of a state A of a species to the state B, where each state is characterized by a different absorption spectrum

$$A_{\lambda 1} \overset{h\nu}{\rightarrow} B_{\lambda 2}$$

If at least one of the characteristic absorption wavelengths of one state is in the visible range, the phenomenon is called photochromism. Photochromic glasses might be applied to windows, sunglasses and optical switches. In these applications, it is desirable to obtain rapid darkening under illumination, rapid and reversible fading on withdrawal of light, and minimum fatigue on cycling the process.

B. Undoped Glasses

Ordinary soda lime glasses and many other silicate glasses are found to be photochromic under shorter than visible wavelength illumination if they are prepared from purest (iron free) materials under excessively reducing conditions[8]. It is amusing to think that in such an environment, ordinary windows so made would darken in sunlight and fade at dusk. The active centers (2.2eV)[16] must be present in the structure of these silicate glasses. The band is promoted by Ti, Ga, Ge, Al, Te, In, Tl, stabilized by Zr^{4+} [25]. The former also promote a 2.7 band.

C. Sensitized Glasses

By the addition of Eu^{2+} or Ce^{3+} using 4f-5d transitions, it has become possible to transfer energy from the available sunlight to the active centers. In this process Eu^{2+} changes to Eu^{3+}, Ce^{3+} to Ce^{4+}, and the glass darkens through the same centers that would darken under shorter wave illumination in the undoped glass[8][16]. These glasses fade too slowly and show too much fatigue for immediate broad use. The active absorption bands are at 3.92 eV (3150 A°) and 3.71 eV (3325 A°).

D. Silver Halide Glasses

It appears that isolation of active centers for a reversible process can be better achieved in ordered zones, such as small silver halide crystals developed in certain glasses by special heat treatment[26]. The glasses are characterized by

1. High solubility for silver.
2. Facility of silver migration (such as in aluminosilicate glasses of Al:Na ratios near one).
3. A crystallization process involving silver, chlorine, and fluorine.
4. A minimum participation of the glass base in the precipitation process, e.g., absence of Ca to avoid CaF_2.
5. (Advantageous but not necessary) electron transfer to silver by copper.

A typical composition[27] (wt%) is: SiO_2, 62.8; Na_2O, 10.0; Al_2O_3, 10.0; B_2O_3, 15.9; Ag, 0.38; Cl, 1.7; F, 2.5; Cu, 0.016. The basic reaction had been investigated in connection with photographic processes[28], but in this case irreversible reactions dominated and were, of course, desired. In these reactions colloidal silver precipitates and the silver can not be reconverted to Ag^+. The control of complex reactions is required in the mastery of these photochromic glasses.

IX. INFRARED ABSORPTION

A. Principles and Topics of Study

The infrared absorption in a solid is determined by the dynamics of motion of the atomic constituents. The energy of the absorption can be predicted with reasonable confidence when the symmetry of the atomic arrangements, the atomic masses, and the force constants are known. This is the case for crystals of known structure and assignments for different crystal classes can be found in the literature. In the case of glasses the result from ligand field optical studies, Moessbauer spectra and magnetic spin resonance are not sufficient so far to predict the absorption frequencies, in particular the technologically important high frequency absorption edges, of glasses. However, the infrared spectrum of the glass will be essentially a smeared out version of the spectrum of crystalline compounds of equivalent composition. Although the evaluation of the phonon process involving the entire lattice is complicated, and indeed forbidding for the more disordered glass, something can be said about the highest frequency modes which determine the technologically important cut-off. These modes are stretching modes involving, as in simple molecules, the atoms of lowest coordination number, largest charge, and shortest cation-anion distance; in silicates the Si-O stretching mode determines the cut-off.

If high transmission up to high wavelengths is desirable, compositions must be developed which are free from, or low in, content of such atoms. In oxide glasses silica free formulations had to be developed. The nonoxide glasses designed around the higher-mass anions S, Se, and Te, however, surpass all oxide glasses in high-wavelength ir transmission. The large theoretical and phenomenological differences between an ordered and disordered solid must nevertheless be admitted. High wavelength absorptions are generally larger in glasses due to disorder characteristics. Starting from a comparison with high point-defect concentration in crystals a theoretical evaluation is offered for instance by Hadni[29]. It involves the small phonon path known to be associated with the low thermal conductivity of glasses that persists to very low temperatures.

B. Data Evaluation

The large extinction coefficient observed in the IR spectroscopy of glasses makes it necessary to work with thin films. But normally the resonant frequency or true absorption maximum cannot be evaluated from the extinction minimum obtained by usual relations derived for such thin films. One must return to the relation ot the complex index in the form:

$$n^{*2} - 1 = 4\pi Nq^2/m\ [\omega_0^2 - \omega^2 + ig\omega]$$

and find that the imaginary part of $n^{*2} = (n-ik)^2$ (equal to 2nk from $n^{*2} = n^2 - 2ink + k^2$) must become a maximum for the resonant frequency. One again has to correct for the local field in the glass and replaces nk by: $3nk/n^2 + 2$ which is plotted against ω to find the true maximum[30].

The important thing is that both n and k can not be obtained from absorption data, and reflection spectra must be measured to obtain structural information from infrared absorption.

It should be noted that Simon writes the complex index $n^* = n - ik_0 = n(1-ik)$ so that his k is k_0/n and his k_0 is our k. Therefore his final formula has $3n^2k$ instead of 3nk. Our nomenclature seems more conventional.

C. Silica

In vitreous silica broad maxima are observed as shown in Fig. 13 (p. 937). Those at 1000 cm^{-1} are assigned to Si-O bond stretching, those at 450 and below to bond bending. The stretching band corresponds to one in quartz and one in cristobalite indicating similar short range order.

D. Alkali Silicate Glasses

The addition of R_2O to SiO_2 introduces nonbridging oxygen ions. Proportionally the 1100 cm^{-1} band decreases and a new band at 950 cm^{-1} appears at about 25% Na_2O (mole); a double band (940, 1075) structure above 33% (Fig. 14). The 950 and 940 bands may be assigned to nonbridging O-Si stretching, an assignment supported by the similarity of bands in Na, Li, K glasses[31]. A low frequency (540) Raman band appearing also above 25 mole% Na_2O may be assigned[32] to bond-bending force at nonbridging oxygen. More recent detail can be found, for instance, in the investigations of Su and associates[33].

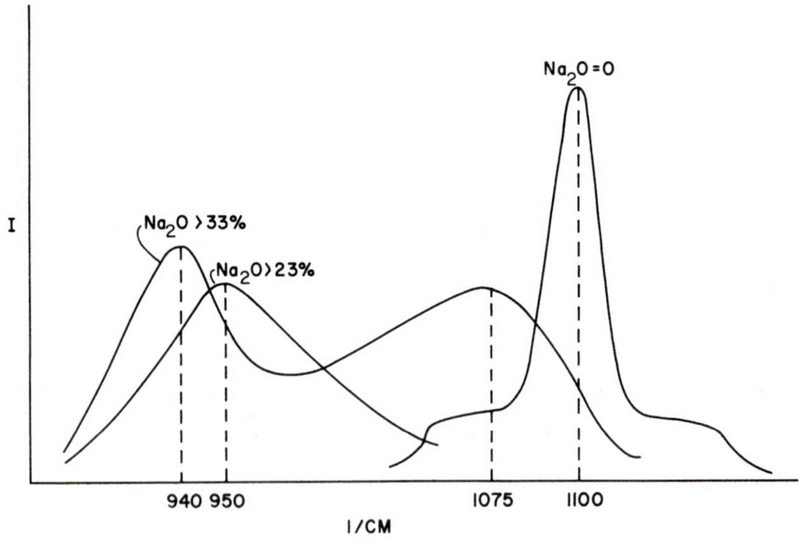

FIG. 14.
Infrared spectrum in Na_2O-SiO_2 glasses. Na_2O contents labeled. Arbitrary units of intensity (I) versus reciprocal wavelength.

While the interpretation of this to mean that crystallites are present[34] has not been accepted, recent observations of phase separation[35] support the abrupt structural change evident from ir-Raman studies, and recommend further use of this tool. The bands, characteristic of nonbridging oxygen, appropriately appear in alkaline earth glasses[36].

E. Borate Glasses

The complexity of the borate structure has been demonstrated and well analyzed in recent spin resonance studies. The appearance of four coordinated boron on adding alkali remains, however, confirmed in all these studies and appears to be manifest in the 1350-1400 band replacing that at 1250 when Na_2O in borate binary glasses exceeds 20%.

F. Germanate Glasses

All bands are similar to those in silica and silicate glasses but shifted to lower frequencies as expected with a lower force constant (Ge being larger than Si). The 1110 band of vitreous SiO_2, e.g., is located at 895 cm^{-1} (GeO_2). The spectrum is related to that of hexagonal crystalline GeO_2, but not to that of the rutile type six coordinated tetragonal variety.

Complex germanate glasses consequently transmit deeper in the infrared and have been applied as infrared windows and lenses in the modest range of the lead sulfide cell (to about 6μ)[37][38][39]. Such glasses may contain 35–45% GeO_2 with substantial amounts of TiO_2, and barium lanthanum, and (or) zinc oxide, with some ZrO_2, ThO_2 and PbO as well as fluorine.

G. Other Oxide Glasses

More transparent in the infrared than silicate glasses are

a. Tellurite glasses

b. Antimonite glasses

c. "Calcium aluminate" glasses

The latter, based on the narrow glass forming field near $CaO \cdot Al_2O_3$, have found some practical application particularly in England[38][40].

The participation of oxygen in the vibrational modes limits these applications to the range of 6μ. Nonoxide glasses or crystals must be used for ranges beyond.

H. Nonoxide Glasses

The appearance of detectors capable of "seeing" colder objects emitting above 10μ stimulated the intensive study of infrared transparencies in this range since 1953. The first important ir transmitting material in this higher range was "arsenic trisulfide" glass (actually not defined as a compound As_2S_3, but an arsenic sulfur glass around the 2:3 As:S ratio). It was discovered by Schultz and Sellak[41], rediscovered by Frerichs[42] and developed as well as extended into many sulfide, selenide, and telluride (chalcogenide) systems by several teams of research[43][44][45][46].

Because of the apparent requirement of large weakly bonded constituent atoms for transparency in this range all these materials are physically weak and soft with poor thermal characteristics.

A reasonable compromise has now been attained with the silicon, and, particularly the germanium chalcogenide systems. The thorough investigation of Hilton and co-workers (see e.g. Ref. [47]) resulted in operative transparencies such as $Ge_{28}\ Sb_{12}\ Se_{60}$ glass (for detail on this and similar glasses see, e.g., Ref. [48]). In 1968, the Academy of Science's Materials Advisory Board convened a panel to reinvestigate the limitations on developing strong, temperature resistent high wavelength ir transparencies with results given in a recent report[49]. In general, the fundamental nature of the limitations was recognized and further studies recommended are mostly in the area of improving the germanium selenide glass types as well as in the area of polycrystalline transparencies synthesized under pressure (see Section IX, I). Work is now in progress by the author on the possibilities of affecting the physical properties of ternary germanium selenide glasses by controlled fine structure (phase separation and cross-linking in time-temperature variations of undoped and doped base glasses).

I. Polycrystalline Hot Pressings

Polycrystalline hot pressings are very promising mechanically, and are strong competitors of IR transmitting glasses. The author worked on this idea around 1955, using MgF_2 crystals. While such materials are now conventional, they do not suffice for the now required transmittance range. Sulfide and selenide polycrystalline hot pressings are now available with transmittances to and beyond 14μ, such as ZnS (to 14.7μ, Eastman Kodak's Irtran 2) or ZnSe (to beyond 20μ, Eastman Kodak's Irtran 4) and more recently CdTe (Irtran 6) and, CdS. (See Refs. [50][51][52] for details on these and other ir materials.

These materials are mechanically and thermally superior, comparing equal IR transmittance; they may have greater limitations in regard to homogeneity and large scale fabrication, but remain important competitors for glass transparencies because of their physical properties.

J. "Water" in Glass

It is much harder to remove water from the glass melt and to prevent its reaction to affect the composition and structure of the final product than one might deduce from its low boiling point. Most conventional glasses contain "water" in the form of hydrogen-oxygen assemblies bonded by hydrogen bonds.

It is possible[53][54] to distinguish three bands depending on the tightness of the bond (tightest when associated with the bridging oxygen of silica or low alkali silicates):

a. 2.75 – 2.95 μm

b. 3.35 – 3.85 μm

c. 4.25 μm (for high alkali (R) glasses involving more than one "nonbridging" (Si-O-R) oxygen per silicon.

These IR absorption bands can be used to characterize the glass structure and, last not least, to analyze for "water" content.

Calibrations by Goetz et al.[55] were used to characterize the H_2O content of glasses in our studies of the influence of H_2O on phase separation. In infrared transmitting oxide glasses the water content must be suppressed by bubbling the melt.

X. ULTRAVIOLET TRANSMISSION OF GLASSES

Very little shall be said here on the uv transmission of glasses. It should be emphasized that published data on this subject to this time must be considered unreliable. The uv absorption edge for all glasses free of coloring additives is generally considered related to the fundamental transition of electrons in the oxygen bonds. However, it has become increasingly evident how large the role of minute concentrations of transition elements, specifically iron, is. It remains true that oxygen, and more so, fluorine are the most favored anions if a high energy transition (low wavelength uv absorption edge) is desired; and that for oxide glasses phosphate glasses tend to transmit to lower wavelengths because the P — O bond is stronger than the Si — O bond. It also remains true that generally nonbridging oxygen introduced by alkali and other "modifiers" in the silicate structure shift the absorption to longer wavelength. However, as particularly Sigel and Ginther[20] showed, the presence of Fe^{3+} at levels below 10 ppm is crucial. The reduction of Fe^{3+} to Fe^{2+} improves the transmission. Quantitative data of past studies on bulk compositions of oxide glasses must be reexamined. Among glasses, in practice, pure SiO_2 and low impurity phosphate glasses are most transparent. In uv optics, however, one must resort to fluoride crystals, particularly LiF which transmits to 180 nm. (For details consult Kreidl and Rood[50] and Schroeder and Neuroth[51].)

XI. ABSORPTION IN LASER MATERIALS

Although the existence of a long decay time metastable emission level is crucial for a solid state laser material, some important requirements are related to absorption processes. Several levels above the metastable emission level should exist and possess strong and broad absorption bands to permit exploitation of a large amount of pumping light, which usually has a continuum type spectrum.

These high levels also should have short life times so that they could rapidly feed the metastable emission level. It is desirable that electrons from the absorption levels go to the emission level by a radiationless process. Furthermore, the material should be free of absorption in the spectral range of emission. In the case of neodymium, for instance, the absorption range is 570 to 600 nm[57]. In spite of favorable emission conditions, glasses are frequently chosen for the broad absorption bands of the laser ion in glass, and, particularly, superior homogeneity in large specimens. The broad absorption of e.g. Nd^{3+} should be, and is found, both in crystals and glass in an area where the host material is thoroughly transparent. For Nd^{3+} the entire level scheme is given in Fig. 15.

In the well-known relations for laser efficiency the value of a derives from the desirable absorption characteristics for the pump light.

$$dN_2 = a\,IN_0 dt - N_2/t_2 \cdot dt \qquad (33)$$

or in the solution

$$E_s = AN_2/aN_0 \cdot T_s/t_2 \cdot (1 - \exp(-T_s/t_2)^{-1} \qquad (34)$$

in which

N_2	Number of excited centers
a	Cross section for pumping light
I	Intensity of pumping light
N_0 ($\gg N_2$)	Total number of particles
t_2	Life time of metastable level
E_s	$I_s T_s$ energy of emission, to start emission
A	Area

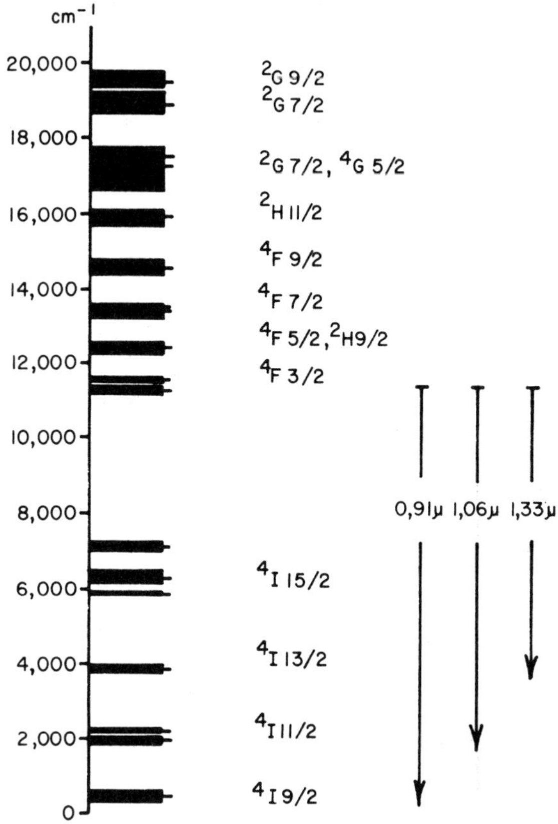

FIG. 15
Level scheme of Nd^{3+}, the ion mostly used in glass laser.

I_s	Intensity of emission, to start emission
T_s	Pumping time at threshold

An interesting accessory role of absorption in laser glass is that of a sensitizer ion capable of absorbing pumping light and transferring electrons to the emitting species. An example is Mn^{2+} in Nd^{3+} glass, with a claimed efficiency increase of 35%[58].

Right now, absorptions, in part impurity induced, in part perhaps the consequences of nonlinearity, causing failure of the glass, are in the foreground of glass laser research.

XII. HIGH ENERGY INDUCED ABSORPTION

The interaction of high energy radiation with solids is a vast topic in itself. The present outline is concerned with changes in optical properties resulting from irradiation. Radiation effects may be structural or electronic. Particle radiation is the typical cause of structural changes, ionic radiation of electronic changes. However, structural changes are invariably accompanied by electronic changes, and high intensity ionic radiation may also cause structural rearrangement. The electronic changes often result in new absorption centers which may have characteristic absorption and (or) emissions in the optical range just like inherent species. Only some examples of such centers are given here. An extensive review will be found in a text by Lell, Kreidl and Hensler[59].

Under heavy neutron bombardment structural changes are radical and one may speak of new materials resulting from it. Crystalline as well as fused silica, e.g., are converted to a new (metamict) phase characterized by new typical properties.

By and large, however, the silicon-oxygen structure is quite resistant to bombardment of medium intensities and a large amount of optical centers has been traced to defects and impurities present before irradiation (precursors).

Intrinsic absorption centers induced by radiation in pure silica are weak and positioned in the far ultraviolet. For the purpose of this presentation it is enough to state that pure silica has, and maintains under quite intensive high energy radiation, high transmittance. The far uv centers appear to represent oxygen and oxygen-silicon vacancies. Color centers induced by such radiation, or found in minerals having radiation histories, are associated with precursor impurities, mostly combinations of aluminum, lithium, and germanium. The fused silica obtained from hydrolysis of pure $SiCl_4$ is particularly radiation resistant. It is very pure but contains a high concentration of "water."

In complex glasses colors resembling those of transition elements are induced. Some of these appear to be electron holes trapped at or near nonbridging oxygen and alkali. They were identified by Stroud, Schreurs, and Tucker[60] and can be suppressed by the addition of transition elements, advantageously by the fairly colorless cerium. The resulting center can be shown to be an electron hole trap characterized by a reaction of the type:

$$Ce^{3+} \overset{h\nu}{\rightarrow} Ce^{3+,\ +}$$

the latter symbol representing[60] a cerium ion of an electronic state similar to that of Ce^{4+} but in the oxygen environment more characteristic of Ce^{3+}.

In practice these conditions lead to the following technological phenomena:

a. Optical glasses darken under high energy radiation.

b. This darkening may be suppressed by Ce^{3+} in "protected" glasses.

c. Color changes might be used to measure radiation doses (dosimeter glasses). Examples are the conversions of cobalt or bismuth to new centers (the latter may respond linearly to 10^9 rad).

d. Radiophotoluminescence refers to the process of creating a luminescent center. An example is silver which, according to the most recent data, perhaps surprisingly, seems to be converted from Ag^{+} to Ag^{2+}, the orange fluorescence of which may be measured down to the millirad range (<.01 rad)[61].

e. Induced thermoluminescence had been observed, although so far crystals only have been utilized in dosimetry.

f. Neutron reaction with lithium gives rise to emission in cerium present in a lithium aluminum silicate glass which permits its use as a scintillation counter.

Other effects are possible. In particular, refraction and optical rotation may change permanently in a radiation field.

REFERENCES

[1]. R. Benumof, Concepts in Physics, Prentice-Hall, Englewood Cliffs, N. J., 1965.

[2]. A. J. Dekker, Solid State Physics, Prentice-Hall, Englewood Cliffs, N. J., 1965, 154-157.

[3]. R. P. Feynmann, R. B. Leighton, and M. Sands, The Feynmann Lectures on Physics I, II, III, Addison-Wesley, Reading, Mass., 1965, 1966.

[4]. R. S. Longhurst, Geometrical and Physical Optics, Wiley, New York, 1964.

[5]. J. R. Hensler, VII International Congress on Glass, CR Short Papers 366, 1965.

[6]. W. A. Weyl, "Coloured Glasses," Soc. Gl. Tech., Sheffield (1951).

[7]. T. Bates, Modern Aspects of the Vitreous State (J. D. MacKenzie, ed.), 1962.

[8]. A. J. Cohen and H. L. Smith, Science, 137, 981 (1962).

[9]. A. Paul and W. Douglas, Phys. Chem. Gl., 6 (6), 207 (1965).

[10]. C. Bamford, Phys. Chem. Gl., 3 (2), 54 (6), 189 (1962); 1 (15), 159, 165 (1960); 2 (5), 163 (1961).

[11]. J. H. Linwood and W. A. Weyl, JOSA, 32, 443 (1942).

[12]. K. Bingham and S. Parke, Phys. Chem. Gl., 6, 224 (1965).

[13]. Y. Tanabe and S. Sugano, J. Phys. Soc. Jap., 9, 753 (1955).

[14]. C. R. Kurkjian and E. A. Sigety, Phys. Chem. Gl., 9 (3), 73-83 (1968).

[15]. A. C. Hirayama, J. G. Castle Sr., and M. Kurijama, Phys. Chem. Gl., 9 (4), 109 (1968).

[16]. A. J. Cohen, VII International Congress on Glass, 1965, CR III-2-111; Cohen and Hapke, Science, 161, 1238 (1968).

[17]. R. G. Tucker, Adv. Gl. Tech., 103 (1967).

[18]. T. Volokhova, Izv. Zav. Fiz., 9, 10 (1965).

[19]. F. N. Steele and R. W. Douglas, Phys. Chem. Gl., 6, 246 (1965).

[20]. G. H. Sigel and R. J. Ginther, Glass Tech., 9 (3), 66-69 (1968).

[21]. W. Vogel, A. Renfeld, and H. Ritschel, Symp. on Colored Gl., Sept. 1967, Prague.

[22]. R. W. Douglas and M. S. Zaman, Phys. Chem. Gl., 10 (4), 125-132 (1969).

[23]. S. D. Stookey, Ind. Eng. Chem., 41, 856 (1949).

[24]. R. Yokota, Advances in Glass Technology, 1962, Plenum, New York, 424.

[25]. E. Swarts and J. Pressau, J. Am. Cer. Soc., 48 (7), 333 (1965).

[26]. W. H. Armistead and S. D. Stookey, Science, 144 (3615), 150 (1964).

[27]. G. P. Smith, VII International Congress on Glass, 1965, CR III-2-108.

[28]. N. R. Nail, et al., J. Phys. Chem. Sol., 9 (3,4) 217 (1959).

[29]. A. Hadni, J. Appl. Opt., 4 (4), 487 (1965).

[30]. I. Simon in Modern Aspects of the Vitreous State (J. D. MacKenzie, ed.) Butterworth, London, 1960.

[31]. P. Jellyman and Procter, Trans. Soc. Glass Tech., 39 T 173 (1955).

[32]. G. Wilmot, MIT Thesis, 1954.

[33]. R. Hanna and G. J. Su, J. Am. Cer. Soc., 47 (12), 597-601 (1964).

[34]. V. A. Florinskaya, Structure of Glass, Cons. Bur. New York, 1953, 55.

[35]. J. Hammel and S. Ohlberg, VII International Congress on Glass, CR I-3.2-34.

[36]. I. Simon and H. McMahon, J. Amer. Cer. Soc., 36, 160 (1953).

[37]. H. Blau, Ohio State Univ. Resch. Found., 1955, Rep #10; U. S. Pat. 2 705 208, 2-1-1955.

[38]. R. Glaze, Cer. Bull., 34, 291 (1955).

[39]. C. W. Cleek, U. S. Pat. 3 022 182, 2-29-1962.

[40]. H. Hafner, J. Amer. Cer. Soc., 41, 315 (1958).

[41]. H. Schultz and L. Sellak, Ann. Phys., 139, 162 (1870).

[42]. R. Frerichs, JOSA, 43, 1153 (1953).

[43]. W. Fraser and J. Jerger, JOSA, 43, 823 (1953).

[44]. J. Jerger, Jr. and R. Sherwood, Contr. Nonr., 4212 (00) (1963).

[45]. C. J. Billian and J. Jerger, Contr. Nonr., 3647 (00) (1963).

[46]. B. T. Kolomiets, Phys. Stat. Sol., 7, 359 (1964).

[47]. A. R. Hilton, Office of Naval Res., Contr. Nonr., 3810 (00) (ARPA #269), 1965.

[48]. C. Jones and H. Hafner, Res. on IR Optical Materials, Final Report Contr AF 1968, 33 (615), 3963.

[49]. Materials Advisory Board, "Infrared Transmitting Materials," Rep. MAB 234 July, 1968 (National Research Council and National Academy of Science/National Academy of Engineering).

[50]. N. J. Kreidl and J. Rood, Appl. Optics and Optical Eng. (R. Kingslake, ed.) Academic, New York, 1965, Vol. 1, Chap. 5.

[51]. H. Schroeder and N. Neuroth, Optik., 26, 381-401 (1967).

[52]. N. Neuroth, Glastech Ber., 41 (6), 243-252 (1968).

[53]. H. Scholze, Glastech Ber., 32, 142, 278, 314, 381 (1959).

[54]. H. Scholze, Glass Ind., 546 (10,11,12) (1966).

[55]. J. Goetz and E. Vosáhlová, Glastech Ber., 41 (2), 47-55 (1968).

[56]. See Chapter 30, this volume.

[57]. L. F. Johnson, J. Appl. Phys., 34, 897-909 (1963).

[58]. N. T. Melamed, Westinghouse Res. Lab. Sci. Paper 65-7C1-LGLAS-P1, June 8, 1965.

[59]. E. Lell, N. J. Kreidl and J. R. Hensler, Progress in Ceramic Science (J. Burke, ed.), Pergamon Press, New York, 1966, Vol. IV.

[60]. J. S. Stroud, J. E. H. Schreurs, and R. F. Tucker, VII International Congress on Glass, to be published, 1965, CR I-3.3-42.

[61]. J. H. Schulman, J. Appl. Phys., 22, 1479 (1951).

Chapter 28

LIGHT SCATTERING FROM GLASS

Joseph J. Hammel

PPG Industries, Inc.
Glass Research Center
Pittsburgh, Pennsylvania

I. INTRODUCTION

Light scattering from glass can be divided into two main categories: (1) Light scattering from single phase glasses; and (2) Light scattering from multiphase glasses. Interpretation of results from some of the earliest light scattering measurements may have been complicated by phase separation. A glass was considered to be of optical quality if it did not contain gaseous inclusions, striations, and was not visually opalescent. With the advent of electron microscopy and small angle scattering, it has become increasingly evident that many glasses which appeared to be of optical quality and homogeneous were phase separated[1],[2],[3]. These phase separated regions will grow when heat treated above the glass transition, substantially increasing the turbidity of the glass. Many of the earliest light scattering measurements may have been from phase separated glasses so that postulates on structure assuming single phase glasses were not applicable.

Also, a brief review of some applications of light scattering in glasses will be given. One use has been to obtain diffusion coefficients from phase separated glasses[4]. In addition Maurer[5],[6],[7] used light scattering very effectively to study the mechanism of nucleation and growth in photosensitive and crystallizable glasses and to study radiation damage in glass.

II. SOME THEORETICAL ASPECTS

A. Rayleigh and Rayleigh-Gans Scattering

When the scattering centers are small compared to the wavelength and are nonabsorbing, Rayleigh theory can be used to interpret light scattering measurements. A useful equation for turbidity, τ, which is readily derived from Rayleigh theory[8] is

$$\tau = \frac{24\pi^3 NV^2}{\lambda^4}\left(\frac{m^2 - 1}{m^2 + 2}\right) \tag{1}$$

where N is the number of particles, V is their volume, m is the ratio of refractive index of the particle to the medium, and λ is the wavelength of light in the medium. This equation assumes single particle scattering, where the particles are nonabsorbing and smaller than 1/10 the wavelength of light. Knowing the number of particles one can obtain their size or knowing their size one can obtain the number.

Einstein[9] predicted that the light scattering from single component liquids is caused by thermal fluctuations. These cause fluctuations in density, hence in refractive index, thus producing light scattering. Using Rayleigh and fluctuation theories and the Lorentz-Lorenz formula[8], Einstein derived the following turbidity equation:

$$\tau = \frac{8\pi^3 kT}{3\lambda^4} (n^2 - 1)^2 \beta \qquad (2)$$

where k is Boltzmann's constant, T is the absolute temperature, n the refractive index of the liquid, and β the compressibility. In deriving Eq. (2), it is assumed that fluctuations in one volume element, v_0, are independent of those in the surrounding volume elements. Ornstein and Zernicke[10] showed that this assumption is valid if the range of the intermolecular forces is of a smaller order of magnitude than the wavelength of light. With Eq. (2) the light scattering from liquids or glasses can be predicted from a knowledge of β, the compressibility.

Using fluctuation theory but considering only compositional fluctuations the following equation can be derived in much the same manner as Eq. (2)[11].

$$\tau = \frac{32\pi^3 n^2 \left(\frac{n^2 + 2}{3}\right)^2 \left(\frac{\partial n}{\partial c}\right)^2 kTM_a c}{3\lambda^4 \rho \left(-\frac{\partial \bar{F}_a}{\partial c}\right)} \tag{3}$$

where c is the concentration, M_a is the molecular weight of component a (solvent), ρ is the density and $\bar{F}_a$ is the partial molar free energy of a, where $\bar{F}_a$ equals $\partial F/\partial n_a$. Compositional fluctuations, which are relatively temperature independent, depend on the magniture of $(-\partial\bar{F}_a/\partial c)$. Generally, the closer a glass composition is to a miscibility gap, the smaller is $(-\partial\bar{F}_a/\partial c)$. At the critical solution temperature, this quantity goes to zero and a very intense scattering or critical opalescence is observed. These are stable fluctuations; inside a miscibility gap the fluctuations become metastable or unstable and phase separation will occur.

In addition, Rayleigh theory predicts dissymmetries, $I_{45°}/I_{135°}$, and depolarization ratios, ρ_h, equal to one. The ρ_h is equal to H_v/H_h where the subscripts v and h represent the incident beam polarizer with its plane of polarization vertical and horizontal, respectively, and H the detector polarizer in the horizontal plane at 90° to the incident beam. These measurements can be used to check whether Rayleigh theory is applicable.

When the scattering centers are larger than 0.1λ but the refractive index difference between the scattering centers and their surroundings is small Rayleigh-Gans theory[12] can be used. This theory no longer considers the scattering source as a simple dipole but takes a vector sum of the amplitudes of the scattered light from all of the scattering elements of the scattering source. Rayleigh-Gans theory has extended the usefulness of light scattering in polymer chemistry so that shape, size, and interaction of large polymers can now be obtained. It predicts dissymmetries greater than one and depolarization ratios, ρ_h, less than one[12] and has been successfully applied to predict light scattering in phase separated glasses[13].

B. Light Scattering from Phase-Separated Glasses

Recently it was found that light scattering from phase separated glasses produced two so-called anomalous effects, namely, an initial inverse eighth-power dependence of turbidity on wavelength[14] and a dissymmetry less than one[6]. No light scattering theory predicted these effects. Voishvillo[15] described these phenomena as arising from regions of positive and negative deviations from the average refractive index of the glass; these regions being approximately the same size and not much smaller than the wavelength of light. Maurer[6], who was the first to observe a dissymmetry less than one, described his results differently. He used a diffusion-controlled growth model originally proposed by Walker and Guinier[16] to explain low angle X-ray scattering from precipitated alloys. In this model the scattering is explained by interference effects between the particle and its surrounding diffusion field.

Goldstein[17] derived a quantitative light scattering theory for a system of particles undergoing diffusion-controlled growth. His theory allows one to obtain diffusion coefficients from experimental light scattering data provided the experimental system fits the model. His model requires: (1) that the growth of the new phase is diffusion-controlled; (2) that the particles behave as Rayleigh scatterers; (3) that in the scattered light only interference effects between particles and their diffusion fields are important; and (4) that compositional changes during phase separation are relatively small and have little or no effect on the diffusion coefficients. It will be shown later that his theory was successfully applied to obtain diffusion coefficients in glasses. Goldstein also analyzed the light scattering from a system undergoing spinodal decomposition[18]. In the case of spinodal decompositon there is no nucleation barrier to phase separation, only a diffusional one. He showed that it would be difficult to distinguish between light scattering from such systems and those undergoing nucleation and growth. The model proposed by Voishvillo for phase separated glasses is more likely to apply to a system undergoing spinodal decomposition than one undergoing nucleation and growth. All three theories predict an initial inverse eighth-power dependence of turbidity on wavelength and a dissymmetry less than one.

III. EXPERIMENTAL PROCEDURES

Light scattering instruments are available in which one can accurately measure depolarizations and obtain intensity measurements over a wide range of angles and wavelengths[19],[20]. Most of the initial measurements were made using a high pressure mercury lamp as a source. With a laser source more precise results on low scattering intensity glasses should be available. For example, a rough calculation shows that a 15 mW, 633 nm laser source produces approximately 1000 times the intensity of the 436 nm mercury line generally used. In addition the 436 nm generally produces fluorescence in most glasses; the 633 nm line would produce little or no fluorescence.

One of the main problems in measurements from polymer solutions is dust. This problem is eliminated in glasses, however, in glass there are the problems of fine gas bubbles, unmelted inclusions and striations due to improper mixing. Optical quality glasses generally eliminate these problems. Therefore, if the means to make optical quality glass are available a serious light scattering study can be made.

IV. RESULTS

A. Single Phase Glasses

The earliest measurements on light scattering from glasses were made by Krishman[21],[22]. He reported depolarization measurements, ρ_h, ranging from 0.91 to 0.68 for 17 optical glasses and ρ_h equal to 0.80 for vitreous silica. From these results he concluded that the scattering centers in optical glasses and vitreous silica are large compared to the wavelength of light, i.e., greater than about 500 Å. In 1948, Rank and Douglas[23] measured depolarization ratios from two optical glasses supplied by Bausch and Lomb and obtained similar results and arrived at the same conclusions. On the basis of Krishnan's results, Mueller[24] postulated a type of nonannealable strain that should be present in all glasses. These strains could develop from frozen-in density or compositional fluctuations existing in the liquid state. As the temperature is lowered below the glass transition to room temperature the fluctuations produce local shearing strains which gradually increase due to temperature contractions. Thus the original hypothesis by Krishnan that the scattering centers are large compared to the wavelength of light used to explain ρ_h less than one would no longer be required.

More recently Goldstein[25] measured the depolarization ratios for vitreous silica and 12 optical glasses and found within experimental error ρ_h equal to unity in all but one glass. This was in direct contrast to the results of Krishnan. Goldstein thoroughly analyzed all possible sources of error in measuring depolarization ratios and concluded that Krishnan's glasses may have had foreign inclusions or inhomogeneities. These inhomogeneities could have been small platinum inclusions or two phase glasses. Whatever the reason for the discrepancies, Goldstein's results are probably the more reliable. His one glass, a borosilicate, in which he found ρ_h equal to 0.92 could be explained on the basis of stable compositional fluctuations large in degree or extent, or the glass could have been phase separated.

From his results Goldstein concluded[26]: (1) that the light scattering he investigated in 11 optical glasses and vitreous silica is the result of frozen-in thermal fluctuations characteristic of the liquid state; (2) that the magnitude of the predicted shear strains produced by these fluctuations is small and has little or no effect on lowering ρ_h; and (3) that the finding of a ρ_h of unity for vitreous silica is evidence against the existence of appreciable regions of cristobalite-like or tridymite-like properties in vitreous silica.

In 1949, Debye and Beuche[27] examined the scattering intensity as a function of scattering angle for two optical glasses. They obtained dissymmetries of 1.22 and 1.18 for the two glasses. Also, from a correlation function derived from theory and experimental data they predicted inhomogeneities of from 500 to 600 Å in these glasses. A more comprehensive study was made by Maurer in 1956[28]. He measured dissymmetries, Rayleigh ratios, R_u, and depolarization ratios, ρ_v, on 16 optical glasses. The results are given in Table I along with values of ρ_h made by Goldstein on the same glasses.

ρ_v is defined as V_h/V_v and gives an indication of the anisotropy of the scattering centers[21]. Goldstein showed that Maurer's values of R_u are in small error since he used the wrong value of R_u for benzene (standard) and a correction factor was applied incorrectly[25].

Only a few of the glasses show any real evidence of dissymmetry of scattering. Values of dissymmetry as large as about 1.03 are within the limits of experimental precision. Glass samples with larger dissymmetries may have contained foreign inclusions. By careful experimentation Maurer eliminated possible contributions to scattering measurements of incipient crystallization or incomplete melting. He also studied the effect on light scattering of heating above the glass transition and on the rate of cooling through the glass transition. The results shown in Fig. 1 are for one of his glasses. The upper curve represents samples of the glass held until equilibrium at various temperatures above the glass transition and air quenched to room temperature. The highest temperature produced the largest scattering. The bottom curve is for samples cooled through the glass transition at different rates. The more rapidly cooled the higher the scattering. He interpreted these results as evidence for frozen-in high temperature fluctuations. In contrast a study by Prod'Homme[29] on two optical glasses showed initial increases in scattering on heating above the glass transition, maxima and subsequent decreases in intensity. The results are shown in Figs. 2 and 3. These curves are very similar to those by Andreev[30]

TABLE I

LIGHT SCATTERING DATA FOR FIFTEEN OPTICAL GLASSES[a]

Type of glass	Dissymmetry I_{45}/I_{135}		Rayleigh's ratio 546 nm ($\times 10^{-6}$cm^{-1})	Depolarization ratios	
	546 nm	436 nm		ρ_v	ρ_h
511-635	1.10	1.13	1.60	0.06	0.92
517-645	1.00	1.00	1.90	0.07	
573-425	1.01	1.00	7.45	0.056	
575-414	1.02	1.02	6.45	0.058	
575-428	1.04	1.05	6.40	0.063	
596-397	1.01	1.01	7.90	0.054	
611-572	1.01	1.01	3.75	0.063	
611-588	1.00	1.00	3.90	0.086	
613-442	1.06	1.06	10.2	0.184	1.02
617-550	1.00	1.00	6.40	0.046	
620-603	1.06	1.04	2.90	0.016	
645-480	1.00	1.00	10.0	0.087	1.01
649-338	1.01	1.01	13.3	0.052	
670-472	1.00	1.00	8.5	0.100	
673-322	1.01	1.01	18.8	0.040	0.99
720-293	1.03	1.01	20.3	0.037	

[a]From R. D. Maurer and M. Goldstein[25][28].

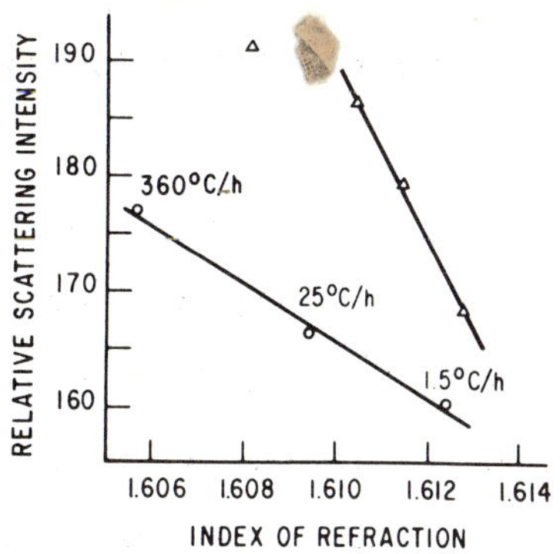

FIG. 1

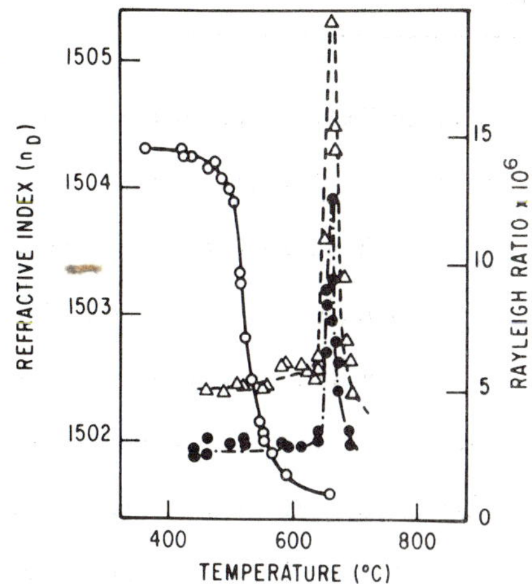

FIG. 2

FIG. 1. Variation of scattering intensity with annealing treatment; Δ-rate annealed sample (from Maurer[28]).

FIG. 2. Light scattering from an optical borosilicate crown with approximately 22% B_2O_3; o-refractive index, •-light scattering at 546 nm, Δ-light scattering at 436 nm (from Prod'Homme, et. al.[29]).

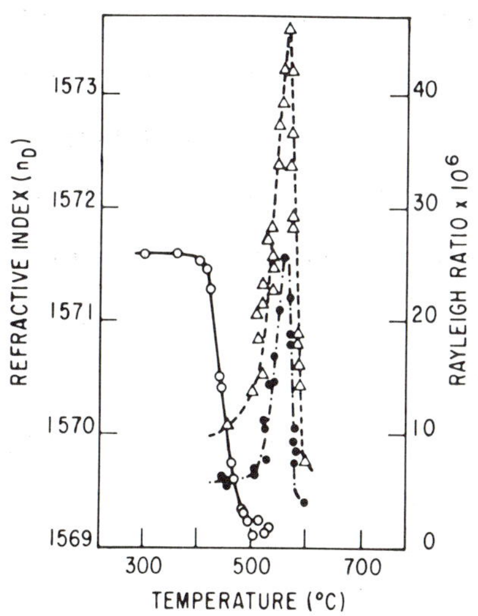

FIG. 3

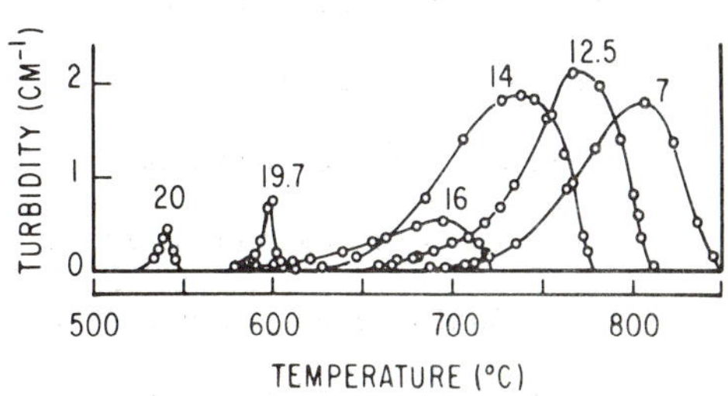

FIG. 4

FIG. 3. Light scattering from an optical light flint with approximately 32% PbO; o-refractive index, •-light scattering at 546 nm, Δ-light scattering at 436 nm, (from Prod'Homme, et al.[29]).

FIG. 4. Light scattering from phase separated soda-silicate glasses with 7 to 20% Na_2O (from Andreev and Ershova[30]).

shown in Fig. 4 for phase separated $Na_2O \cdot SiO_2$ glasses. Prod'Homme's curves are similar in shape to those of Andreev's indicating Prod'Homme may have been looking at phase separation.

In summary, the depolarization measurements by Goldstein and the dissymmetry measurements and heating and cooling experiments by Maurer are strong evidence that the light scattering from optical glasses (single phase) are the result of density and compositional fluctuations, both of which are small compared to the wavelength, i.e., less than 500 Å. Also, the shear strains produced by frozen-in high temperature fluctuations are generally small indicating small fluctuations in either degree or extent. The τ can be calculated using Eq. (2) and an average value of 2.5×10^{-12} cm^2/dyne for the room temperature compressibility of glass. The value for τ of 1.7×10^{-6} cm^{-1} is divided by $16\pi/3$, since τ equals $(16\pi/3)R_u$, and compared with the Rayleigh ratios measured by Maurer. The measured values are from 20 to 200 times larger. This excess scattering is chiefly the result of composition fluctuations. Therefore, it is possible to estimate $(-\partial\bar{F}/\partial C)$ from Eq. (3). Generally, $(-\partial\bar{F}/\partial C)$ is relatively temperature independent, but when a glass approaches a miscibility gap either through temperature lowering or changing composition $(-\partial\bar{F}/\partial C)$ becomes smaller and fluctuations larger. Miscibility gaps are quite extensive in silicate glasses.

Cation clustering has been established in glasses and a relationship between this clustering and $(-\partial\bar{F}/\partial C)$ may exist. It is interesting to speculate whether a relationship also exists between glass strength and composition fluctuations. The ultimate strength of glass may indeed be affected by the magnitude of $(-\partial\bar{F}/\partial C)$.

B. Phase Separated Glasses

In Figs. 5 and 6 relative scattering intensities versus scattering angles are shown for a phase separated glass after various heat treatment times. Prior to the results on phase separated glasses if dissymmetries in light scattering were observed, they were greater than one, e.g., polymers, colloids, etc. Dissymmetries less than one are unique to phase separated glasses. The samples heat treated for longer times have higher scattering intensities and also show maxima (Fig. 6). Goldstein's theory does not predict maxima. These maxima are probably caused by particles which no longer behave as Rayleigh scatterers, i.e., the particles are larger than 1/10λ.

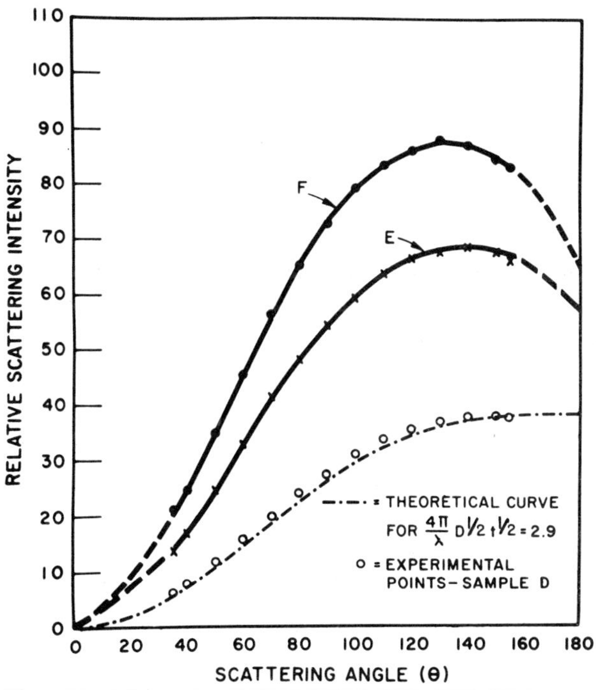

FIG. 5
Light scattering curves for a $22CaO-14Al_2O_3-10B_2O_3-54SiO_2$ glass heat treated at 867°C for o-15, x-20 and ●-30 minutes. The theoretical curve was from the Goldstein model. (From Hammel and Ohlberg[13].)

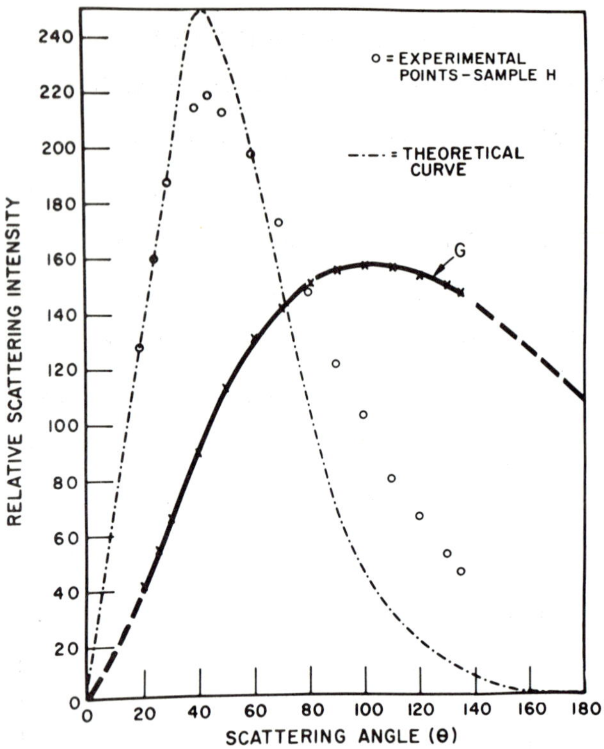

FIG. 6
Light scattering curves for a $22CaO\text{-}14Al_2O_3\text{-}10B_2O_3\text{-}54SiO_2$ glass heat treated at 840°C for x-16 and 0-150 minutes. The theoretical curve was from the Hammel and Ohlberg model[13].

Using Rayleigh-Gans (an extension of Rayleigh theory to large size particles) and diffusion-controlled growth theories, Hammel and Ohlberg[13] showed maxima will occur. Excellent agreement was found between Goldstein's theory and results at early growth times and good agreement between Hammel and Ohlberg's model at longer growth times.

In Fig. 7 scattering curves are calculated from Goldstein's theory for various values of a parameter, C, equal to $(4\pi/\lambda)D^{1/2}t^{1/2}$. An experimental C value is obtained by taking early growth data and fitting it to the best theoretical curve. In Figs. 8 and 9 experi-

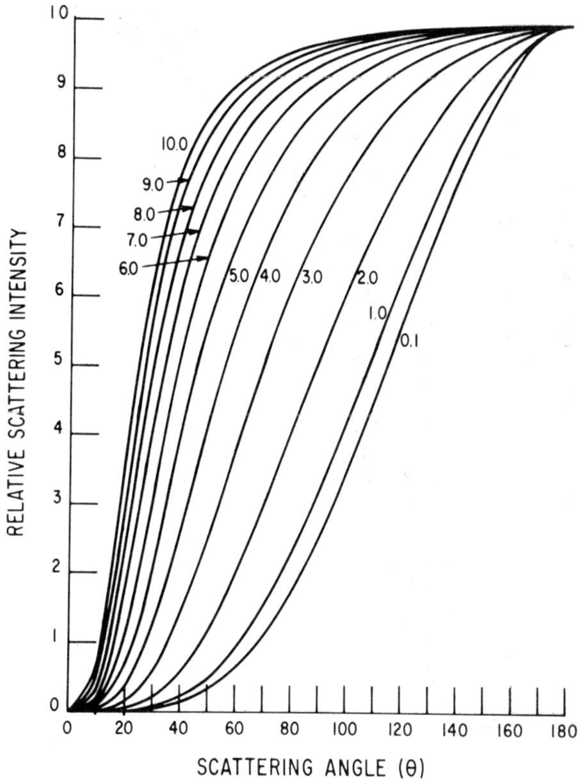

FIG. 7
Calculated curves from Goldstein's diffusion-controlled growth model[4].

mental data and calculated curves for samples of two glasses heat treated at various times and temperatures are shown. The fit plotted to the theoretical curve is very good over all angles; values of C thus obtained are plotted. Knowing the time of heat treatment, a value of D_{1s} is readily obtained. The results for three glasses along with diffusion coefficients (obtained from viscosity data using the Stokes-Einstein relation and a value of 3 Å for a_0) are shown in Table II. The diffusion coefficients, D_{1s}, for glasses A and B are 200 to 500 times larger than those obtained from viscosity, whereas, the diffusion coefficients, D_{1s}, for glass C are only 5 to 10 times

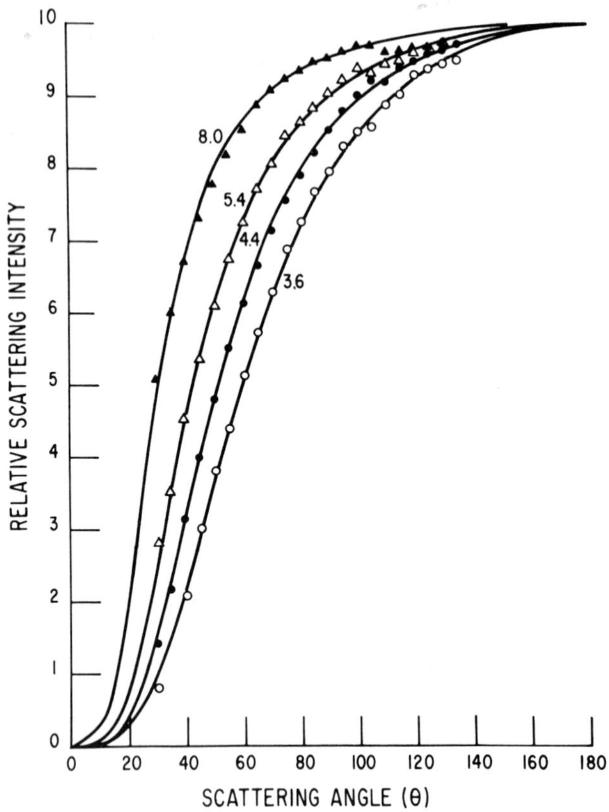

FIG. 8

Calculated curves and light scattering data for a $13Na_2O \cdot 11CaO \cdot 76SiO_2$ glass; O-60 minutes at 635°C, ●-75 minutes at 645°C, ▲-90 minutes at 654, Δ-90 minutes at 674°C (from Hammel, et al.[4]).

larger. On Table III are the temperature ranges and activation energies obtained from viscosity and light scattering data. The activation energies obtained from light scattering are consistently 25% smaller.

In all three glasses the diffusion coefficients and activation energies obtained from light scattering are associated with the growth of silica-rich particles from homogeneous media. It is evident from the results that this growth is not related to viscous forces

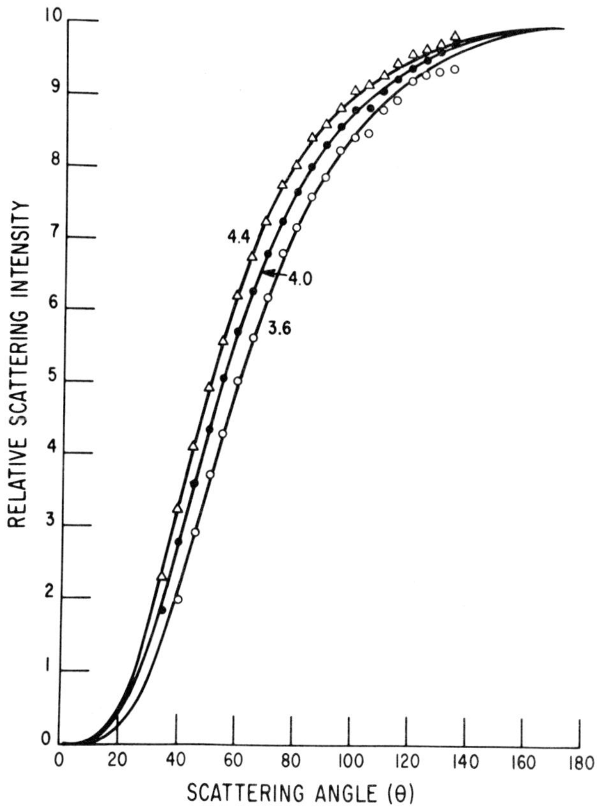

FIG. 9
Calculated curves and light scattering data for a $13Na_2O \cdot 11CaO \cdot 76SiO_2$ glass; O-60 minutes at 635°C, ●-90 minutes at 635°C, Δ-120 minutes at 635°C (from Hammel, et al.[4]).

and that the energy involved is lower. For glasses A and B a model is envisioned in which the growth process involves the diffusion of sodium, calcium, and oxygen away from the silica-rich particles and a local rearrangement of Si-O bonds at the surface of the particle, the diffusion of oxygen being rate controlling. With glass C the diffusion of network formers would also be involved which might explain the relatively small differences in diffusion coefficients obtained from viscosity and light scattering data.

TABLE II

DIFFUSION COEFFICIENTS FROM LIGHT SCATTERING

Temperature (°C)	Viscosity (Poises)	Diffusion Coeff. (D_V)	Diffusion Coeff. (D_{Is})
Glass A	*19.5 Na_2O • 80.5 SiO_2*		
545	3.39×10^{10}	1.18×10^{-17}	
555	1.74×10^{10}	2.32×10^{-17}	8.10×10^{-15}
565	9.34×10^{9}	4.38×10^{-17}	1.33×10^{-14}
575	4.90×10^{9}	8.45×10^{-17}	2.47×10^{-14}
585	2.82×10^{9}	1.48×10^{-16}	
Glass B	*13 Na_2O • 11 CaO • 76 SiO_2*		
615	3.90×10^{10}	1.12×10^{-17}	5.52×10^{-15}
635	1.05×10^{10}	4.25×10^{-17}	1.47×10^{-14}
645	5.52×10^{9}	8.20×10^{-17}	2.50×10^{-14}
654	3.10×10^{9}	1.47×10^{-16}	3.45×10^{-14}
674	9.55×10^{8}	4.86×10^{-16}	7.61×10^{-14}
Glass C	*22 CaO • 10 B_2O_3 • 14 Al_2O_3 • 54 SiO_2*		
804	3.76×10^{8}	1.41×10^{-15}	1.38×10^{-14}
814	2.07×10^{8}	2.56×10^{-15}	2.28×10^{-14}
824	1.17×10^{8}	4.60×10^{-15}	3.65×10^{-14}
834	6.73×10^{7}	8.03×10^{-15}	5.68×10^{-14}
844	4.00×10^{7}	13.7×10^{-15}	7.78×10^{-14}
854	2.41×10^{7}	22.9×10^{-15}	12.2×10^{-14}

TABLE III

ACTIVATION ENERGIES FOR THREE SILICATE GLASSES

Glass	Temperature range (°C)	Activation Energies	
		E_V	E_{Is}
19.5 Na_2O · 80.5 SiO_2	545-585	90	66
13 Na_2O · 11 CaO · 76 SiO_2	615-674	105	75
22 CaO · 10 B_2O_3 · 14 Al_2O_3 · 54 SiO_2	804-854	136	104

C. Some Applications of Light Scattering from Glasses

Maurer[5] took advantage of the high viscosity of silicate glasses near their glass transition and their transparency to study the mechanism of nucleation and growth of colloidal gold in photosensitive glasses. When the particles are small, light scattering is proportional to the number and square of the volume, whereas, optical absorption is proportional to the number and the first power of the volume. This allowed him to determine both the number and the size of the particles after various precipitation times by the simultaneous measurements of light scattering and optical absorption. According to Maurer, Figs. 10 and 11 are evidence for a fixed number of nucleation sites in photosensitive glasses at a given dosage of X rays. The higher the exposures the larger the number of nucleation sites. These were exposures prior to heat treatment above the glass transition. The slopes of the light scattering and absorption curves are equal to 3 and 3/2, respectively, which is evidence for diffusion-controlled growth and no increase in a fixed number of nucleation sites. When antimony, which is added to reduce gold ions, is added in increasing amounts the mechanism shifts from a fixed number of sites to one in which the number of sites increases during precipitation. This was deduced from light scattering and optical absorption curves in which the slopes were equal to 4 and 5/2, respectively, which according to diffusion-controlled theory would be indicative of an increase in nucleation sites with precipitating time. Also, making a few assumptions he was able to estimate nucleation rates and obtain diffusion coefficients[5].

Metal particles were found to catalyze the crystallization of glass and were used to make some of the first glass-ceramics.

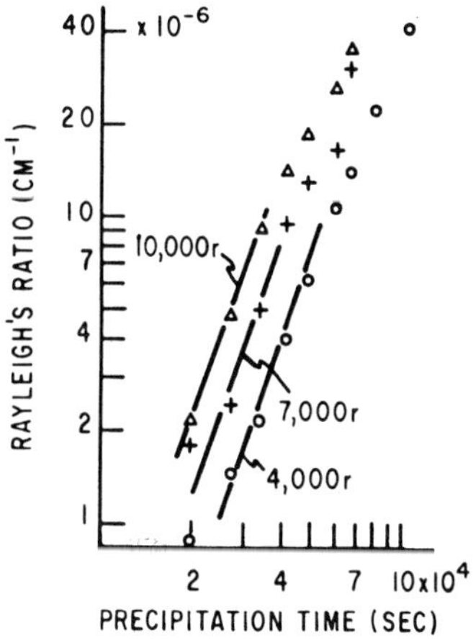

FIG. 10

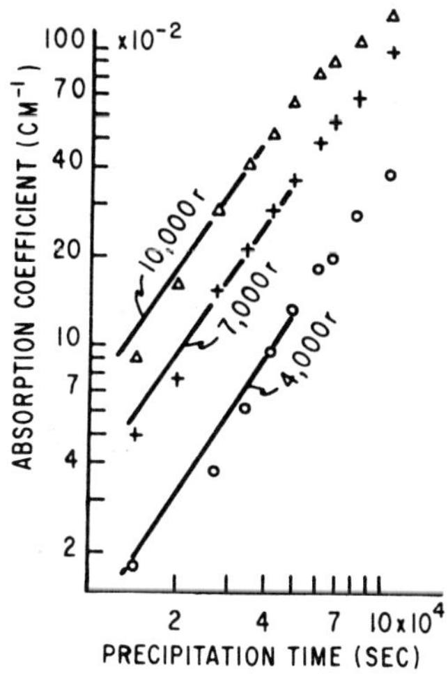

FIG. 11

FIG. 10
Light scattering at 436 nm versus precipitation time at various dosage levels for a photosensitive glass (from Maurer[32]).

FIG. 11
Absorption coefficient at 436 nm versus precipitation time at various dosage levels for a photosensitive glass (from Maurer[32]).

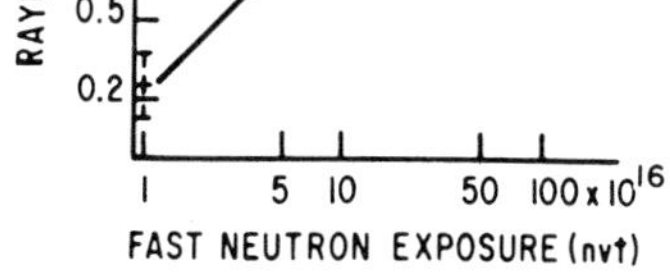

FIG. 12
Scattered light intensity with incident unpolarized light versus neutron exposure (from Maurer[7]).

Stookey[31] notes that very small metallic particles were ineffective in catalyzing crystallization. Maurer[32] in his light scattering and optical absorption studies found that colloidal particles must grow to at least 80 Å before they can be effective nuclei. Approximate calculations by him showed that when a metal particle is less than 80 Å, strain in the crystal embryo caused by distortion in fitting the metal catalyst may inhibit nucleation.

Figure 12 shows a linear increase in scattering as a function of fast neutron exposure for a sample of fused silica. The straight line has a slope of unity indicating that the damaged centers are scattering independently. At the highest level of exposure large dissymmetries of scattered polarized light were observed. According to Maurer's analysis the results indicated damaged regions rodlike in shape and approximately 1800 Å long[7]. Also, a strained region was produced which was uniform along the rod. Using a more rigorous analysis, Goldstein[33] concurs with Maurer's conclusion that strains surround the damaged volumes. However, he does not agree with the interpretation of 1800 Å rodlike damaged regions. Goldstein believes Maurer's results could be interpreted as a "wide separation of damaged regions produced by a single primary recoil."

ACKNOWLEDGMENT

The author wishes to thank H. Golob for her critical reading of the manuscript.

REFERENCES

[1]. S. M. Ohlberg, J. J. Hammel, and H. R. Golob, *J. Am. Ceram. Soc.*, 48, 178 (1965).

[2]. J. J. Hammel, "Experimental Evidence for Spinodal Decomposition in Glasses of the $Na_2O \cdot SiO_2$ System," Proc. Intern. Congress on Glass (1965).

[3]. J. W. Cahn and R. J. Charles, Phys. Chem. Glasses, 6, 181 (1965).

[4]. J. J. Hammel, J. Mickey, and H. R. Golob, J. Colloid and Interface Sci., 27, 329 (1968).

[5]. R. D. Maurer, J. Appl. Phys., 29, 1 (1958).

[6]. R. D. Maurer, J. Appl. Phys., 33, 2132 (1962).

[7]. R. D. Maurer, J. Phys. Chem. Solids, 17, 44 (1960).

[8]. H. C. Van De Hulst, Light Scattering by Small Particles, Wiley, New York, 1957, pp. 63-84.

[9]. G. Oster, Chem. Revs., 43, 319 (1948).

[10]. L. S. Ornstein and F. Zernicke, Physik. Z., 27, 261 (1926).

[11]. B. H. Zimm, R. S. Stein, and Paul Doty, Light Scattering From Dilute Polymer Solutions, Gordon and Breach, New York (1964), pp. 7-36.

[12]. H. C. Van De Hulst, Light Scattering by Small Particles, Wiley, New York, 1957, pp. 85-102.

[13]. J. J. Hammel and S. M. Ohlberg, J. Appl. Phys., 36, 1442 (1965).

[14]. A. I. Kolyadin, Proc. of the Third All-Union Conf. on the Glassy State, Leningrad, 1959, 202; N. S. Andreev, V. I. Aver'yanov, and N. A. Vioshvillo, Proc. of the Third All-Union Conf. on the Glassy State, Leningrad, 1959, 205.

[15]. N. A. Voishvillo, Opt. i Spectroskopiya, 12, 412 (1962). [English transl: Opt. Spectry., 12, 225 (1962)].

[16]. C. Walker and A. Guinier, Acta Met., 1, 568 (1953).

[17]. M. Goldstein, J. Appl. Phys., 34, 1928 (1963).

[18]. M. Goldstein, J. Am. Ceram. Soc., 48, 126 (1965).

[19]. B. A. Brice, M. Halwer, and R. Speiser, J. Opt. Soc. Am., 40, 768 (1950).

[20]. B. R. Jennings and H. Plummer, Brit, J. Appl. Phys., 1, 1201 (1968).

[21]. R. S. Krishnan, Proc. Indian Acad. Sci., 3A, 211 (1936).

[22]. R. S. Krishnan and P. V. Rao, Proc. Indian Acad. Sci., 20A, 109 (1944).

[23]. D. H. Rank and A. E. Douglas, J. Opt. Soc. Am., 38, 966 (1948).

[24]. H. Mueller, Proc. Roy Soc. (London), A166, 425 (1938).

[25]. M. Goldstein, J. Appl. Phys., 30, 493 (1959).

[26]. M. Goldstein, J. Appl. Phys., 30, 501 (1959).

[27]. P. Debye and A. M. Bueche, J. Appl. Phys., 20, 518 (1949).

[28]. R. Maurer, J. Chem. Phys., 25, 1206 (1956).

[29]. L. Prod'Homme and C. Vacherand, Verres et Ref., 20, 354 (1966).

[30]. N. S. Andreev and T. I. Ershova, Sov. Phys.-Doklady, 12, 149 (1967).

[31]. S. D. Stookey, Ind. Eng. Chem., 45, 115 (1953).

[32]. R. D. Maurer, J. Chem. Phys., 31, 444 (1959).

[33]. M. Goldstein, J. Appl. Phys., 33, 3377 (1962).

Chapter 29

ELECTRO-OPTICAL AND MAGNETO-OPTICAL EFFECTS

Y. R. Shen

Department of Physics
University of California
and
Inorganic Materials Research Division
Lawrence Radiation Laboratory
Berkeley, California

I. INTRODUCTION

The properties of a material may vary with applied electric and magnetic fields. In particular, the fields induce a change in the optical dielectric constants ε_{ij} or the refractive indexes n_{ij}, and consequently affect light propagation in the material. Such induced optical effects are known as electro-optical* (induced by electric

*The linear and quadratic electro-optical effects are commonly known as the Pockels effect and the Kerr effect, respectively[1].

field only) and magneto-optical* (induced by magnetic field only) effects. Thus, for example, a dc electric field can induce linear birefringence and a dc magnetic field can induce both linear and circular birefringences in a medium. The electro- and magneto-optical effects make possible many useful optical devices. Scientifically, measurements of electro- and magneto-optical effects often yield valuable information about the structure and electronic properties of a material.

In the presence of the applied dc fields $\mathbf{E}^{dc}$ and $\mathbf{H}^{dc}$, the optical dielectric constants of a medium can be written as

$$\varepsilon_{ij}\left(\mathbf{E}^{dc},\mathbf{H}^{dc}\right) = \varepsilon^0_{ij} + \Delta\varepsilon_{ij}\left(\mathbf{E}^{dc},\mathbf{H}^{dc}\right) \tag{1}$$

Two questions arise in the discussion of electro- and magneto-optical effects: (1) For a given medium, what do we know about the tensor $\Delta\boldsymbol{\varepsilon}$? (2) Given $\Delta\boldsymbol{\varepsilon}$, how does it affect light propagation in the medium? In principle, the second question can always be answered by solving the Maxwell equations. Here, we simply refer to the literatures on the subject[3]. In this review, we shall limit ourselves to the discussion of the first question.

The dielectric constants of Eq. (1) can usually be expanded into power series of $\mathbf{E}^{dc}$ and $\mathbf{H}^{dc}$.

$$\varepsilon_{ij} = \varepsilon^0_{ij} + 4\pi\Big(\chi^{(2e)}_{ijk}\,E^{dc}_k + \chi^{(2b)}_{ijk}\,H^{dc}_k + \chi^{(3ee)}_{ijk\ell}\,E^{dc}_k E^{dc}_\ell + \chi^{(3eb)}_{ijk\ell}\,E^{dc}_k H^{dc}_\ell + \chi^{(3bb)}_{ijk\ell}\,H^{dc}_k H^{dc}_\ell + \cdots\Big) \tag{2}$$

*Investigation of magneto-optical effects was initiated by Faraday[2].

where χ's are coefficients independent of the fields*, and the convention that a tensor product is expressed by repeated subindexes is used. For simplicity, we shall assume in this paper that ε_{ij} depends weakly on the applied fields, and, therefore, only the leading terms in Eq. (2) give observable effects. In some special cases, one can also include the full expression of ε_{ij} in the description without too much complication.

Our problem now is to find χ's in Eq. (2) for a given medium. Because of the existing symmetry of the medium, many components of the χ tensors are vanishing and only a few independent. The problem can then be separated into two parts: first, to find from the existing symmetry of a given medium the independent nonvanishing components of the χ tensors and then, to find the magnitudes of these nonvanishing components. While the second part of the problem depends on the microscopic properties of the medium, the first part can be solved by a simple thermodynamic treatment as we shall now briefly describe.

Again, for simplicity, we shall assume that the medium is nondissipative.† Let us consider a quasi-monochromatic light which is represented by the field

$$\mathbf{E}(t) = \mathcal{E}(t)\,\exp(i\omega t) + \mathcal{E}^{*}(t)\,\exp(i\omega t) \qquad (3)$$

where $|\partial\mathcal{E}/dt| \ll |\omega\mathcal{E}|$. If the light is propagating in a medium with a

---

*We assume here the electric-dipole approximation.

†The following symmetry argument is also valid for absorbing media.

dielectric constant $\varepsilon(\omega)$, then the time-averaged energy density stored in the medium is given by[4]

$$\bar{U} = (1/4\pi)\left[E_i^*\left(\partial\omega\varepsilon_{ij}/\partial\omega\right) E_j + |H|^2\right] \tag{4}$$

which becomes, with the expression of ε_{ij} in Eq. (2)

$$\begin{aligned}\bar{U} = (1/4\pi)\Big[&E_i^*\left(\partial\omega\varepsilon^0_{ij}/\partial\omega\right) E_j + |H|^2 + 4\pi E_i^*(\partial/\partial\omega)\omega \\ &\times \Big(\chi^{(2e)}_{ijk} E^{dc}_k + \chi^{(2b)}_{ijk} H^{dc}_k + \chi^{(3ee)}_{ijk\ell} E^{dc}_k E^{dc}_\ell + \chi^{(3eb)}_{ijk\ell} E^{dc}_k H^{dc}_\ell \\ &+ \chi^{(3bb)}_{ijk\ell} H^{dc}_k H^{dc}_\ell + \cdots\Big) E_j\Big]\end{aligned} \tag{5}$$

Suppose we now rotate the crystal about a certain axis by a certain angle. This energy density should change accordingly from $\bar{U}$ to $\bar{U}'$. If, however, the rotation happens to be a symmetry operation in the symmetry group of the medium, then the medium would appear unchanged after the rotation, and therefore $\bar{U} = \bar{U}'$. This symmetry argument applies to each term of Eq. (5) independently, and can be used to find the independent components of χ.

Let us consider as an example the $\chi^{(2e)}$ term which is responsible for the linear electro-optical effect. We write

$$\bar{U}_e = E_i^*\left(\partial\omega\chi^{(2e)}_{ijk}/\partial\omega\right) E_j E^{dc}_k \tag{6}$$

If S_{ij} represents a symmetry operation, then we would have

$$\bar{U}'_e = E^*_{i'} S^*_{i'i}\left(\partial\omega\chi^{(2e)}_{ijk}/\partial\omega\right) S_{jj'} S_{kk'} E_{j'} E^{dc}_{k'} = \bar{U}_e \tag{7}$$

and, hence,

$$S^{*}_{i'i}\chi^{(2e)}_{ijk}S_{jj'}S_{kk'} = \chi^{(2e)}_{i'j'k'} \tag{8}$$

we have n such relations if there are n symmetry operations in the symmetry group of the medium. In addition, we have, from general principle of symmetry for a nondissipative medium,

$$\varepsilon_{ij} = \varepsilon^{*}_{ji} \tag{9}$$

These relations make some components of $\chi(2e)$ vanish and others depend on one another. An immediate consequence of Eq. (7) or (8) is that all components of $\chi(2e)$ should be zero for crystals with inversion symmetry. As a simple illustration, let us consider the zinc blende crystals, such as GaAs, which belong to the class T_d. These crystals remain invariant under 180° rotations about the three four-fold axes. From Eq. (7) or (8), we find

$$\chi^{(2e)}_{iii} = -\chi^{(2e)}_{iii} = 0, \quad \chi^{(2e)}_{ijj} = -\chi^{(2e)}_{ijj} = 0, \quad \chi^{(2e)}_{iij} = -\chi^{(2e)}_{iij} = 0, \quad i \neq j \tag{10}$$

and $\chi^{(2e)}_{ijk}$ with $i \neq j \neq k$ are the only nonvanishing components. The crystals also remain invariant under mirror reflections about the three diagonal planes. This symmetry and Eq. (9) give

$$\chi^{(2e)}_{xyz} = \chi^{(2e)}_{yzx} = \chi^{(2e)}_{zxy} = \chi^{(2e)}_{yxz} = \chi^{(2e)}_{zyx} = \chi^{(2e)}_{xzy} \tag{11}$$

and, consequently, we have only one independent component of $\chi^{(2e)}$ for crystals in the class T_d. It can be seen that the linear electro-optical effect and the piezoelectric effect follow the same symmetry considerations. We can, therefore, obtain the linear electric-optical tensor $\chi^{(2e)}$ for a crystal from the piezoelectric tensor listed in many texts[5].

The quadratic electro-optical tensor $\chi^{(3ee)}$ for a medium can be determined similarly. For example, for an isotropic medium, the nonvanishing components of $\chi^{(3ee)}$ are $\chi^{(3ee)}_{iiii}$, $\chi^{(3ee)}_{ijji}$, and $\chi^{(3ee)}_{iijj}$ with $i \neq j$. The symmetry considerations for the terms involving magnetic field in Eq. (5) are slightly different, since the magnetic field, being a pseudovector, is invariant under inversion. For a pseudovector, the transformation matrix for a pure rotation is the same as that for a rotation plus an inversion. Therefore, we should have

$$S^*_{i'i}\chi^{(2b)}_{ijk}S_{jj'}S^M_{kk'} = \chi^{(2b)}_{i'j'k'}, \text{ etc.} \tag{12}$$

where $S^M_{kk'}$ are transformation matrices for symmetry operations associated with a pseudovector. As a consequence $\chi^{(2b)}$, unlike $\chi^{(2e)}$, does not vanish in a medium with inversion symmetry. For example, in an isotropic medium or a cubic system, the nonvanishing components of $\chi^{(2b)}$ are $\chi^{(2b)}_{xyz} = \chi^{(2b)}_{yzx} = \chi^{(2b)}_{zxy} = \chi^{(2b)*}_{yxz} = \chi^{(2b)*}_{zyx} = \chi^{(2b)*}_{xzy}$. In addition to Eqs. (9) and (12), one can also show from general principle of symmetry that[6]:

$$\varepsilon_{ij}\left(\mathbf{H}^{dc}\right) = \varepsilon_{ji}\left(-\mathbf{H}^{dc}\right) \tag{13}$$

As a result, the real part of ε_{ij} should be an even function of $\mathbf{H}^{dc}$ and the imaginary part of ε_{ij} an odd function of H^{dc}. Physically, the imaginary part of ε_{ij} induced by the magnetic field leads to induced circular birefringence or the Faraday effect[1], and the real part leads to induced linear birefringence or the Voigt effect[7].

Therefore, we can conclude that in a nondissipative medium, all the nonvanishing components of $\chi^{(2b)}_{ijk}$ and $\chi^{(3eb)}_{ijk\ell}$ are imaginary with $i \neq j$ and those of $\chi^{(2e)}_{ijk}$, $\chi^{(3ee)}_{ijk\ell}$, and $\chi^{(3bb)}_{ijk\ell}$ real.

We note in passing that the same energy density $\bar{U}$ of Eq. (5) also governs the inverse electro- and magneto-optical effects. If we consider $\bar{U}$ as the stored energy density of the dc fields, then the corresponding dc polarization $\mathbf{P}^{dc}$ and magnetization $\mathbf{M}^{dc}$ are[9]

$$P_k^{dc} = \mathcal{E}_i^*(\partial/\partial\omega)\omega\left[\chi^{(2e)}_{ijk} + \chi^{(3eb)}_{ijk\ell}\, H_\ell^{dc} + 2\chi^{(3ee)}_{ijk\ell}\, E_\ell^{dc}\right]\mathcal{E}_j \tag{14}$$

$$M_k^{dc} = \mathcal{E}_i(\partial/\partial\omega)\omega\left[\chi^{(2b)}_{ijk} + \chi^{(3eb)}_{ijk\ell}\, E_\ell^{dc} + 2\chi^{(3bb)}_{ijk\ell}\, H_\ell^{dc}\right]\mathcal{E}_j \tag{15}$$

Both $\mathbf{P}^{dc}$ and $\mathbf{M}^{dc}$ are quadratic functions of the optical field. These equations show that dc polarization and magnetization can be induced by an optical field. The inverse electro-optical effect, known as optical rectification, has been observed in electro-optical crystals $(\chi^{(2e)} \neq 0)$[9]. The inverse Faraday effect, which comes from $\chi^{(2b)}$, has also been observed in both paramagnetic and diamagnetic materials[10]. Calculations show that the inverse Voigt effect, arising from the $\chi^{(3bb)}$ term, should also be observable[11][12].

We have seen how symmetry considerations can help us determine the forms of the χ tensors for a given medium. We now come to the central theme of this review—how to find the magnitudes of these χ coefficients*. In the following sections, we shall discuss electro-optical and magneto-optical effects separately. We shall first

*In general, the electro-optical effects can be divided into true and false effects. The false effect arises from the piezo-optical effect in which the stress in the medium is induced by the applied electric field. See, for example, Ref. [13]. We shall consider in this paper only the true electro-optical effects.

outline the procedure for deriving the exact microscopic expressions for χ. We shall then use a two-level or two-band model to simplify the expressions greatly, so that quantitative estimates of χ become feasible.

II. LINEAR ELECTRO-OPTICAL EFFECT

In general, both electrons and ions in a medium contribute to electro-optical effects. At optical frequencies much larger than the frequencies of lattice vibration, the ionic contribution comes in because the dc electric field disturbs the lattice, but the effect is usually negligible. We can limit our discussion to the electronic contribution only.

The microscopic expression for $\chi^{(2e)}_{ijk}$ can be obtained from straightforward perturbation calculation. In the electric-dipole approximation, the Hamiltonian of the system is given by

$$H = H_0 - e\mathbf{r} \cdot \left(\mathbf{E}^{dc} + \mathbf{E}\, e^{-iwt} + \mathbf{E}^*\, e^{iwt}\right) \tag{16}$$

where H_0 is the Hamiltonian for the unperturbed material system, $\mathbf{r}$ is the electron coordinate with respect to the position of the associated molecule or the unit cell, and the symbol for summation over electrons is omitted for simplicity. The first-order perturbation theory with $\mathbf{E}^{dc} = 0$ leads to the usual expression for the linear dielectric constant[14]

$$\varepsilon_{ij} = 1 + 4\pi\, \chi_{ij}$$

$$\chi_{ij} = Ne^2 \sum_{n,n'} \left[\frac{(r_i)_{nn'}(r_j)_{n'n}}{\hbar(\omega_{n'n} - \omega)} + \frac{(r_j)_{nn'}(r_i)_{n'n}}{\hbar(\omega_{n'n} + \omega)}\right] \rho_n^0 \tag{17}$$

where

$$(r_i)_{nn'} = \langle n|r_i|n'\rangle$$

$$H_0|n\rangle = E_n|n\rangle$$

$$\hbar\omega_{n'n} = E_{n'} - E_n,$$

N is the density of molecules* or unit cells, and ρ_n^0 is the population in the $|n\rangle$ state at thermal equilibrium. Application of a dc electric field on the system modifies the linear susceptibility χ_{ij} through changes in the energies, the wave functions, and the populations of the electronic states. In most electro-optical materials, the states are either fully occupied or empty at normal temperature, and the change in ρ_n^0 is negligible. Then, the next-order perturbation calculation yields the microscopic expression for $\chi_{ikj}^{(2e)}$ [15]-[17]

$$\chi_{ijk}^{(2e)} = \left(Ne^3/\hbar^2\right) \sum_{g,n,n'} \left\{ \frac{\omega_{ng}(r_i)_{gn}(r_k)_{nn'}(r_j)_{n'g}}{\left(\omega^2 - \omega_{ng}^2\right)\omega_{n'g}} + \frac{\left(\omega^2 + \omega_{ng}\omega_{n'g}\right)(r_i)_{gn}(r_j)_{nn'}(r_k)_{n'g}}{\left(\omega^2 - \omega_{ng}^2\right)\left(\omega^2 - \omega_{n'g}^2\right)} + \frac{\omega_{n'g}(r_j)_{gn}(r_i)_{nn'}(r_k)_{n'g}}{\left(\omega^2 - \omega_{n'g}^2\right)\omega_{ng}} \right\} \qquad (18)$$

where $|g\rangle$ are the fully occupied ground states and $|n\rangle$ and $|n'\rangle$ are the empty excited states.

---

*In general, one should sum over different types of molecules present in the medium.

It is obvious from the above expression that without detailed knowledge about energies and wave functions of all the electronics states, one cannot possibly evaluate $\chi_{ijk}^{(2e)}$. Unfortunately, our knowledge about excited electronic states is very limited. To find the magnitudes of $\chi_{ijk}^{(2e)}$, we must resort to some approximation to circumvent this difficulty. The usual approximation is to assume an effective mean transition frequency $\bar{\omega}_{ng}$ for all transitions. It is equivalent to approximating the complex system by a simple two-level system. The approximation is reasonable if the integrated oscillator strength for transitions in a fairly narrow range of frequencies dominates, and if the frequency ω is far away from any resonance absorption. By pulling the dispersion factors outside the summation sign, we can then write[17][18]

$$\sum_{n,n'} \left(r_i\right)_{gn} \left(r_j\right)_{nn'} \left(r_k\right)_{n'g} = \left(\tilde{r}_i \tilde{r}_j \tilde{r}_k\right)_{gg} \tag{19}$$

where $\tilde{\mathbf{r}} = \mathbf{r} - (\mathrm{r})_{gg}$. Consequently, Eq. (18) reduces to the form[17]

$$\chi_{ijk}^{(2e)} = \left(Ne^3/\hbar^2\right)\left[\left(3\omega^2 - \bar{\omega}_{ng}^2\right)/\left(\omega^2 - \bar{\omega}_{ng}^2\right)\right] \sum_g \left(\tilde{r}_i \tilde{r}_j \tilde{r}_k\right)_{gg} \tag{20}$$

Then, for evaluating $\chi_{ikj}^{(2e)}$, we only need to know $\bar{\omega}_{ng}$ and the wave functions for the ground states $|g\rangle$.

To find $\bar{\omega}_{ng}$, we assume that the same mean frequency can be used to reduce χ_{ii}, so that Eq. (17) gives

$$\chi_{ii} = \left[2Ne^2\bar{\omega}_{ng}/\hbar\left(\bar{\omega}_{ng}^2 - \omega^2\right)\right] \sum_g \left(\tilde{r}_i \tilde{r}_i\right)_{gg} \tag{21}$$

The above expression describes reasonably well the dispersion of the linear susceptibility for many materials. We can, therefore, deduce $\bar{\omega}_{ng}$ from measurements of refractive indexes at several frequencies.

The ground-state wave functions or electron distribution depend critically on the microscopic structure of the medium. However, in most cases, they can be approximated by simple linear combinations of atomic orbitals or other basic functions such as the Bloch functions. Thus, for example, in a III—V compound, the valence electrons are mainly responsible for the optical transitions. The valence states, therefore, correspond to the ground states $|g\rangle$ in Eqs. (20) and (21). The molecular model of a III—V compound $(A_{III}B_V)$ assumes that the eight valence electrons around each A_{III} atom are paired up to form four tetrahedral bonds connecting the A_{III} atom to the four adjacent B_V atoms as shown in Fig. 1. From

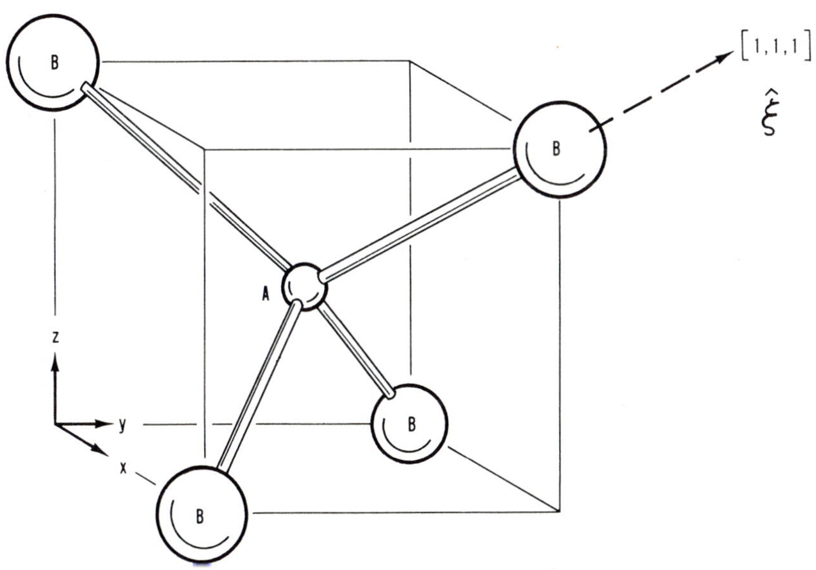

FIG. 1
Four equivalent tetrahedral bonds connecting the A(III) atom to the four adjacent B(V) atoms.

the theory of molecular orbitals[19], we can then describe the ground-state wave functions of these valence electrons approximately by the linear combinations of sp^3 hybridized orbitals centered at A_{III} and B_V and pointing in the [1,1,1] directions.

$$\psi_{III-V} = (\lambda\psi_{III} + \psi_V)/(1 + \lambda^2)$$

$$\psi_{III} = \frac{1}{2}\left(\phi_{ns} + \sqrt{3}\ \phi_{np}\xi\right)$$

$$\psi_V = \frac{1}{2}\left(\phi_{n's} - \sqrt{3}\ \phi_{n'p}\xi\right) \tag{22}$$

where ϕ_{ns}, $\phi_{np}\xi$, etc., are hydrogenlike atomic wave functions, ψ_{III} describes an orbital centered at A_{III} and pointing in the bond direction $\hat{\xi}$, ψ_V is centered at B_V and pointing in the $-\hat{\xi}$ direction, and λ is a parameter which determines the ionicity or polarity of the bonds.* Thus, knowing the ground-state wave functions, we can readily evaluate $\chi^{(2e)}_{ijk}$ from Eq. (20) by summing over contributions from the eight valence electrons per each A_{III} atom[20],[23]. The factor N in Eq. (20) is now the density of A_{III} atoms, and the overlapping of two bond orbitals has been neglected.

A somewhat different approach has been taken by Levine[24] using Phillips' theory of the chemical bond. Again assume that we can replace all the transition frequencies in a semiconductor or

---

*The parameter λ can be determined from a variational calculation. Phillips and Van Vechten have, however, expressed λ in terms of a mean energy gap $\bar{E}_g$ and a charge-transfer energy C_ξ using a two-band model. They show that $\chi^{(2e)}_{ijk}$ can also be written in terms of $\bar{E}_g$ and C_ξ, which can be calculated from Eq. (26). See Ref. [20].

insulator by the mean frequency $\bar{\omega}_{ng}$. Then, the Thomas-Reiche-Kuhn sum rule becomes[23]

$$\sum_n \frac{2m}{\hbar} \omega_{ng} |(r_i)_{ng}|^2 = \frac{2m}{\hbar} \bar{\omega}_{ng} (r_i r_i)_{gg} = 1 \tag{23}$$

Along a crystal axis, Eq. (21) reduces to the form

$$\chi_{ii} = \hbar^2\omega_p^2/(\bar{E}_g^2 - \hbar^2\omega^2) \tag{24}$$

where $\omega_p^2 = Ne^2/m$ is the plasma frequency with N being the density of valence electrons, and $\bar{E}_g = \hbar\bar{\omega}_{ng}$ is the effective energy gap, which is usually larger than the true energy gap. In the limit of zero frequency, we obtain $\chi_{ii} = \hbar^2\omega_p^2/\bar{E}_g^2$, which has been derived by Penn from band-theory calculation[26]. Application of a dc electric field should modify χ_{ii} through $\bar{E}_g$. We can express χ_{ii} in terms of the bond polarizability $\alpha_{\xi\xi}$ along the bond direction ξ, assuming that the bond polarizability in the direction perpendicular to $\hat{\xi}$ is negligible,

$$\chi_{ii} = \sum_\xi T_{i\xi} N_\xi \alpha_{\xi\xi} \tag{25}$$

where $T_{i\xi}$ is a geometric factor arising from coordinate transformation between $\hat{\xi}$ and $\hat{i}$, and N_ξ is the density of valence electrons associated with bonds along $\hat{\xi}$. If all bonds are alike but pointing in different directions, then we have

$$\chi_{ii} = F_{ii}(\hat{\xi})\chi_{\xi\xi} \tag{25'}$$

with $F_{ii}(\hat{\xi}) = \sum_\xi T_{i\xi} N_\xi/N$. This is true in many simple crystals. From a two-band model, Phillips has shown that[25]:

$$\bar{E}_g = [E_h^2 + C_\xi^2]^{(1/2)} \tag{26}$$

$$C_\xi = f\, e^2[(Z_A/r_A) - (Z_B/r_B)]$$

where E_h is the homopolar effective energy gap determined mainly by the lattice constant, C_ξ is the charge-transfer energy, Z_A and r_A are respectively the number of valence electrons and the covalent radius of the A atom in the ξ bond, and f represents the effects of Thomas-Fermi screening ($f \simeq 0.12$)[24]. From Eq. (26), one can now calculate $\bar{E}_q$ numerically for a given crystal.

We are, however, interested in finding $\chi^{(2e)}$. As Levine has pointed out[22], an applied dc electric field should displace the bond charge $q = (^2/\varepsilon_h)e$ by a distance Δr along the bond direction, where ε_h is the homopolar part of the static dielectric constant[25]. This produces a dc dipole moment so that

$$\chi_{\xi\xi}^{dc} \Delta E_\xi^{dc} = Nq\Delta r \tag{27}$$

On the other hand, it leads to a change in the charge-transfer energy C_ξ.

$$\Delta C_\xi \simeq -fe^2 (Z_A + Z_B)\, \Delta r/r^2 \tag{28}$$

We then have, from Eqs. (24)-(28),

$$\chi_{\xi\xi\xi}^{(2e)} = \partial\chi_{\xi\xi}/\partial E_\xi^{dc} = (\partial\chi_{\xi\xi}/\partial C_\xi)(\partial C_\xi/\partial E_\xi^{dc})$$

$$= (1/F_{ii}^2)(8fe^2(Z_A + Z_B)\; C_\xi/\hbar^2 qd^2 N\omega_p^2)\; \chi_{ii}^2 \chi_{ii}^{dc} \tag{29}$$

where $d = r_A + r_B \simeq 2r$. To find $\chi_{ijk}^{(2e)}$, we should sum over contributions from all the bonds.

$$\chi_{ijk}^{(2e)} = \sum_{\xi} g_{ijk}(\hat{\xi})\, \chi_{\xi\xi\xi}^{(2e)} \tag{30}$$

which reduces to

$$\chi_{ijk}^{(2e)} = G_{ijk}(\hat{\xi})\, \chi_{\xi\xi\xi}^{(2e)} \tag{30'}$$

when all bonds are alike. This geometric factor $G_{ijk}(\hat{\xi})$ is zero for those $\chi_{ijk}^{(2e)}$ components which must vanish as required by symmetry. When all the quantities in the above expression for $\chi_{ijk}^{(2e)}$ are known for a given crystal, we can then calculate $\chi_{ijk}^{(2e)}$ numerically. Using this model, Levine[22] has calculated $\chi_{ijk}^{(2e)}$ for many crystals of zincblende and wurtzite structures. The agreement with experimental data is excellent, considering the many approximations involved in the calculation.

From Eqs. (29) and (30'), we find

$$\frac{\chi_{ijk}^{(2e)}}{\chi_{ii}^{2}\chi_{ii}^{dc}} = \Delta_{ijk}$$

$$\Delta_{ijk} = \left(\frac{G_{ijk}}{F_{ii}^{2}}\right)\left(\frac{8fe^{2}(Z_A + Z_B)}{h^{2}qd^{2}N\omega_p^{2}}\right) C_{\xi} \tag{31}$$

This relation was originally suggested by Miller[26] who pointed out that Δ is of the same order of magnitude for a large variety of materials. For all III—V and II—VI compounds, the factor Δ_{ijk}/C_{ξ} should be nearly a constant. The agreement of this assertion with experimental data is again excellent[22]. For large $\chi_{ijk}^{(2e)}$, we must therefore have large linear susceptibility (or small $\bar{E}_g$ as seen from

Eq. (24)) and large charge-transfer energy C_ξ compatible with small $\bar{E}_g$.

In the above discussion, the local-field correction has been neglected. This is clearly nonnegligible, however, since the valence electrons are fairly localized as seen from the bond picture. If the observed macroscopic χ_{ii} and χ_{ii}^{dc} are used in the expression for $\chi_{ijk}^{(2e)}$, then the local-field effect is automatically taken into account.

III. QUADRATIC ELECTRO-OPTICAL EFFECT

We can also derive the microscopic expression for $\chi^{(3ee)}$ by carrying out the next higher-order perturbation calculation[18], but again it will be of little use for actual evaluation of $\chi^{(3ee)}$. For the latter purpose, we must use the approximations discussed in the last section.

Using Levine's approach, we find from Eqs. (24)-(28)

$$\chi_{\xi\xi\xi\xi}^{(3ee)} = \frac{1}{2}\frac{\partial^2\chi_{\xi\xi}}{\partial\left(E_\xi^{dc}\right)^2}$$

$$\simeq \left(\frac{1}{F_{ii}}\right)\left(\frac{\chi_{ii}^2}{\hbar^2\omega_p^{\,2}}\right)\left(1 - \frac{4C_\xi\chi_{ii}}{\hbar^2\omega_p^{\,2}}\right)\left(\frac{\Delta C_\xi}{\Delta E_\xi^{dc}}\right)^2$$

$$= \left(\frac{1}{F_{ii}}\right)^3\left(\frac{4fe^2\left(Z_A + Z_B\right)}{Nqd^2\hbar^2\omega_p^{\,2}}\right)\left(1 - \frac{4C_\xi^2\chi_{ii}}{\hbar^2\omega_p^{\,2}}\right)\chi_{ii}^2\left(\chi_{ii}^{dc}\right)^2 \tag{32}$$

where we have neglected the $\partial^2 C_\xi/\partial\left(E_\xi^{dc}\right)^2$ term, which is usually small in most crystals[22]. We then obtain

$$\chi_{ijk\ell}^{(3ee)} = G_{ijk\ell}\,(\hat{\xi})\,\chi_{\xi\xi\xi\xi}^{(3ee)} \tag{33}$$

where $G_{ijk\ell}$ is, again, a geometric factor coming from coordinate transformation. Numerical values of $\chi^{(3ee)}_{ijk\ell}$ calculated from the above expression for several crystals also agree well with experimental data[22]. We can also define

$$\Delta_{ijk\ell} = \frac{\chi^{(3ee)}_{ijk\ell}}{\chi^2_{ii}\left(\chi^{dc}_{ii}\right)^2} \tag{34}$$

but now, for the group of semiconductors, $\Delta_{ijk\ell}$ is roughly proportional to $\left(1 - 4C^2_\xi\chi_{ii}/\hbar^2\omega^2_p\right)$.

For materials in which molecules can rotate and move around, there is also contribution to the quadratic electro-optical effect from molecular reorientation and redistribution by the dc electric field (or in the quantum picture, population redistribution induced by the field in the ground states). This, in fact, constitutes the major part of the observed quadratic electro-optical effect or Kerr effect in liquid media. Because of the limited scope of this review, we shall not go into detailed theoretical discussion on this subject here, but refer the readers directly to the existing literature[27].

IV. MAGNETO-OPTICAL EFFECTS

We now consider the effect of a dc magnetic field on the linear dielectric constant ε_{ij} or the linear susceptibility χ_{ij}. From Eq. (17), we recognize that the field dependence of χ_{ij} comes in through the wave functions of $|n\rangle$ and $|n'\rangle$, the transition frequencies $\omega_{n'n}$, and the population factors ρ^0_n. In a diamagnetic medium, the states are either fully occupied or empty, and ρ^0_n are unaffected by the dc magnetic field. This change in χ_{ij} induced by field dependence in wave functions and energies is known as the diamagnetic effect[28].

In a paramagnetic medium, however, the populations of the ground states do change as a result of Zeeman perturbation. This population redistribution, which is responsible for paramagnetization, also leads to a change in χ_{ij}, known as the paramagnetic effect[29]. The paramagnetic effect, when present, usually dominates over the diamagnetic counterpart.

Let us first discuss the linear magneto-optical effect. We choose the crystal axes as the coordinate axes. Then, from symmetry argument (Section I), $\chi_{ijk}^{(2b)}$ with $i \neq j$ are imaginary and are the only nonvanishing components. The microscopic expressions for χ_{ij} and $\chi_{ikj}^{(2b)}$ with $i \neq j$ can be written as[30][31]:

$$\begin{aligned}
\chi_{ij} &= \left(\frac{1}{2}\right)\left(\chi_{ij} - \chi_{ij}^{*}\right) \\
&= Ne^{2}\omega \sum_{g} \rho_{g}^{0} \sum_{n} \left[\frac{|(r_{+})_{gn}|^{2}}{(\omega_{ng}^{+})^{2} - \omega^{2}} - \frac{|(r_{-})_{gn}|^{2}}{(\omega_{ng}^{-})^{2} - \omega^{2}}\right] \\
&= \left(Ne^{2}/\omega\right) \sum_{g} \rho_{g}^{0} \sum_{n} \left[\frac{(\omega_{ng}^{+})^{2}|(r_{+})_{gn}|}{(\omega_{ng}^{+})^{2} - \omega^{2}} - \frac{(\omega_{ng}^{-})^{2}|(r_{-})_{gn}|^{2}}{(\bar{\omega}_{ng})^{2} - \omega^{2}}\right]
\end{aligned} \tag{35}$$

$$\begin{aligned}
\chi_{ijk}^{(2b)} &= \partial\chi_{ij}/\partial H_{k} \\
&= Ne^{2}\omega \sum_{g} \left\{\rho_{g}^{0}\left(\partial/\partial H_{k}\right) \sum_{n} \left[\frac{|(r_{+})_{gn}|^{2}}{(\omega_{ng}^{+})^{2} - \omega^{2}} - \frac{|(r_{-})_{gn}|^{2}}{(\bar{\omega}_{ng})^{2} - \omega^{2}}\right]\right. \\
&\quad \left. + \left(\partial\rho_{g}^{0}/\partial H_{k}\right) \sum_{n} \left[\frac{|(r_{+})_{gn}|}{(\omega_{ng}^{+})^{2} - \omega^{2}} - \frac{|(r_{-})_{gn}|^{2}}{(\omega_{ng}^{-})^{2} - \omega^{2}}\right]\right\}
\end{aligned} \tag{36}$$

where $r_{\pm} = \left(r_{i} \pm r_{j}\right)/\sqrt{2}$ and $\omega_{ng}^{\pm} = \omega_{ng} \pm g\mu_{B}H_{k}$.

The above expression for $\chi^{(2b)}_{ijk}$ is again too complicated for numerical evaluation. We are tempted to use the same approximation as used in the discussion of electro-optical effects. However, if $\omega^{\pm}_{ng}$ are replaced by $\bar{\omega}_{ng}$ in Eqs. (35) and (36), both χ_{ij} and $\chi^{(2b)}_{ijk}$ become zero.* This is the result of cancellation between $\sum_g |(r_+)_{gn}|^2$ and $\sum_n |(r_-)_{gn}|^2$. This cancellation is imperfect when the dispersion factors are included in the summation. For order-of-magnitude estimate, we approximate the diamagnetic part of $\chi^{(2b)}_{ikj}$ as

$$\left(\chi^{(2b)}_{ijk}\right)_D \simeq Ne^2\omega \sum_g \rho^0_g \sum_n \left[\left(\frac{-2\omega_{ng}}{(\omega^2_{ng} - \omega^2)^2}\right)\left(|(r_+)_{gn}|^2 + |(r_-)_{gn}|^2\right)\right.$$
$$\times \left(\frac{g\mu_B}{h}\right) + \sum_{n'} \left(\frac{1}{(\omega^2_{ng} - \omega^2)}\right)\left((r_+)_{gn}(r_-)_{n'g} + (r_-)_{gn}(r_+)_{n'g}\right)$$
$$\left.\times \left(\frac{g\mu_B}{h\omega_{n'n}}\right)\right] \simeq \chi_{ii}\left[\left(\frac{-2\omega\bar{g}\mu_B}{h(\bar{\omega}^2_{ng} - \omega^2)}\right) + \left(\frac{\omega\bar{g}\mu_B}{h\bar{\omega}_{n'n}\bar{\omega}_{ng}}\right)\right] \tag{37}$$

The paramagnetic part of $\chi^{(2b)}_{ikj}$, however, depends critically on the cancellation between transitions of right and left circular polarizations. We use the approximation

$$\left(\chi^{(2b)}_{ijk}\right)_P \simeq Ne^2\omega \sum_g \left(\frac{\partial\rho^0_g}{\partial H_K}\right) C_g(\omega)\left(\frac{1}{(\bar{\omega}^2_{ng} - \omega^2)}\right)(r_+r_-)_{gg}$$
$$\simeq \chi_{ii}\left(\frac{\omega}{\bar{\omega}_{ng}}\right) \sum_g C_g\left(\frac{\partial\rho^0_g}{\partial H_K}\right) \tag{38}$$

---

*In semiconductors with free carriers, this sum rule is somewhat modified (see Ref. [31]).

where $C_g(\omega)$ represents the cancellation effect. It is zero for $\omega << \omega_{ng}$ [31], but approaches the order of unity when ω is close to a strong resonance. We assume here that $C_g = \bar{\omega}_{ng}\Delta/(\bar{\omega}_{ng}^2 - \omega^2)$ for $\omega > \Delta$, where Δ is the half-width of the strongest absorption band. Eq. (38) then becomes

$$\left(\chi_{ijk}^{(2b)}\right)_P \simeq \chi_{ii}\left[\frac{\omega\Delta}{\left(\bar{\omega}_{ng}^2 - \omega^2\right)}\right] \sum_g \left(\frac{\partial\rho_g^0}{\partial H_K}\right) \tag{39}$$

Clearly, the approximation here is poor, and we should be satisfied with an order-of-magnitude estimate only.

Let us calculate the diamagnetic $\chi_{ikj}^{(2b)}$ of pure CdS as an example. We have $\chi_{ii} = 0.34$ and $\hbar\omega_{ng} \simeq 7$ ev $\simeq \hbar\omega_{n'n}$ [25]. Then, at $\hbar\omega = 1.1$ eV, we find from Eq. (37) with $\bar{g} = 2$, $\left(\chi_{ikj}^{(2b)}\right)_D \simeq 2 \times 10^{-10}$/oe. The Verdet constant for the Faraday rotation is given by

$$V = (2\pi\omega/c)\, \chi_{ikj}^{(2b)} \tag{40}$$

We then have $V \simeq 7 \times 10^{-5}$ rad cm^{-1} oe for CdS. The observed Verdet constant[35] for CdS at $\hbar\omega = 1.1$ eV is 3×10^{-5} rad cm^{-1} oe^{-1}. As another example, we consider the case of CaF_2 doped with paramagnetic $E\mu^{2+}$ ions. At low temperatures, the paramagnetic term $\left(\chi_{ikj}^{(2b)}\right)_P$ dominates. Since the absorption edges in both $E\mu^{2+}$ and CdS appear near 4000 Å, we assume the same values of χ_{ii} and $\hbar\bar{\omega}_{ng}$ for $E\mu^{2+}$ as for CdS. The ground states of $E\mu^{2+}$ are the $^8S_{7/2}$ multiplet. We, therefore, have $\sum_g \left(\partial\rho_g^0/\partial H\right) \simeq 2\mu_B/kT$. The half-width of a strong band is usually around a few tenths of eV. We set $\Delta \simeq 0.2$ eV, corresponding to a band width of 3000 cm^{-1}. At T = 4.2°K and $h\omega = 2.26$ eV, we find, from Eq. (39), $\left(\chi_{ikj}^{(2b)}\right)_P \simeq 10^{-7}$/oe. Then, the

Verdet constant for CaF_2:1% Eu^{2+} is $V \simeq 7 \times 10^{-4}$ rad cm^{-1} oe^{-1}. The observed value is $V \simeq 5 \times 10^{-4}$ rad cm^{-1} oe^{-1} [36]. Therefore, in order to have large paramagnetic Verdet constant, we must not only find a crystal with large Zeeman perturbation on the ground-state populations, but should also have the frequency ω close to a strong absorption band so that C_g is large.

In general, one can show that under limited conditions, the paramagnetic part of the susceptibility χ_{ij} is proportional to the magnetization (M) [29][30].

$$\left(\chi_{ij}\right)_P = A \sum_m m\rho_m^0 = A(M) \tag{41}$$

The magnetization, and hence $\left(\chi_{ij}\right)_P$, saturates in the presence of strong field. This is usually the case in ferromagnets. The paramagnetic rotation is always very strong in ferromagnets, but is stronger if ω is close to a strong absorption band in materials where only a few narrow, strong, well-separated bands exist.

We can also find the microscopic expression for $\chi_{ijk\ell}^{(3bb)}$ from $\partial^2\chi_{ij}/2\partial H_k \partial H_\ell$. It is easy to see that for ω sufficiently far away from absorption bands the diamagnetic part of $\chi_{ik\ell i}^{(3bb)}$ is roughly $g\mu_B/\hbar\bar{\omega}_{ng}$ times smaller than $\left(\chi_{ikj}^{(2b)}\right)_D$, and the paramagnetic part of $\chi_{ik\ell i}^{(3bb)}$ is $g\mu_B/kT\ C_g$ times smaller than $\left(\chi_{ikj}^{(2b)}\right)_P$. In general, the paramagnetic Voigt effect is given by $\left(\chi_{ii}(H^2) - \chi_{ii}(0)\right)_P$, which becomes comparable in order of magnitude with $\chi_{ij}(H)$ when the magnetization becomes saturated.

In liquid media, molecules can also be realigned and redistributed by the applied dc magnetic field. Molecular reorientation and redistribution are partially responsible for the observed quadratic magneto-optical effect in liquid. The linear birefringence induced by a magnetic field in liquid is known as the Cotton-Mouton effect. Theoretical treatment on the Cotton-Mouton effect should follow closely to that on the Kerr effect in liquid.

V. CONCLUSION

While symmetry arguments are useful in identifying the nonvanishing independent components of the χ tensors for electro- and magneto-optical effects, a microscopic theory is required to estimate the magnitudes of these nonvanishing components. Calculations based on a simple two-level or two-band models show the usefulness of these approximations. The good agreement obtained between theory and experiment for the electro-optical coefficients has been particularly encouraging. The same calculations also predict the nonlinear optical coefficients successfully[19],[20],[22]-[24]. The approximations used to calculate the Verdet constant for magneto-optical rotation are rather crude, but they do yield order-of-magnitude agreement with the experimental results. The quantitative calculations discussed here break down in the limit where optical absorption becomes appreciable.

ACKNOWLEDGMENT

This work was done under the auspices of the U. S. Atomic Energy Commission.

REFERENCES

[1]. F. Pockels, Aghandl. Ges. Wiss. Gottingen, 39, 1 (1893); Lehrbuch der Kristalloptik Teubner, Leipzig, 1906; J. Kerr, Phil. Mag., 4, 50, 337, 5346 (1875).

[2]. M. Faraday, Phil. Trans., 136, 1 (1846).

[3]. See, for example, Handbuch der Physik, Vol. XXV/1 (S. Flügge, ed.) Springer-Verlag, Berlin, 1961.

[4]. See, for example, L. D. Landau and E. M. Lifshitz, Electrodynamics in Continuous Media, Addison-Wesley Publishing Co., Inc., Reading, Mass., 1959, p. 252.

[5]. See, for example, J. F. Nye, Physical Properties of Crystals, Oxford University Press, New York, 1957, p. 123.

[6]. See, for example, L. D. Landau and E. M. Lifshitz, Electrodynamics in Continuous Media, Addison-Wesley Publishing Co., Inc., Reading, Mass., 1959, p. 331.

[7]. W. Voigt, Magneto- and Electro optik, B. G. Teubner, Leipzig, 1908.

[8]. L. D. Landau and E. M. Lifshitz, Electrodynamics for Continuous Media, Addison-Wesley Publishing Co., Inc., Reading, Mass., 1959, p. 49.

[9]. M. Bass, P. A. Franken, J. F. Ward, and G. Weinreich, Phys. Rev. Letters, 9, 446 (1962).

[10]. P. S. Pershan, J. P. van der Ziel, and L. D. Malmstrom, Proceedings of the Physics of Quantum Electronics Conference, Puerto Rico, 1965 (P. L. Kelley, B. Lox, and P. E. Tennewald, eds.) McGraw-Hill Book Co., New York, 1966, p. 3.

[11]. Y. R. Shen and N. Bloembergen, Bull. Am. Phys. Soc., 9, 292 (1964).

[12]. P. S. Pershen, J. P. van der Ziel, and L. D. Malmstrom, Phys. Rev., 143, 574 (1966).

[13]. I. S. Zheluder, Usp. Fiz. Nauk, 88, 253 (1966); (translation: Sov. Phy. Uspekhi, 9, 97 (1966).

[14]. H. A. Kramers and W. Heisenberg, Z. Physik, 31, 681 (1925); M. Born, W. Heisenberg, and P. Jordon, Z. Physik, 35, 570 (1926).

[15]. J. Armstrong, N. Bloembergen, J. Ducuing, and P. S. Pershan, Phys. Rev., 127, 1918 (1962).

[16]. N. Bloembergen and Y. R. Shen, Phys. Rev., 133, A37 (1964).

[17]. F. N. H. Robinson, Bell System Tech. J., 46, 913 (1967).

[18]. S. S. Jha and N. Bloembergen, Phys. Rev., 171, 891 (1968).

[19]. C. Coulson, L. Ridei, and D. Stocker, Proc. Roy. Soc. (London) 270A, 357 (1962).

[20]. J. C. Phillips and J. A. Van Vechten, Phys. Rev., 183, 709 (1969).

[21]. C. Flytzanis and J. Ducuing, Phys. Rev., 178, 1218 (1969).

[22]. B. F. Levine, Phys. Rev. Letters, 22, 787 (1969).

[23]. See, for example, J. M. Ziman, Principles of the Theory of Solids, Cambridge University Press, 1964, p. 224.

[24]. D. R. Penn, Phys. Rev., 128, 2093 (1962).

[25]. J. C. Phillips, Phys. Rev., 166, 832 (1968); Phys. Rev. Letters, 20, 550 (1968).

[26]. R. C. Miller, Appl. Phys. Letters, 5, 17 (1964).

[27]. C. V. Raman and K. S. Krishnan, Phil. Mag., 3, 724 (1927); R. W. Hellworth, Phys. Rev., 152, 156 (1966).

[28]. R. Serber, Phys. Rev., 41, 489 (1932).

[29]. J. H. Van Vleck and M. H. Hebb, Phys. Rev., 46, 17 (1934).

[30]. Y. R. Shen, Phys. Rev., 133, A511 (1964).

[31]. H. S. Bennett and E. A. Stern, Phys. Rev., 137, A448 (1965).

[32]. M. Balkanski and J. J. Hopfield, Phys. Stat. Sol., 2, 623 (1962).

[33]. Y. R. Shen and N. Bloembergen, Phys. Rev., 133, A515 (1964).

Chapter 30

THE INFLUENCE OF THE COMPOSITION ON THE GAIN OF Nd-DOPED GLASSES

C. F. Rapp

Owens-Illinois, Inc.
Toledo, Ohio

I. INTRODUCTION

The laser is essentially an optical amplifier which is based on the principle that if an atom or molecule can exist in two states of different energy, an electromagnetic wave can stimulate the atom to make a transition between these states. Normal absorption is the result of exciting the molecule from the lower to the higher energy state. If the molecule was originally in the higher energy state, the incident radiation would stimulate the molecule to make a transition to a lower energy state and emit a quantum of light[1].

In solid state lasers, the active ions or atoms are dispersed in a host matrix. The host has an effect on the lasing behavior since the electronic transitions of the active ion are influenced by the local environment of the site. For example, crystalline hosts have well-ordered structures and active ions such as neodymium substitute into well-defined sites in the crystalline structure. The glass host has a more random structure and the neodymium ions substitute into random positions in the glass structure. Thus, in a glass host, due to the variety of positions, the various neodymium ions would have slightly different wavelengths for absorption and emission. This results in broader absorption and emission lines for the neodymium ion in a glass host than in a crystalline host. Figure 1 contrasts the neodymium emission line in a glass and crystalline host.

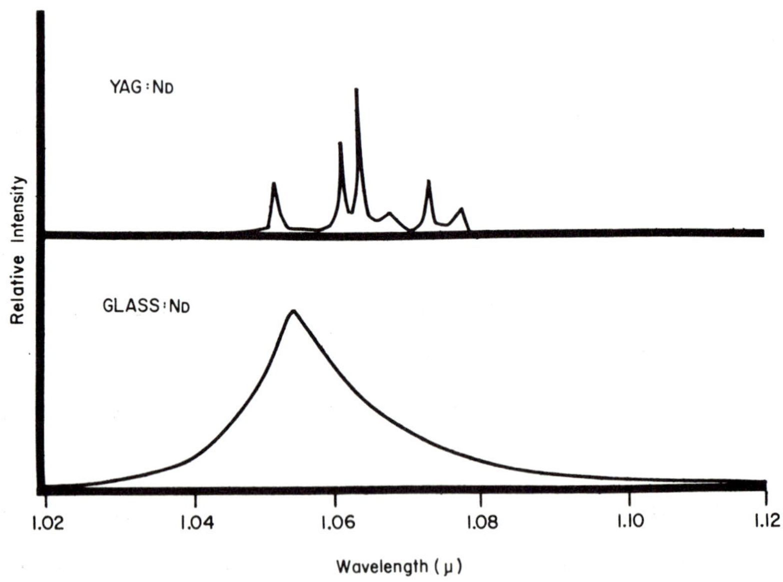

FIG. 1

Comparison of neodymium emission line in a glass with crystalline host.

It is important to remember that the term "glass" encompasses a great variety of materials having different composition, properties, and structures. In selecting a glass composition for use as a laser host there are essentially four factors to be considered. These are summarized in Table I. First, it must be possible to produce extremely high quality glass from the composition with no stria, i.e., index of refraction variation or inhomogeneities. Second, the glass should have a high quantum efficiency for the conversion of absorbed pump light into fluorescent light. Third, there should be broad absorption bands to absorb as much pump light as possible. Fourth, the glass should have a high gain coefficient to extract as much of the stored energy as possible.

In attempting to develop a laser with the optimum combination of the last three items listed in Table I, the task is complicated by the fact that there are no stoichiometric rules for the composition as there are for crystalline compounds, and almost any cation can be incorporated into the composition regardless of the charge or size. As a result of this, there is almost an unlimited number of glass compositions which can be made and tested. Since this would lead to an unending search for the best composition for a particular use, one of the major goals in glass research is to obtain very simple trends in the properties as they relate to the composition so that the best composition can be predicted rather than just happen by chance. The importance of this becomes quite obvious when one considers that many commerical glasses may have as many as eight or ten different oxides present in the glass.

The objectives of the present investigation were to establish the cause of the variability in the lasing properties and to develop sufficient data to allow predictions of the optimum composition to be used as a glass host.

II. EXPERIMENTAL PROCEDURE

A. Melting

Selected Nd-containing alkali and alkali aluminosilicates were melted in Pt-10Rh crucibles in an electric furnace for times up to 48 hrs with continuous stirring. The melting temperatures were generally between 1350° and 1500°C, depending on the composition. The glasses were annealed in the crucible.

B. Evaluation

The fluorescent decay time was obtained by exciting polished plates (7/8 inch x 7/8 inch x 1/4 inch) with a 10 μsec flash of a xenon flash lamp. The fluorescence of the excited sample was detected with an RCA 7102 photomultiplier and displayed on an oscilloscope with a time base. A 1.06 μ narrow band pass filter was placed between the sample and the detector. The resulting trace was then used to calculate the decay time.

III. RESULTS AND DISCUSSION

In the evaluation of a neodymium-doped laser glass composition, the fluorescence decay time is one of the most important factors, since the gain of the laser glass is a direct function of the decay time. The simplest expression for the gain of a laser is essentially the same as the Beer-Lambert law for absorption. The gain is equal to $e^{\alpha(N_2-N_1)l}$ where α is the gain coefficient, N_2-N_1 is the population inversion, and l is the length of the laser rod. It can be shown that the gain coefficient, α, is proportional to the Einstein

B_{21} coefficient which is a measure of the probability for a stimulated transition. The Einstein B_{21} coefficient is, of course, directly proportional to the A_{21} coefficient and, therefore, inversely proportional to the fluorescence decay time. From this, then, it is not surprising to find a direct correlation between the lasing efficiency of a glass and its fluorescence decay time. Table II demonstrates such a correlation between the measured fluorescence decay times of several Nd-containing glasses and their corresponding lasing efficiencies. It can be seen here that the lasing efficiency is almost directly proportional to the inverse of the fluorescence decay time.

TABLE I

LASER GLASS REQUIREMENTS

1. good optical quality
2. a high quantum efficiency
3. broad absorption bands
4. a high gain coefficient

TABLE II

RELATION OF FLUORESCENCE DECAY TO LASING EFFICIENCY

Glass number	τ (μsec)	$\frac{1}{\tau} \times 10^3$	Efficiency at 363 J in	Slope efficiency
1	300	3.33	1.39%	1.66%
2	330	3.03	1.40	1.71
3	488	2.05	0.90	1.19
4	702	1.42	0.61	0.85

Within the neodymium ion itself, the fluorescent transition is a transition within the f orbital and is, therefore, a forbidden transition. This gives rise to the relatively long fluorescence decay times. A number of effects can remove part of this forbiddenness but of concern here is the effect of an asymmetric ligand field upon the ion.

If the neodymium ion were incorporated into a pure fused silica host, the ion would have an octahedral coordination with six oxygen ions as represented in Fig. 2. Each of these coordinating oxygen ions would be identical giving the neodymium ion a center of symmetry. If, however, some other cation such as an alkali ion were added to the glass, the six oxygen ions coordinated with the neodymium would not all be the same since some would be bound to the alkali and some would not. There would now be no center of symmetry for the neodymium ion and the asymmetrical ligand field would cause a slight hybridization between the f orbitals of the neodymium ion and some other p or d orbitals of the ion. This would, in part, remove the forbiddenness of the transition and a shorter fluorescence decay time

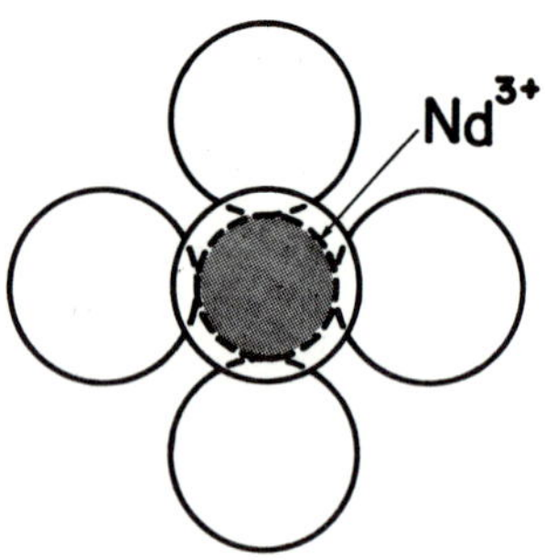

FIG. 2

Neodymium ion having an octahedral coordination with six oxygen ions after incorporation into pure fused silica host.

would result. This general effect is shown in Fig. 3 which is a plot of the fluorescence decay time versus the percent potassium oxide in a silica base glass. As would be expected, the decay time decreases, or the transition probability increases, as the amount of potassium in the glass is increased.

If one assumes that the degree of hybridization in the neodymium ion is dependent on the amount by which these "odd" oxygen ions are affected by the added cation, then the probability for a transition would be directly dependent upon the external field which the added cation exerted on the oxygen ions. This external field would simply be the electrostatic force of the cation. The electrostatic force "f" can be calculated by $f = \varepsilon\varepsilon'/Kr^2$, where K is the dielectric constant for a vacuum, ε is the charge of the oxygen ion, ε' is the charge of the cation, and r is the radius of the cation. In the case of the alkali silicate glasses, ε, ε', and K are all constants so that there should be a direct relationship between the transition probability of the neodymium ion and $1/r^2$ for the alkali ion in the glass. Figure 4 shows a plot of this for the simple alkali silicates. The five glasses shown all contain 70 mole% SiO_2, 2.5 mole% Al_2O_3, 0.5 mole% Nd_2O_3, and 27.5 mole% alkali. Figure 5 shows this same plot of the five simple alkali silicate glasses along with glasses of various mixtures of alkali totaling 27.5 mole%. It can be easily seen from this plot that there is no "mixed alkali" effect but rather only a simple averaging effect which gives glasses with transition probabilities lying between those of the simple lithium silicate glass, which has the greatest transition probability and therefore the highest gain, and the cesium silicate, which has the lowest transition probability and therefore the lowest gain.

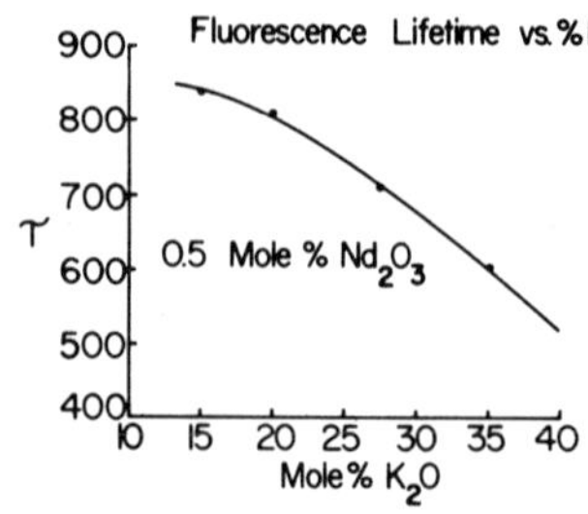

FIG. 3

Fluorescence decay time versus percentage potassium oxide in a silica base glass.

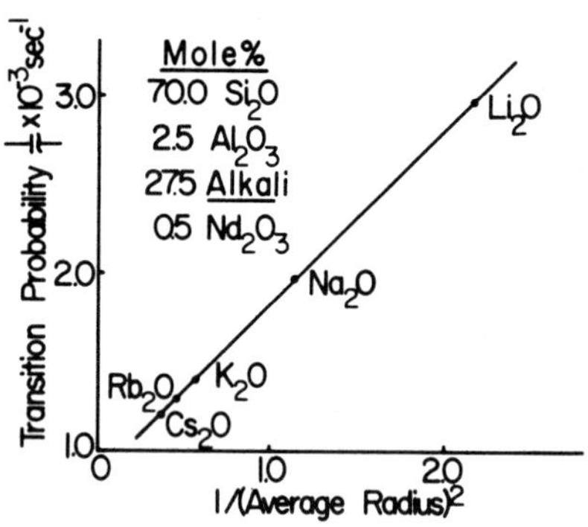

FIG. 4

Relationship between transition probability of the neodymium ion and $1/r^2$ for the alkalai ion in a glass.

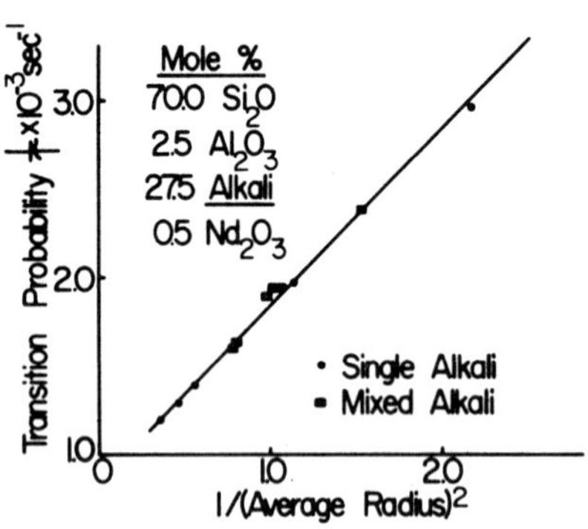

FIG. 5

Relationship between transition probability of the neodymium ion and $1/(\text{Average Radius})^2$ for the mixed alkalai ion in a glass.

The effect of adding alkaline earth oxides to the glass is shown in Fig. 6. The compositions of these additional glasses are all 60 mole% SiO_2, 2.5 mole% Al_2O_3, 27.5 mole% Li_2O, 0.5 mole% Nd_2O_3, and 10 mole% alkaline earth oxide. The plot is extended slightly to the higher transition probability region, because the charge on the alkaline earth oxides of 2 results in a higher field strength than that of an alkali of similar radii. Since the electrostatic force is proportional to the charge and inversely proportional to the radius squared, the charge of two was accounted for in the plot by treating it as a charge of one but decreasing the effective diameter by the $\sqrt{2}$ or 0.707 times the diameter. The plot is displaced upwards from the simple alkali plot because the addition of the alkaline earth oxides to the glass gives a greater number of cations affecting the neodymium. This same effect of shifting the plot up or down is observed in the simple alkali silicate systems when the alkali is increased or decreased.

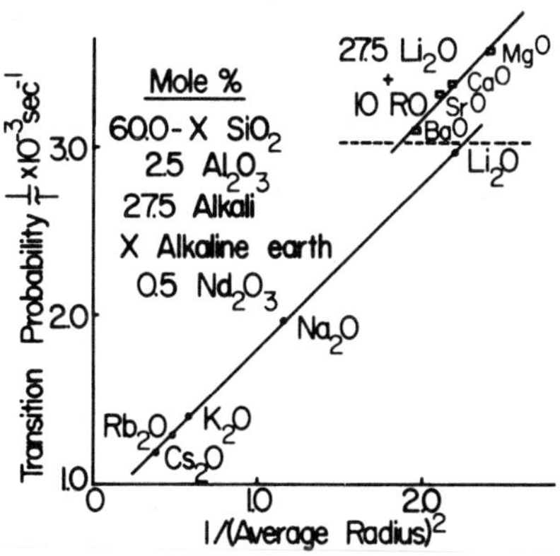

FIG. 6

Relationship showing the effect on the transition probability with the addition of alkaline earth oxides in a glass.

The importance of this relationship of the relative gain to the alkali or akaline earth contained in the glass is that almost all the differences in the fluorescent properties of all the silicate laser glasses arise from the differences in the alkali or alkaline earth incorporated in the glass. Although there are many other oxides included in the glasses such as those of B, Zn, Cd, La, Ce, or Sb, these are generally present in small amounts and their importance arises in the melting and forming, the chemical durability, or the solarization of the glass and usually play a secondary effect on the fluorescence.

IV. SUMMARY

This paper has examined various compositional factors and how they affect the fluorescence properties. This data can be used to explain why the various glass laser hosts that have been developed do not have equally good lasing properties. In some cases, greater attention has been paid to factors affecting the formability of the glass rather than to the factors affecting the lasing parameters.

REFERENCES

[1]. B. A. Lengyel, Introduction to Laser Physics, Chapt. 1, John Wiley, New York, 1966.

Chapter 31

SOLID STATE REACTIONS IN THE PREPARATION OF ZIRCON STAINS

Richard A. Eppler

Ceramics Group
Glideen-Durkee Division
SCM Corporation
Baltimore, Maryland

I. INTRODUCTION

There are a number of methods to obtain color in a vitreous material, such as a glass or a glaze. In the first place, certain transition metal ions can be smelted into a glass when it is made. Some examples include blue glasses containing cobalt and green glasses containing chromium. The chemistry of these colored glasses has been treated by Weyl[1].

A second method to obtain color in a vitreous system is to induce the precipitation of a crystal[2]. Certain oxides such as tin oxide, zirconium dioxide, and titanium dioxide dissolve in a vitreous material at high temperature. If the temperature is reduced, however, the solubility is likewise greatly reduced and precipitation occurs[3]. This method is widely used for opacification, but is less often used with coloring oxides.

The third method to obtain color in a vitreous matrix is to disperse in that matrix an insoluble crystal which is colored. The color of the crystal is then imparted to the matrix. This method is the one most commonly used in introducing color to vitreous coatings such as glazes for use on wall tile and other whiteware bodies.

Most of the crystals used for ceramic colors are oxides. The one major exception is the family of CdS-CdSe red pigments. The various oxide pigments can be classified as shown in Fig. 1[4]. In some cases, pigments are made from crystals which are themselves colored. Examples include cobalt aluminate blue, Victoria green, and Naples yellow. Solid solutions of colored materials, such as solutions of cobalt aluminate and cobalt chromate, are also used.

OXIDE PIGMENTS

- Colored crystals
  - True Colored Compounds
    - $CoAl_2O_4$
    - $3\ CaO \cdot Cr_2O_3 \cdot 3\ SiO_2$ (Victoria Green)
    - $Pb_2Sb_2O_7$ (Naples Yellow)
  - Solid solutions
    - $Co(Al_xCr_{2-x})O_4$
- Colorless crystals
  - Doped crystals
    - $ZrSiO_4$:V
    - $ZrSiO_4$:Fe
    - $ZrSiO_4$:Pr
    - Al_2O_3:Cr
  - Mordant pigments
    - ZrO_2:V
    - SnO_2:V

FIG. 1
Classification of oxide pigments.

The use of colored crystals is severely limited by the simultaneous requirement of intense color and acceptable values of the other properties required of a ceramic pigment (see Section II). These requirements can be optimized separately by the use of colorless host crystals to which a chromophore is added in small amount during manufacture. In most cases the dopant is trapped in the lattice of the host. Examples of this type of ceramic color include the various zircon stains and chrome-alumina pinks. In some cases, however, the dopant is believed to be "adsorbed" on the surface of the host. These colors, which include tin-vanadium and zirconia-vanadium yellows are called mordant stains.

These latter ceramic colors can be further classified on the basis of the nature of the chromophore[2]. The most common technique is to add as the dopant one of the transition metal oxides (Ti, V, Mn, Fe, Co, Ni, and Cu) or one of the rare earth oxides. These oxides are themselves colored due to the presence of the partially filled 3d or 4f shells in the atomic structure of the cation. Several other effects are also used. In some cases, the host crystal permits the stabilization of an unusual valence state of the dopant ion. The effect is particularly important in the case of the zircon stains, where the tetravalent state is imposed. Thus the addition of vanadium produces the blue color characteristic of V^{+4}[5]. Alternatively, the addition of praseodymium produces a yellow color. This yellow color is characteristic of the very unusual Pr^{+4} state.

When the host crystal cation can have more than one valence state, charge compensation effects can occur. For example, in the tin-vanadium yellow, the ability of tin ions to be reduced to the

+2 state permits the stabilization of pentavalent vanadium in the tetravalent tin oxide structure.

In some cases, the geometry of the host crystal distorts the dopant ion so that unique colors result. Probably the best known example of this effect is the ruby, where the alumina lattice distorts the chromium ion from a green to a red color. The same effect occurs with the chrome-alumina pink stains.

II. PROPERTIES DESIRED IN A CERAMIC COLOR

To be suitable as a ceramic color, a material must possess a number of properties. Among these properties two are preeminent. The first is high intensity of color. Related to the intensity is the diversity of colors obtainable with similar materials (for example, different stains with the same host crystal). The latter is important because most users mix two or more colors to obtain the final shade they desire.

The other important property of a ceramic color is stability under the high temperatures and corrosive environments encountered in the firing of glazes and enamels. The rates of solution in the molten vitreous material should be very low in spite of the very fine particle sizes ($\sim 10\ \mu$) which are used. Neither should gases be given off as a result of the contact of the ceramic color with the molten vitreous material. These requirements limit ceramic colors to those materials which are fully reacted and relatively inert.

For these same reasons, colors that are useful at one temperature of firing may not be useful at higher temperatures. For example, at the firing temperatures of porcelain enamels (750-850°C), stains based on the spinel structure (cobalt aluminates and chromates, for example) are widely used. At the higher temperatures often used in the whitewares industry (1000-1300°C) requirements for resistance to glaze attack have in many cases led to the adoption of the more stable colors based upon zirconia or zircon.

Another desirable property for a ceramic color is a high refractive index[2]. For example, valuable ceramic stains are based upon spinels ($n \cong 1.8$) and upon zircon ($n \cong 1.9$), but no valuable stains are based upon apatite ($n \cong 1.6$) even though the lattice of apatite is as versatile for making ionic substitutions as that of spinel.

For all these reasons, ceramic colors based upon zircon as the host lattice have found increasing acceptance for those applications involving firing over 1000°C. The first such stain[5]-[6] to be introduced commercially contains vanadium as the dopant, and produces a turquoise blue. Yellow stains containing praseodymium[7] and pink stains containing iron[8]-[9] are also produced commercially. Various other ions have also been suggested[10].

With the exception of strong red shades, a full pallette of colors is attainable by using one or more of these colors. Their resistance to dissolution is superior to almost all other ceramic colors, permitting use at the higher (1000-1300°C) firing temperatures.

III. PREPARATION OF ZIRCON STAINS

A. Composition

Zircon stains are usually made by calcining a mixture of silica, zirconia, a dopant ion which produces the color, and, usually, a mineralizer, or low melting material which catalyzes the reaction[11]-[12]

A number of studies[12]-[14] have shown that the dopant must be present at the time the zircon is formed from its constituents; the dopant cannot be diffused into zircon with or without the presence of mineralizers. The zirconia/silica ratio[12]-[15] is normally close to the stoichiometry of the reaction:

$$ZrO_2 + SiO_2 \rightarrow ZrSiO_4 \quad (1)$$

although Seabright[5]-[7] claims very broad ranges in this property. For the case of the vanadium doped blue stains, it has been shown[15] that the dopant substitutes for zirconia in the zircon lattice. X-ray diffraction has shown that the final product consists of only one phase, zircon[11]. Some compositions which are illustrative of these principles are shown in Table I.

TABLE I

ILLUSTRATIVE COMPOSITIONS

Composition	A	B	C
Zirconia	62.8	61.76	59.3
Silica	31.1	30.12	28.2
Ammonium metavanadate	3.82	--	--
Praseodymium oxide	--	4.87	--
Yellow iron hydroxide	--	--	10.0
Sodium silicofluoride	--	--	2.5
Sodium fluoride	2.26	3.25	--
Color of stain	Blue	Yellow	Pink
Reference	[12]	[12]	[9]

The materials which are called mineralizers are one or more of various halides and or silicohalides, which are added to the batch in concentrations of 1-5%[16]. Although a weak blue stain can be made without mineralizers[14]-[17], they are required for incorporation of more than 2% vanadium in the zircon lattice[15]. The iron containing pink stain cannot be made without a mineralizer[16].

B. Processing

The compositions, as in Table I, are normally compounded from conventional technical grade ceramic sources. They are weighed out and mixed dry and are then micronized on a laboratory size hammer mill.

Crucibles, which may be fire clay, porcelain, fused silica, etc., are used as containers during firing.

Selection of the proper temperature for firing involves a balance between kinetics and volatility effects. On the one hand, some of the oxides which serve as color centers are volatile at temperatures of 700-1000°C (e.g. vanadium)[2]. On the other hand, as is well known, the rates of most reactions increase rapidly with the temperature. Generally, temperatures in the range 800-1100°C are used[5]-[9].

The atmosphere in the furnace during firing can be either air or oxygen, but not nitrogen or hydrogen[15]. This observation indicates that oxygen atoms in the atmosphere are used in the reaction.

IV. MECHANISM OF FORMATION OF ZIRCON STAINS

Study of the mechanism of formation of these materials should provide insight into the mechanisms of transport whereby the zirconia, silica, and the dopant are brought together. The role of the mineralizer can also be considered. A useful technique for studies of mechanism is the marker experiment[18]-[21], several variations of which will now be discussed.

A. Experiments upon Polycrystalline Samples

In these experiments, which are described more fully elsewhere[16], one of the host reactants (ZrO_2 or SiO_2) is placed in the bottom of a crucible (see Fig. 2a). On top of this reactant are placed some 0.020 inch diameter platinum wires, to serve as inert markers. On top of the wires is placed the other reactant. The dopant and the mineralizers can be mixed with either of the host reactants, or with both. All nine combinations of the latter materials have been studied in some cases in order to better understand the role of these materials[16]. The crucible is then fired in accordance with the normal stain-making schedule.

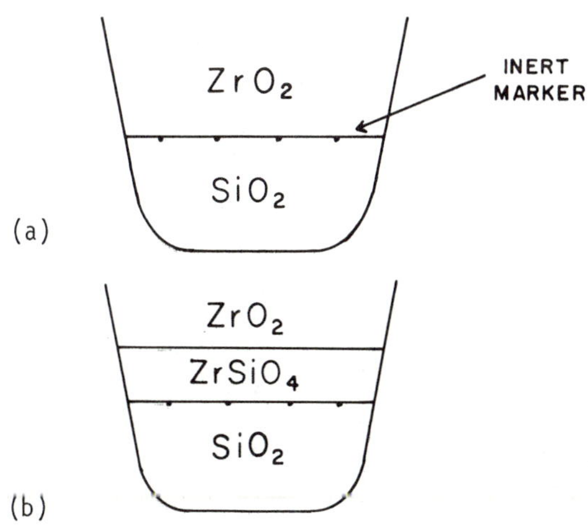

FIG. 2
Schematic drawing of the marker experiment: (a) before reaction and (b) after reaction.

Figure 3 is a view of an experiment[16] in which vanadium was the dopant, NaCl and NaF the mineralizers. Both were distributed throughout the reactants. On the left is a silica layer. It is white and is quartz by X-ray diffraction. There is no evidence of any product on this side of the markers which can be seen on top of the zirconia layer on the right.

The interface of this zirconia layer is a deep blue and is the product zircon by X-ray diffraction. Underneath this interface, which is 1-2 mm thickness, is unreacted baddeleyite, which is an off-white.

The fact that the product is confined to the zirconia layer suggests that silicon is the only mobile cation. Furthermore, Fig. 4

FIG. 3
Marker experiment on a vanadium doped zircon stain; interface between the ZrO_2 layer (blue) and the SiO_2 layer (white).

FIG. 4
Marker experiment on a vanadium doped zircon stain; side view—SiO_2 layer over ZrO_2 layer.

which is a side view of a white silica layer and a blue zirconia layer in the same experiment shows that the blue stain appears around the entire periphery of the ZrO_2 layer, not only at the interface with the SiO_2 layer. Schmalzried[19] has shown that such a distribution of the product is characteristic of a reaction wherein one of the cations is transported through the vapor state.

Figure 5 shows another experiment[16] in which iron was the dopant and Na_2SiF_6, $BaSiF_6$, and $MgSiF_6$ were the mineralizers. Iron was found distributed throughout the sample, while the mineralizers were left in the silica layer. The silica layer is on the left. The interface of the silica layer was dark greenish brown and heavily sintered. However, it did not differ from the body of the

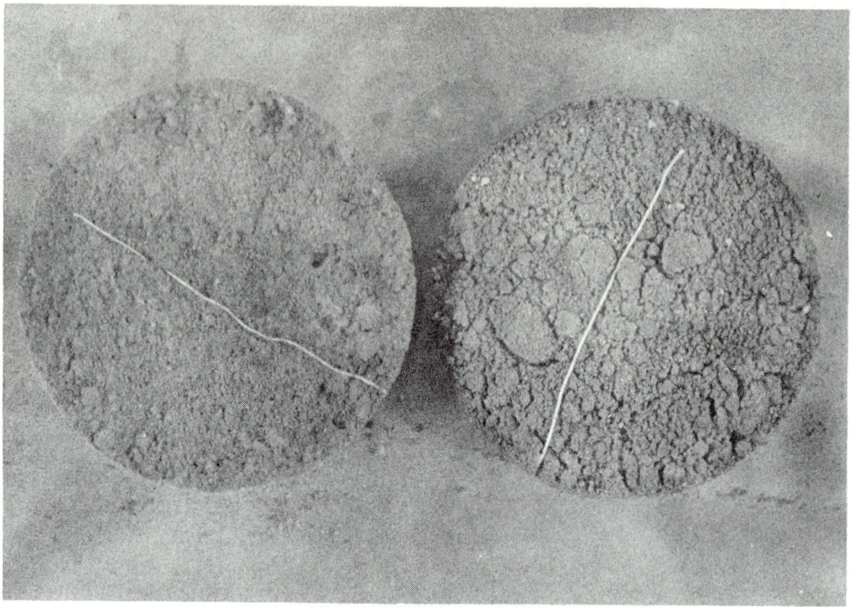

FIG. 5
Marker experiment on an iron doped zircon stain; interface between the ZrO_2 layer (light brown) and the SiO_2 layer (dark brown).

layer in crystalline phases, which were cristobalite and quartz. The dark color was probably due to solid solution of some of the iron in the silica, and to residual hematite.

The zirconia layer, which is on the right, has an interface which is brown. It is the product zircon. Underneath the product layer is a light purple body of unreacted baddeleyite.

These experiments suggest that the silicon ions are transported through the vapor state[16]. If so, it is likely that the silica reacts with the mineralizer to form a silicon tetrahalide.

$$SiO_2 + 4MX \rightarrow SiX_4\ (g) + 2M_2O \qquad (2)$$

The silicon tetrahalide is a volatile species which transports the silicon to the interface with the product layer. Two possibilities have been suggested[19] for the process of transport of the silicon across the product layer. The first of these suggestions, described here as the diffusion model, postulates that the silicon tetrahalide decomposes on the surface of the product layer depositing silicon and electrons in imperfections in the latter layer:

$$SiX_4 \rightarrow Si^{+4} + 4e^- + 2X_2 \quad (3)$$

A flux of electrons and silicon ions then passes across the product layer by a diffusion process, permitting the following reaction at the interface with the unreacted zirconia:

$$Si^{+4} + 4e^- + ZrO_2 + O_2 \rightarrow ZrSiO_4 \quad (4)$$

The other alternative, which will be called the vapor transport model, postulates that the silicon tetrahalide as well as oxygen can pass through pores in the product layer, so that the final reaction occurs completely at the interface with the unreacted zirconia:

$$SiX_4 + ZrO_2 + O_2 \rightarrow ZrSiO_4 + 2X_2 \quad (5)$$

In order to attempt to distinguish between these two mechanisms, a single crystal experiment was performed. Recent studies[21]-[23] have shown the desirability of confirming results in polycrystalline materials with work on single crystals.

B. Single Crystal Studies

The experimental arrangement for the single crystal study is shown in Fig. 6. A mixture of silica, mineralizers, and ammonium metavanadate were placed in the bottom of a porcelain crucible. The platinum markers were placed at the interface, on top of which was

— SINGLE CRYSTAL EXPERIMENT

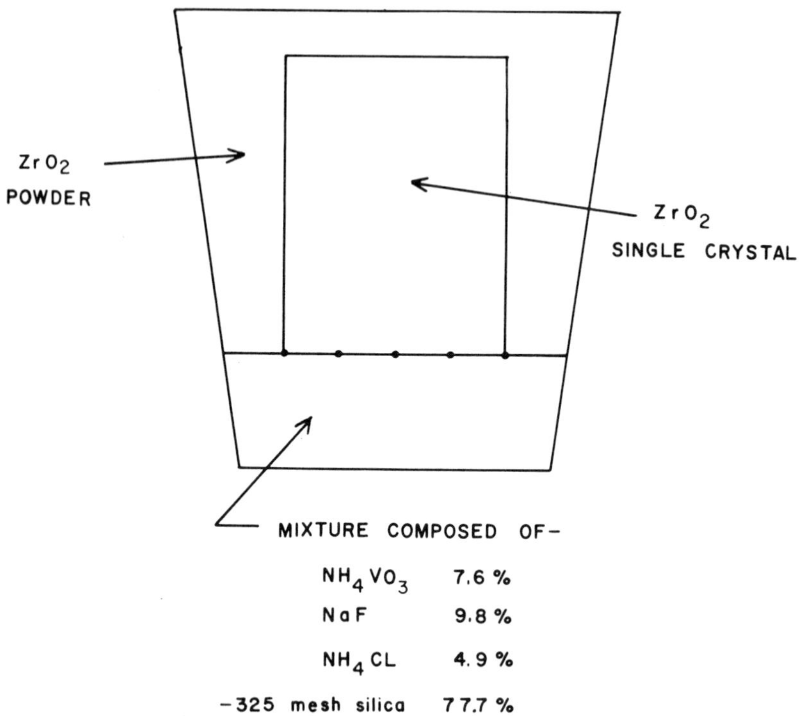

FIG. 6
Single crystal experimental setup.

placed a single crystal of zirconium dioxide, obtained from Semi-Elements, Inc., Saxonburg, Pa. Powdered zirconium dioxide was packed around the single crystal to hold it in place during firing. The crucible was then fired at 900°C for 4 hrs.

After firing, the single crystal was carefully removed for examination (see Fig. 7). A blue reaction layer is found on the single crystal surface which was exposed to the silica-vanadia-mineralizer mixture. The rest of the crystal was undisturbed. A similar blue layer was found in the powdered zirconia surrounding the single crystal. On the other hand, there was no evidence of any reaction product in the powdered silica-vanadia-mineralizer layer.

FIG. 7
The ZrO_2 single crystal after reaction with silica and vanadia.

In order to facilitate examination of the single crystal with the microscope, one corner of the crystal, perpendicular to the reaction layer, was ground flat so that the reaction layer could be examined with the microscope. Some results are shown in Fig. 8 and Fig. 9. Figure 8 shows one corner of the ground off area. To the left is the undistrubed crystal. To the right is the reaction layer.

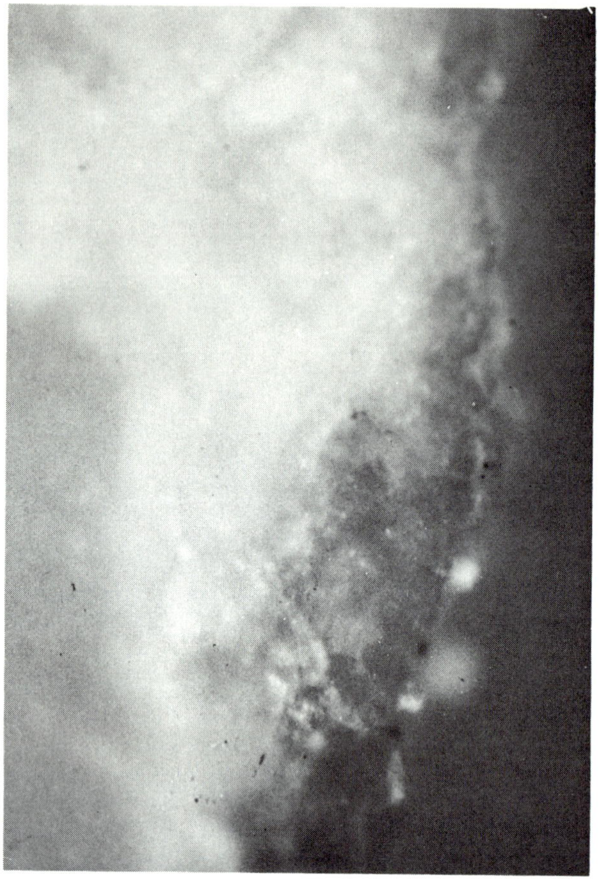

FIG. 8
Single crystal experiment, micrograph of a corner of the polished area—⊥ to the reaction direction.

At the top of the picture is the intersection of the reaction layer with the side of the crystal. It is found that the penetration of the blue color is uneven. Note the white area at the corner of the ground off area in the top of the picture. Note also that the reaction layer appears altered relative to the rest of the undisturbed crystal, but is not obviously polycrystalline. Perhaps the observed alteration is due to stresses caused by differences in the specific volume between zircon and zirconia. Figure 9 shows an area adjacent

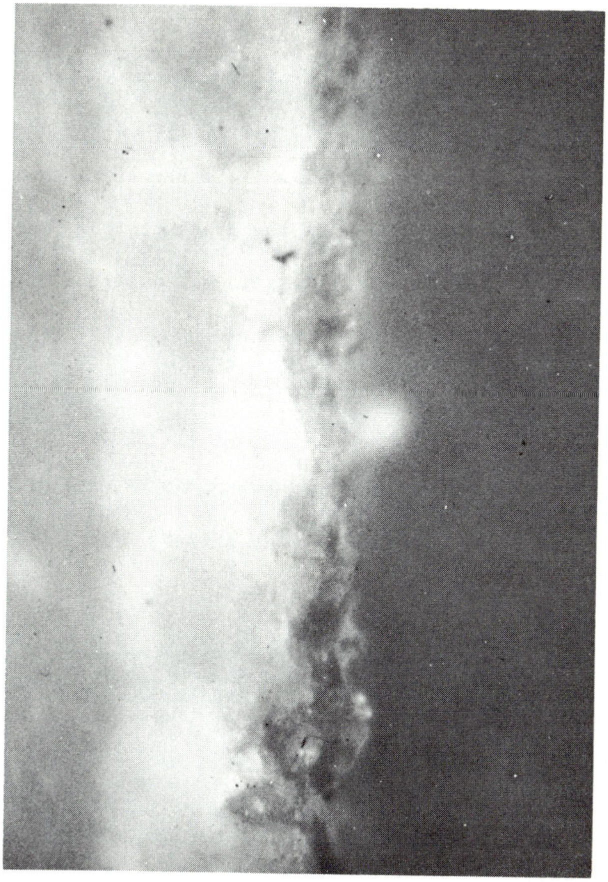

FIG. 9
Single crystal experiment, micrograph of the edge of the polished area—⊥ to the reaction direction.

to that in Fig. 8. It also illustrates that there is a white alteration layer between the blue reaction layer and the undistrubed crystal.

Returning to Fig. 8, note in the upper left hand corner of the picture that the reaction layer appears deeper than in the rest of the picture. This region was adjacent to powdered zirconia during the reaction process and illustrates the important point that the depth of the reaction layer in the powdered material was approximately twice the depth of the reaction layer in the single crystal.

Figure 10 is a photograph looking down onto the reaction layer of the single crystal, from the direction of the silica-vanadia-mineralizer mixture. The irregular white material is probably silica particles adhered to the surface of the crystal. The faceted particles may be crystals of the product nucleated on the zirconia surface, or else they may be facets of the surface of the crystal itself.

The same corner of the single crystal shown in Fig. 9 was examined by an electron probe at Biodynamics Research Corp. Rockville, Md. The α X-ray image for silicon corresponds exactly with the development of the blue product layer. This sharp delineation is

FIG. 10
Single crystal experiment, micrograph of the reaction surface.

strong evidence that transport of the silicon ions is the rate limiting step in producing the stain.

On the other hand, the vanadium K α X-ray image shows that vanadium has penetrated much deeper into the zirconia crystal. The vanadium concentration is higher in the product layer than elsewhere, but it is nevertheless appreciable at some distance from the product layer.

C. Reaction Kinetics

Still further information concerning the mechanism of the transport across the product layer can be obtained from a study of the thickness of the product layer as a function of time, under isothermal conditions[16]. If the diffusion mechanism is operative the mathematical formalism developed by Schmalzried[18] applies, and:

$$(\Delta x)^2 = kvt \tag{6}$$

where Δx = thickness of the product layer; t = time; v = specific volume of the product; and k = the rate constant.

To test this equation for an iron doped stain, a series of marker experiments using the same materials as the experiment shown in Fig. 5 were set up in 1.3 ml crucibles, and were fired at 1120°C for 1, 2, 4, 8, and 16 hr, respectively.

Without unloading, each crucible was mounted in a plastic, sectioned and polished for microscopic examination. Using a built-in reticle, the thicknesses of the product layers were measured. The results are plotted in Fig. 11. A very satisfactory straight line is obtained, which is evidence that a diffusion process is operable. From the slope of the line, the reaction rate constant is found by Eq. (6) to be equal to 4.3×10^{-10} moles/cm second.

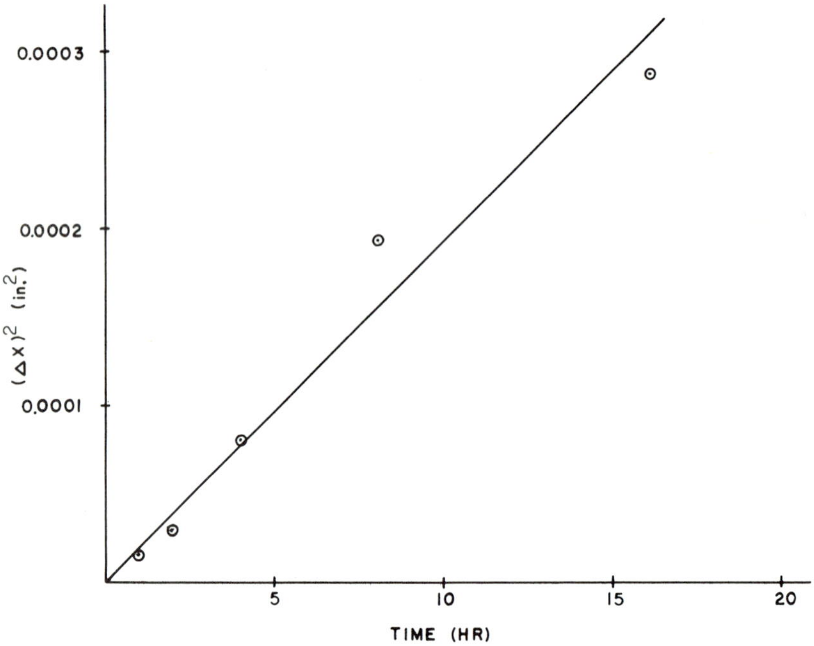

FIG. 11
Reaction layer thickness as a function of firing time.

Schmalzried[18] has also shown that the reaction rate constant can be related to the appropriate diffusion coefficient by:

$$k = c_A \bar{D}_A \left[\frac{n\Delta G}{RT}\right] \tag{7}$$

provided that a diffusion mechanism is operative. In this equation the subscript A refers to the rate limiting species which the electron probe data indicate is the silicon ions. The number n depends upon the nature of the defects which predominate in the product layer and are effective in facilitating ionic transport; also upon the valences of the several cations[23]. A method for calculation of n has been derived by Schmalzried[24] from the application of the principles of stoichiometry and thermodynamics to the applicable defect chemistry.

Because both silicon ions and zirconium ions have valence 4; and because they react in the ratio 1:1, n = 1 for practically any reasonable selection of defect pairs in the product layer. The free energy of the reaction can be estimated from standard free energy data and heat capacity data on zircon[25], zirconia[26], and silica[26] and is found to be -16.8 kcal/mole. Thus for motion of silicon ions: $D_{Si} \simeq 3 \times 10^{-9}$ cm^2/second at 1120°C, if a diffusion model applies.

D. Discussion

The results of the study of the thickness of the reaction layers as a function of time support the diffusion model (Eqs. 3 and 4) for the transport of silicon ions through the product layer. The square of the layer thickness is found to vary linearly with time.

On the other hand, it has been shown[16] that the structure of the product layer is very open, with large interconnected interstices. This observation would appear to make the vapor transport model possible.

In the study of the single crystal it was found that the thickness of the reaction layer in powdered zirconia was approximately twice that in the single crystal. This indicates that perhaps the most reasonable conclusion is that both processes occur simultaneously. That is, as the silicon tetrahalide reaches and passes through the product layer it reacts in substantial amount with that product layer, resulting in the formation of silicon ions, and electrons, which then proceed to the zirconia interface by a diffusion process. On the other hand, a percentage of the silicon tetrahalide does manage to pass entirely through the product layer, and reacts directly with the zirconia.

V. SUMMARY

Zircon is used as the host for a family of ceramic colors because it can be doped to give suitable colors and it is resistant to dissolution in molten glass. These colors are prepared by calcining a mixture of zirconia, silica, a chromophore, and one or more mineralizers. The initial step in the reaction is believed to occur between the silica and the mineralizer, producing a volatile species, probably a silicon tetrahalide. The volatile species transports the silicon ions to the surface of the product layer.

Two simultaneous processes occur in transporting the silicon ions across the product layer to the interface with the other reactant, ZrO_2. In part, the silicon tetrahalides react with the product layer, placing silicon ions and electrons in defects therein. A flux of the latter can then diffuse through the product layer to the interface with the zirconia. Simultaneously some of the silicon tetrahalide passes through interstices in the product layer to react directly with the zirconia.

REFERENCES

[1]. W. A. Weyl, _Coloured Glasses_, Society of Glass Technology, Sheffield, England, 1951.

[2]. L. Tcheichvili and W. A. Weyl, "The Synthesis of Ceramic Pigments," _Glass Ind._, 44(1) 24-5, 49, (2) 82-3, 86, 104, 107-8, (3) 145-8, 173, 175, (4) 208-9, 234-5, 1963.

[3]. R. A. Eppler and W. A. McLeran, Jr., "Kinetics of Opacification of a Titania Opacified Cover Coat Enamel," _J. Am. Cer. Soc._, 50(3): 152-6, 1967; R. W. Eppler, "Crystallization and Phase

Transformation in Titania Opacified Porcelain Enamels I," J. Am. Cer. Soc., 52(2): 89-94, 1969; R. A. Eppler, "Crystallization and Phase Transformation in Titania Opacified Porcelain Enamels I," J. Am. Cer. Soc., 52(2): pp. 94-99, 1969.

[4]. W. D. J. Evans, "Ceramic Pigments: A Structural Approach," Trans. Brit. Cer. Soc., 67: 397-419, 1968.

[5]. C. A. Seabright, Ceramic Pigments, U. S. Patent 2,441,447, May 11, 1948.

[6]. C. A. Seabright, Blue Pigments for Ceramic Ware, U. S. Patent 3,025,178, Mar. 13, 1962.

[7]. C. A. Seabright, Yellow Pigments for Ceramic Ware, U. S. Patent 2,992,123, July 11, 1961; C. A. Seabright, Yellow Ceramic Pigments, U. S. Patent 3,012,898, Dec. 12, 1961.

[8]. C. A. Seabright, Iron Ceramic Pigment, U. S. Patent 3,166,430, Jan. 19, 1965.

[9]. J. E. Marquis and R. E. Carpenter, Pink Ceramic Stain and Process, U. S. Patent 3,189,475, June 15, 1965.

[10]. J. P. Bruneau, Praseodymium, Cerium Yellow, U. S. Patent 3,168,410, Feb. 2, 1965; F. T. Booth, Improvements in Ceramic Stains, Brit. Patent 983,175; H. D. Jamieson, Ceramic Glaze Pigment, U. S. Patent 3,046,150, July 24, 1962; J. K. Olby, Zirconium Stain for Ceramic Glazes, U. S. Patent 3,171,753, Mar. 2, 1965; J. K. Olby, Vanadium-Zirconium Stains for Ceramic Glazes, U. S. Patent 3,058,838, October 16, 1962; W. A. Weyl, Ceramic Pigments, U. S. Patent 2,875,086, Feb. 24, 1959.

[11]. C. A. Seabright and H. C. Draker, "Ceramic Stains from Zirconium and Vanadium Oxides," Bull. Am. Cer. Soc., 40(1) 1-4, 1961.

[12]. F. T. Booth and G. N. Peel, "The Preparation and Properties of Some Zirconium Stains," Trans. Brit. Cer. Soc., 61(7) 359-400, 1962.

[13]. K. Shaw, "Color Chemists Get Down to Fundamentals;" Ceramics, 17(207) 22-24, 1966; 17(208) 33-4, 6, 1966.

[14]. V. I. Matkovich and P. M. Corbett, "Formation of Zircon from ZrO_2 and SiO_2 in the Presence of V_2O_5," J. Am. Cer. Soc., 44(3) 128-30, 1961.

[15]. T. Demiray, D. K. Nath, and F. A. Hummel, "The Zircon-Vanadium Blue Pigment," submitted to J. Am. Cer. Soc.

[16]. R. A. Eppler, "The Mechanism of Formation of Zircon Stains," presented at the Fall Meeting, Whitewares Division, American Ceramic Society, for abstract see Am. Cer. Soc. Bull., 48(8); 819, 1969.

[17]. A. S. Bystrikov and B. S. Cherepanov, "X-ray Study of the Formation of Zircon in the System SiO_2-ZrO_2-V_2O_5," Russ. J. Inorg. Chem., 9(5) 654-6, 1964.

[18]. H. Schmalzried, "Solid State Reactions," Angew, Chem. Intern. Ed. Engl., 2(5) 251-4, 1963.

[19]. H. Schmalzried, "Solid State Reactions," Ber. Deut. Keram. Ges., 42(1) 11-22, 1965.

[20]. J. A. Hedvall, Solid State Chemistry, Whence, Where and Whither, Elsevier, Amsterdam, 1967.

[21]. C. Kooy, Material Transport in Solid State Reactions, Pure and Applied Chem., 9(3) 441-59, 1964.

[22]. H. Schmalzried, "Chemical Reactions Between Crystalline Solids," in Reactivity of Solids (J. W. Mitchell, R. C. De Vries, R. W.

Roberts, and P. Cannon eds.), Wiley-Interscience, New York, 1969.

[23]. R. E. Carter, "Mechanism of Solid State Reaction Between Magnesium Oxide and Aluminum Oxide and Between Magnesium Oxide and Ferric Oxide," J. Am. Cer. Soc., 44(3), 116-20, 1961.

[24]. H. Schmalzried, "Point Defects in Ternary Ionic Crystals," Prog. Solid State Chem., 2:265-303, 1965.

[25]. Mr. Donald Wagman, National Bureau of Standards, Gaithersburg, Maryland, private communication. To be published in National Bureau of Standards publications, series 270, issue 5.

[26]. --, Handbook of Chemistry and Physics, 45th ed., Chemical Rubber Publishing Company, Cleveland, Ohio, 1964; A. Goldsmith, T. E. Waterman, and H. J. Hirschhorn, Handbook of Thermophysical Properties of Solid Materials, the Macmillan Co., N. Y., 1961.

Chapter 32

COMPUTER COLOR CONTROL FOR CERAMIC WALL TILE

William K. Culbreth, Jr.
Florida Tile Industries, Inc.
Lakeland, Florida

In the use of ceramic wall tile, the problems of color control are most emphasized. The nature of a tile wall is such that a large area is made up of similar, distinct parts. These parts form a regular and unvarying pattern. Any variation in color is amplified when the differences are adjacent to one another. An ordinary installation is made up of many shapes. In addition to the flat wall tile, there are the trim pieces; caps, bases, and angles. The nature of the manufacturing technique dictates that each of these items be produced at different times, and in most cases, on different equipment. The American tile market demands a few installations which show a random variation in color; however, the majority of the installations are required to be perfectly uniform in color. In times past, the American public was satisfied to have their bathroom walls in one color and the trim in another. This is no longer the case. The overwhelming majority of the installations today use the same color for wall, trim and angles.

The tile manufacturer engages in two kinds of schemes to satisfy these needs. The first is a before-the-fact scheme in which the manufacturer attempts to produce a perfectly uniform product. The second is an after-the-fact scheme, whereby the manufacturer does the best he can with the not so perfect product he has produced. This after-the-fact scheme involves the use of shade sorting techniques in which all the tile which look alike are given a code number. In most instances, this code number is a part of a system by which a knowledgeable person can decide whether two lots of tile with different numbers are close in color, or are distinctly different. The manufacturer will then attempt to schedule his shipments so as to minimize the color variation as seen by his customers. It is very easy to see the problems of high inventory, poor delivery, and dissatisfied customers that could result from inadequate color control in the manufacturing phase.

How then does the manufacturer go about minimizing his color problems? First of course is the glaze formulation. This involves both the selection of raw materials and the chemistry of the glaze. Even then a perfect glaze formula will not insure perfectly uniform color. Process control is another key. Process control is involved first with the maintenance of a uniform supply of raw material, then with formulation control, with glaze preparation, glaze application, and with firing. Each of these areas can and do effect color uniformity.

The particular aspect of control that we are concerned with in this paper is that of formulation control. In formulation control, we are concerned with the materials which go into a glaze and how they behave. In the context of this discussion, the formula should be taken to mean the makeup of the various stains used to achieve the color and does not refer to the makeup of the glaze itself. We are going to assume that all problems of color variation with which we are dealing, are caused by errors or variations in the formulation. In some respects, this is a ridiculous assumption. However, this assumption appears more valid when you consider that controls have been established on all the other factors which can contribute to color problems. Therefore, if all these other factors are under control, how else can you reasonably improve the situation except by adjustments in the formula?

The situation can easily be pictured as it would occur in the manufacturing plant. A batch of glaze has been prepared; the raw materials, including the ceramic stains, have been weighed and charged in a ball mill. Proper amounts of water have been added and the batch has been ground to the proper fineness. A sample has been taken from the batch and applied to a tile and fired in controlled conditions. After firing, this sample tile is compared to a standard and found not to match. Well, what can be done at this point? The obvious answer is that the color must be changed by either adding some more stain to the batch or by diluting with an uncolored base. The traditional approach for correcting a glaze color has been for a skilled and experienced technician to visually judge what is wrong with the color and to estimate which stains and how much are necessary to properly adjust the glaze. This man's batting average is as important to a tile company as the star hitter's batting average is to a baseball team. Modern industry, being what it is, should not risk a batting slump by such a person or the loss of him to another team. Neither can they afford keeping a second and third string man on the bench. It is, therefore, important to develop a system which is less risky.

The scientific description of colors can be done in any one of several different ways. The color description which is the most useful in color formulation and batch correction work is the spectrophotometric curve, which shows the reflectance value of a color across the visable spectrum; that is, from 400 to 700 nanometers. Each color has its own reflectance profile. Two objects having identical reflectance profiles would be perfect color matches under all conditions. It would be a reasonable objective then to minimize the difference between the reflectance curves of a production glaze and a standard sample, thereby assuring a good match.

Striving for this objective, a system has been devised whereby the difference between the standard color and the test color is minimized by simulating mathematically the addition and subtraction of stains to the test batch. The Kubelko-Munk analysis of absorption and scattering coefficients supplies us with the following relationships which are used in a simulation system:

$$K/S = \frac{(1 - R)^2}{2R}$$

$$K/S_{\text{Color}} = K/S_{\text{stain 1}} + K/S_{\text{stain 2}} + \ldots K/S_{\text{stain n}}$$

$$K/S_{\text{stain}} = M_{\text{stain}} \times C_{\text{stain}}$$

$$\Delta K/S_{\text{Color}} = (M_1 \times C_1) + (M_2 \times C_2) + (M_3 \times C_3) \ldots (M_n \times C_n)$$

$$R = 1 + K/S - (K/S^2 + 2\ K/S)^{\frac{1}{2}}$$

Where

K/S is the ratio of absorption and scattering coefficients.

R is the fraction of light reflected at a specific wavelength.

C is the concentration of a stain or pigment.

M is the slope of the relationship between K/S and concentration.

In the simulation system, the difference between the test and the standard are first calculated. Then, a small increase in one of the pigments is simulated. The effect of this increase on the reflectances of the test batch is calculated and the difference between the standard and the computed reflectance is compared to the original difference. If the new difference is less than the original, further increases in concentration are made. This process will be repeated over and over again until there is no further improvement in the match. If no improvement in the match can be made by increasing the stain concentration, we try decreasing the concentration. As before, this reduction in concentration is repeated until there is no improvement. We then move to the next pigment in the formula and perform a similar operation. Therefore, the process is an attempt to reduce the difference by increasing stain and then by decreasing stain. This process is repeated for all the stains in the formula. This simulation requires a great deal of calculations consequently it is profitable to utilize a time-sharing computer to do this work. Stored in the computer's memory is all of the reflectance data on the colors in production. Also stored in the computer's memory is a set of instructions concerned with doing all the calculations which have just been described. Now, each time that a glaze batch is offshade, we can supply the name of the color that it is supposed to be and the reflectance curve of what it is in actuality. With this information, the computer can execute the instructions which it has in memory and supply us with the changes in stain concentration required to correct the glaze batch.

The use of the time-sharing computer, rather than a service bureau or an in-house computer, is simply a matter of convenience and economics. The nature of a color correction problem calls for an answer as soon as possible. With a time-sharing computer, you can have instant access to the computer from any location in the plant that you desire to install a teletype terminal. Any number of different locations can have access to the program and computer files. The time-sharing system provides access to an extremely powerful computer by users who cannot afford and in fact have no use for a fulltime in-house computer.

How then is this system actually used? Whenever a new batch of glaze is prepared, a sample of this glaze is applied to a tile and fired in the normal fashion. When the tile comes from the kiln, it is inspected by a color technician. This person either approves the match and the glaze goes into production, or else the batch is rejected. In the event of a rejection, the simulation system comes into play. The technician reads 16 data points on an abridged spectrophotometer, thus determining the spectrophotometric curve. This entire process takes approximately two minutes. The technician then proceeds to enter his problem into the computer by the use of a teletype. He types the following information: first, the batch number, simply an identifying number; second, the color name that he is trying to match, such as regency blue, seafoam green; third, the

16 data points he has read from the spectrophotometer; and fourth, the specific gravity of the glaze. The computer then performs the necessary calculations and finally supplies him with an answer such as "add 10 gallons of white glaze at a specific gravity of 1.91, 2.03 pounds of stain 1, and 1.23 pounds of stain 2." This final printout from the computer is all the technician needs to correct the glaze batch.

We have talked about the problems we have with color uniformity and we have presented a system which is supposed to help us with these problems. This is a system which certainly sounds impressive to a production oriented manager. It uses computer input and output. It uses the time-sharing concept, which is certainly one of the latest things in computer technology and it uses simulation technique which also sounds very up-to-date. What effect has all this had on our actual operation? It has had three main effects. First, there are faster matches. Whenever there is a problem with a glaze, there is now much less time between when the glaze is prepared and when it can actually be used in production. This allows us to reduce our glaze-in process inventory and to operate on a tighter production schedule. The second effect is that fewer man hours are required in making this correction. The third effect is that we have been able to institute tighter process and quality controls in manufacturing These tighter controls have resulted in lower production costs, primarily through reduced inspection. It has resulted in lower finished goods inventory requirements and in more uniform product for the customer.

X. CERAMICS AND GLASSES IN COMMUNICATIONS

Chapter 33

CERAMICS AND GLASSES: SOME USES IN THE COMMUNICATIONS INDUSTRY

D. G. Thomas

Bell Telephone Laboratories
Murray Hill, New Jersey

I. INTRODUCTION

This conference on which this volume is based was concerned chiefly with electronic phenomena in ceramics and optical materials. Emphasis has been laid upon understanding the physics and chemistry of these substances. From the point of view of the modern industrialist it would be hard to overestimate the potential importance of such a body of knowledge. These materials are of widespread use, and as their applications become more sophisticated so does an understanding of their fundamental processes and properties become more critical. Components in modern communications systems must have high reliability and must be made in good yield. Reliability, especially, becomes more important as the complexity of the system increases. Reliability and yield depend upon understanding, for without it corrective action when trouble occurs is very hard to take. As an example certain effects have been described recently in glasses which hold promise as leading to electronic switches or memory elements. It seems unlikely, however, that these devices will gain widespread acceptance until their mode of operation is clearly understood, and recognized to be one that is intrinsically stable and reproducible.

In this talk two topics will be discussed. The first considers the application of ceramics to the Bell System Beam Lead-Sealed Junction integrated circuit technology. Here ceramic substrates are used to support thin film interconnection patterns and passive circuit elements, between active silicon chips. Such applications are becoming increasingly widespread and are by no means confined to the particular technology described here. The second topic is concerned with a more remote technology, one in which glass fibers may be used as transmission media for light beams. As is well known, light beams have the ability to carry enormous amounts of information and so a very broadband communications system can be conceived which might use such fibers.

The main message which runs through both these discussions is that benefits would accrue from improved chemical control and processing. However, new physical phenomena may well become relevant as the art develops. For instance, in both the thin film and the glass fiber fields, it is possible to contemplate that electrically and optically active elements may be incorporated into the structures. The successful attainment of such objectives would certainly require understanding of physical principles, and these would then lead on to structure design, and to device fabrication by chemical methods.

II. THE ROLE OF CERAMICS IN THE BELL SYSTEM'S USE OF INTEGRATED CIRCUITS

After the tremendous success of the transistor in revolutionizing the electronics industry, it is now clear that the next major advance in performance and cost reduction will come as many transistorlike devices are built onto one small piece of silicon, producing an integrated circuit. This is accomplished using the so-called planar technique in which windows are etched, using photolithographic techniques, in a protective layer of SiO_2. Diffusion of impurities takes place through these windows, and these areas also serve as contact points. A problem has been how to connect the silicon integrated circuits (or "chips"), which are usually less than 0.1 inch on a side, to the outside world. In addition if, for reliability, the chips have to be encased in hermetically sealed cans, how can economic advances be made if the can costs more than the chip? For the needs of the Bell System high reliability is absolutely essential, yet it must be combined with economy so that the overall result represents an economic advantage over existing art. For this purpose a materials system has been devised which includes thin layers of insulators deposited over the critical junction regions, which allows the silicon chips to be used without the necessity of a sealed can. The electrical junctions are now "sealed." In addition, the electrical lead problem is dealt with by fabricating the chips with tiny gold "beam leads" at the perimeters, which serve the function of providing both electrical connections and mechanical support for the chips. On the chips, the beams are connected by metallization patterns to the active silicon contact regions. Altogether this has been referred to as the "Beam Lead-Sealed Junction" technology. The chips are finally protected from gross water drops by a thin layer of plastic.

Outside the chips the beam leads must be joined to interconnection patterns, and to other passive electrical elements such as resistors and capacitors. It is at this point that ceramics enter the picture. A thin metallic film containing tantalum and other metals is deposited onto a flat ceramic substrate and patterns are etched in this film using photolithographic techniques. With suitable processing, resistors and capacitors can be fabricated from the tantalum film. Finally the beam leaded chips are bonded to the metallization pattern. Figure 1 illustrates a beam leaded chip and a tantalum circuit to which some chips have been bonded. Again the

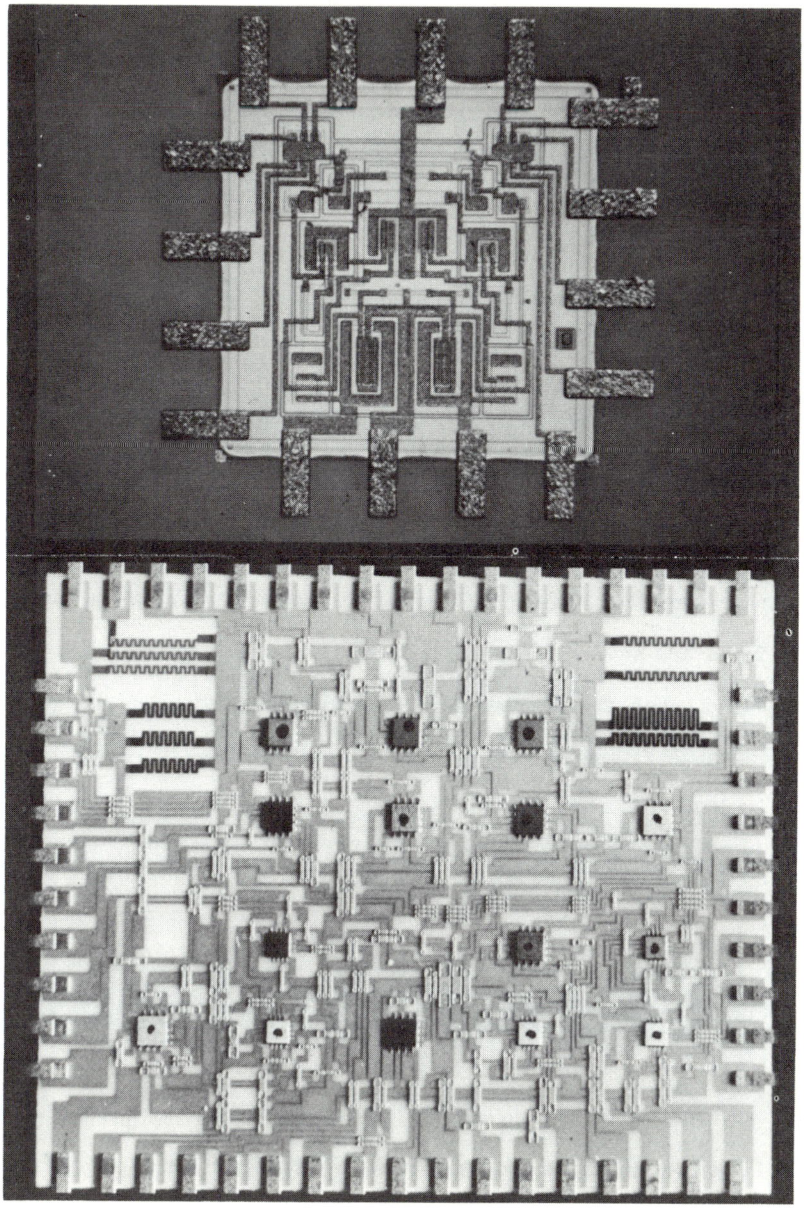

FIG. 1
On the bottom is shown an alumina substrate 2.8 × 3.3 cm which is used in local telephone switching equipment. It supports interconnection patterns with resistors and crossovers. In addition, there are 15 beam leaded silicon chips bonded to the interconnection pattern. On the top a magnified view of a silicon chip is shown with the circuit side up. The beam leads project over the edge of the chip.

tantalum circuit is usually covered with a protective plastic layer. Typically the ceramic substrate circuits have an area of a few square inches, and may contain half a dozen silicon integrated circuits. The ceramic boards may be mounted directly in telephone apparatus, for instance in the modern TOUCH-TONE[R] handsets, or may be mounted on printed circuit boards to form part of a more complex system.

This combination of beam lead-sealed junction silicon devices with tantalum thin film structures mounted on ceramic forms an extremely reliable and adaptive materials system. It is very attractive for telephone use, and it is rapidly being adopted for that purpose.

A. Properties Required of Ceramics for Electronic Use

There are several conditions which a material must meet if it is to be successfully used as a substrate. Some of these are outlined in Table I, and can be discussed in the following terms.

1. Surfaces must be smooth. One reason for this is that as the complexity increases, very fine interconnection lines become necessary. In one circuit currently under development there must be continuity over 150 inches of lines, 0.002 inch in width, with 0.002 inch between the lines. For this purpose, surface smoothness of about 5 microinches peak to valley is desirable. It must be emphasized too, that while the average microsmoothness is important, it is also vital that there should not be occasional large defects, for these cannot be tolerated by long thin line patterns. This draws attention to the importance of care in the preparation of raw material and to cleanliness in manufacture, so that large particles of the raw material or of some foreign substance are rigidly excluded.

TABLE I

SUBSTRATE REQUIREMENTS

1. Surface must be smooth

   5 Microinches peak to valley for interconnections
   1 Microinch peak to valley for capacitors

2. Must be chemically stable

   Alumina resists hf, most glasses do not

3. Must have high thermal conductivity

   High alumina ceramics are 25 times better than most glasses

4. Strong, easily cut, good dimensional tolerances

Even greater demands are made if tantalum circuits are made which include capacitors. These capacitors are made by anodizing tantalum to form an insulating oxide of the order of 2000 Å in thickness. A counter-electrode is made, usually by the evaporation of gold. In order to obtain tolerable yields an extremely smooth substrate surface is necessary, for otherwise high stress points are formed which lead to short circuits through the very thin oxide layer. It has been found that sharp discontinuities greater than 1 μ inch (250 Å), peak to valley, are harmful. Drawn glass or fused silica may be smooth enough for this purpose but for other reasons are not satisfactory. At present Al_2O_3 substrates are generally favored, but the as-prepared surfaces are not exceptionally smooth and the very best ceramics can only allow capacitors to be made in low yield; there is nevertheless hope that this situation will improve as time goes on. At present it is usual to employ glazed alumina when capacitors are made; however, one cannot use a ceramic glazed overall chiefly because if beam leads are bonded to a metallization layer laid down on glaze, there is a tendency for the glaze to crack. So the current practice is to spot glaze just where capacitors are to be made.

2. The substrates must resist attack by etches used to delineate the metallization patterns, and also by the conditions which obtain during the anodizing processes. These etches may include hydrofluoric acid so that soft glass substrates are not acceptable.

3. An important consideration is high thermal conductivity of the substrates. Heat must be removed from the silicon chips, and also from elements such as resistors. The thermal conductivities of glasses including quartz are in the range 0.002 to 0.004 cal/cm/second/°C, while that of high alumina ceramic is near 0.1 cal/cm/second/°C at 25°C, some 25 times greater. This, and its chemical inertness, are among the chief reasons for using pure Al_2O_3 as the substrate for critical applications. (BeO has a thermal conductivity of 0.5 cal/cm/second/°C at 25°C and some day may come into use.)

The importance of thermal conductivity, and also of chemical purity, are illustrated by the aging data of over-stressed tantalum nitride resistors shown in Fig. 2. The dc load here applied of 60 W/inch raises the temperature of the resistors for the top three curves to above 250°C, at which temperature Na^+ ion migration becomes possible in the glass. The Na^+ ions are very detrimental to the resistors which rapidly degrade; for substrates with less Na^+ there is less degradation. For the 7059 glass which is very low in Na^+, it is believed that the aging mechanism is the high temperature

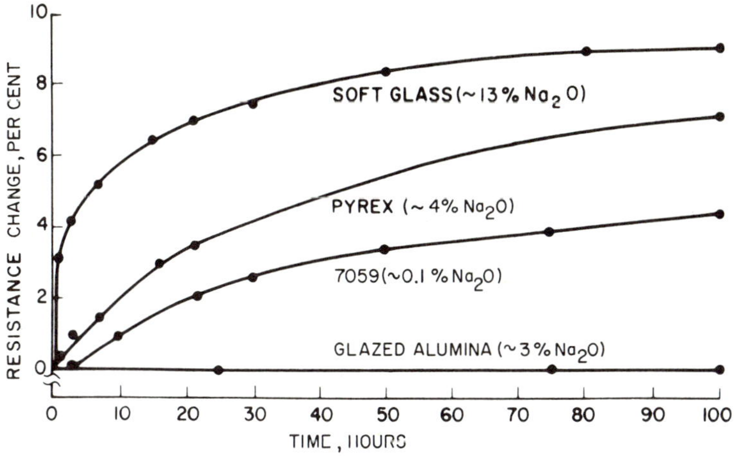

FIG. 2
The aging of thin film resistors laid down on different substrates. The resistors are overstressed by heat dissipation, and degradation is indicated by increase of resistance. Glazed alumina is a good heat conductor and so allows very little temperature increase and very little degradation.

oxidation of the resistor. When glazed Al_2O_3 is used, with its much higher thermal conductivity, the samples are cooler and there is no appreciable degradation, as seen in the lowest curve. Infrared and other direct measurements have shown that on overload the circuit elements can run 100°C hotter on glass than on alumina.

4. Some other advantages offered by high grade alumina, that is, material of greater than 99.5% purity, are as follows. Alumina is much stronger than glass and it can be readily cut, machined, and polished so that high dimensional tolerances can be maintained. For some purposes it is desirable to have accurately positioned holes in the substrates and this is not easy with glass, but drilling, and particularly laser drilling, can be readily applied to alumina. The advantage of laser drilling is that holes can be made in fired ceramic, rather than in the green or partially fired material for which there will be subsequent shrinkage. Indeed it was laser machining that

removed one of the barriers in the way of widespread use of alumina for tantalum circuits, namely, the requirements for easy separation of the circuits after several had been made together on a large ceramic piece in a batch fabrication process. One technique that is now used is to focus a continuous laser beam onto a ceramic surface, not melting it, but rather creating thermal stresses which produce a small fracture. This controlled fracture process allows one to cut ceramic to almost any desired shape, as shown in Fig. 3. Using a power of 100 W from CO_2-N_2-He laser, it is possible to controllably fracture high alumina ceramic 0.025 inch thick at a rate of 120 inches/minute[1].

At present, therefore, high alumina ceramic is the material of choice and in the next section some recent work in the preparation of this material is described.

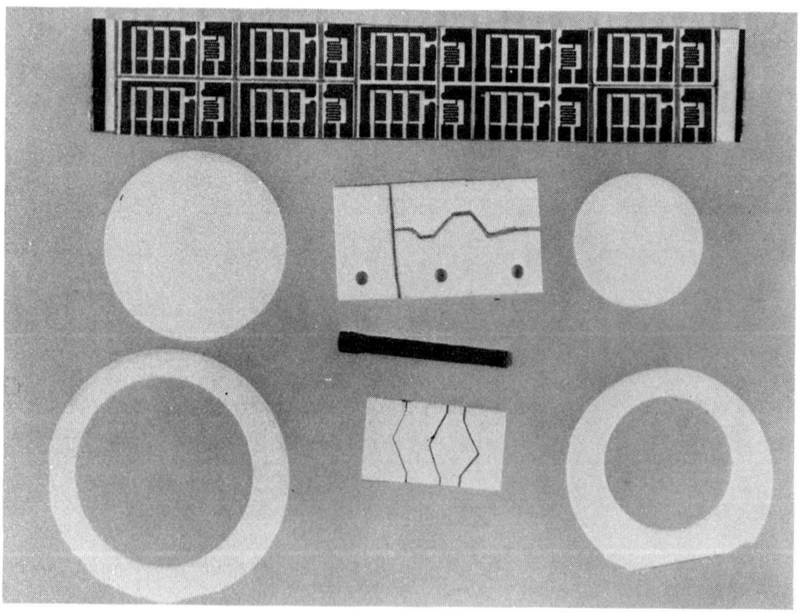

FIG. 3
Various ceramic shapes cut out by laser machining.

B. Recent Progress in the Preparation of High Al_2O_3 Ceramics

There are currently several investigations under way in the Bell Telephone Laboratories and in the Western Electric Co. aimed at improved processing and understanding of high Al_2O_3 ceramic suitable for the substrates just described.

1. Superstrate

Perhaps the most straightforward approach is that of the Western Electric Engineering Center at Princeton, N. J. There, a "doctor blading" method is used to produce a material called "Superstrate." It starts with commercial dry-ground alumina powder which is further ground in an organic medium until the surface area is about 16 m^2/gm. After firing at 1450°C a density of about 3.88 gm/cc is obtained. This is the average density. The surface density down to a depth of a few microns is greater than this, and a very smooth surface results. In fact, using a Talysurf machine to measure surface roughness a "center line average" of 2-3 microinches has been obtained for this material. This is better than can be obtained commercially as is shown by the two microphotographs in Fig. 4. It is encouraging that tantalum capacitors can be made directly on this surface; however, the yields are not very high indicating that surface irregularities are still troublesome. Because the surface is denser than the bulk, the surface properties are not improved by grinding or polishing.

2. Raw Materials from Freeze Drying

An extremely important factor in the preparation of any ceramic is the raw material [2]. This is particularly true of modern ceramics for electronic use. These materials, unlike older materials based on clays, do not form a glass phase in their manufacture which can serve

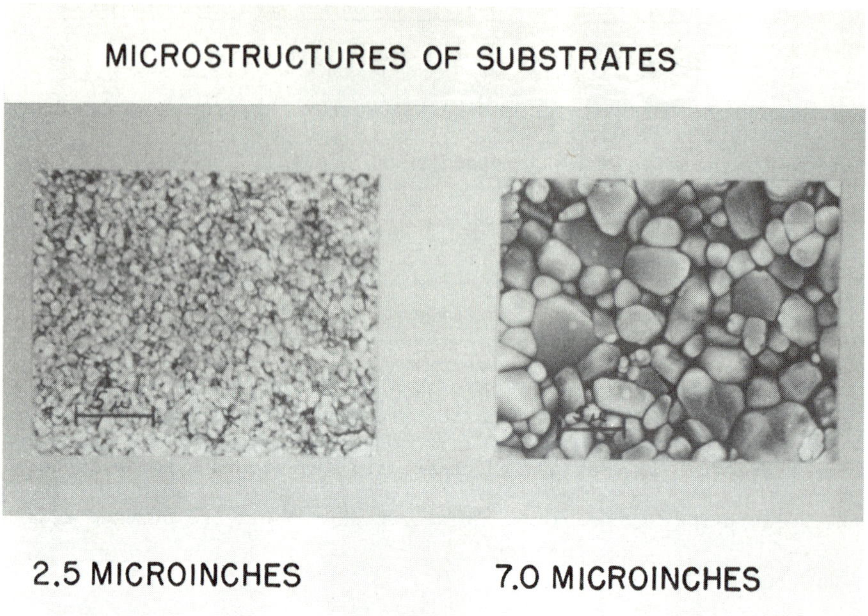

FIG. 4
On the left the surface of "superstrate," as compared with a good commercial ceramic surface. The "center line average" surface roughness is indicated in microinches.

to bind together many particles of variable composition and size. This means that there is at no point a liquid phase, and transport of matter occurs solely by solid state diffusion. Thus, control of the starting materials and understanding and control of the subsequent sintering process is of great concern[3].

One method of raw material preparation which holds promise for the future is the cryochemical or freeze drying process described by Schnettler, Montforte, and Rhodes[4]. In this method a solution containing the cation of the ceramic desired [for instance $Al_2(SO_4)_3$ solution to produce Al_2O_3], is forced through an orifice 10 mils in diameter, which is under the surface of a stirred bath of hexane cooled by a bath of acetone and dry ice. Rapid freezing occurs with extensive nucleation, and a large number of remarkably uniform spherical particles fall to the bottom of the freezing liquid. The spheres are kept cold so that no melting occurs and the water is subsequently removed by sublimation in a vacuum. The dry particles, still spherical in shape, are then calcined to produce the oxide, and again spherical particles result with diameters very similar to the original sphere diameter. These particles are easy to handle and are dust free; they are very porous and friable and may be easily ground up to give a very uniform powder. Figure 5 shows on the left the frozen particles of $Al_2(SO_4)_3$ solution, and on the right particles of Al_2O_3 made from them. The rapid freezing means that even if the solution contains several components these will be precipitated out extremely uniformly. On drying, a solid will result which will correspond precisely to the original composition of the solution.

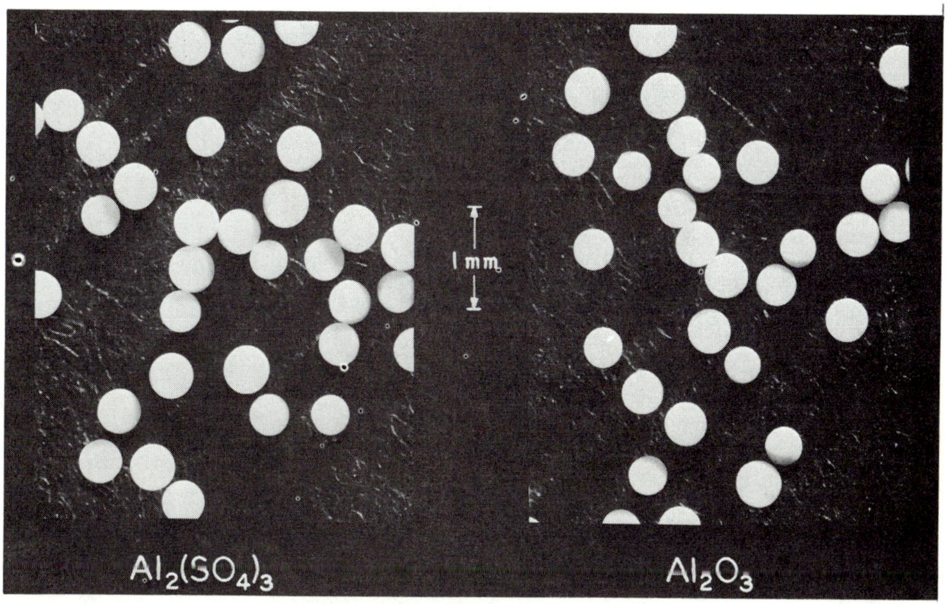

FIG. 5
Freeze dried spheres of $Al_2(SO_4)_3$ on the left, and spheres of Al_2O_3 resulting from calcining on the right.

The uniform mixing that necessarily occurs in the solid means that solid state reactions will take place at unusually low temperatures. Thus, magnesium aluminate spinel is formed from a freeze-dried solid, $MgAl_2(SO_4)_4$, at a temperature below the decomposition temperature of either $MgSO_4$ or $Al_2(SO_4)_3$, and 400°C below the temperature of reaction of conventionally mixed oxides.

Johnson and Schnettler have recently completed a study of the formation of Al_2O_3 by the freeze-drying method and have microscopically followed the topology of the process. Figure 6 shows a

FIG. 6
Freeze dried $Al_2(SO_4)_3$ particle.

scanning electron micrograph (SEM) of a freeze-dried particle of $Al_2(SO_4)_3$. Note the radial structure preserved from the radial freezing process. Figures 7 and 8 show transmission electron micrographs of the resulting Al_2O_3 after varying degrees of calcining. Eight hours at 1200°C give crystallite sizes of 1500 ± 400 Å; after 17.5 hr at 1400°C the crystallites have a size of 4500 ± 1500 Å. The particles tend to form in chains and these chains can be traced all the way back to the initial radial freezing pattern. The

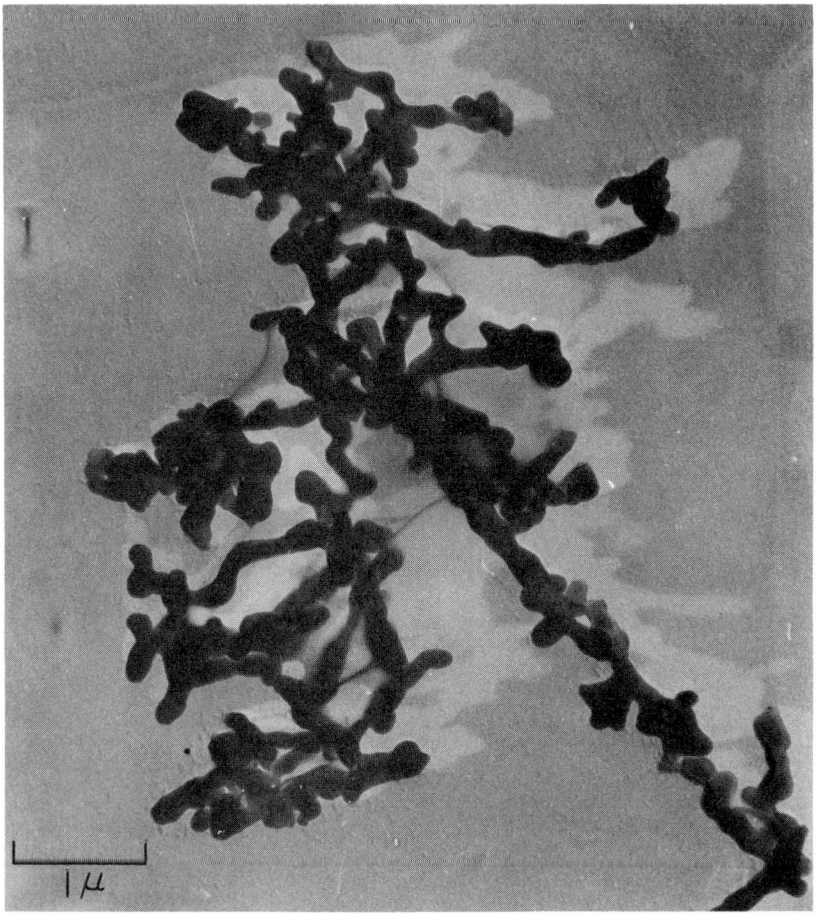

FIG. 7
Al_2O_3 (8 hr at 1200°C) from freeze dried $Al_2(SO_4)_3$.

FIG. 8
Al_2O_3 for 17.5 hr at 1400°C.

particles in the chains are found to grow in diameter as a power of time between 1/5 and 1/7, which indicates that by the time measurements can be made the particles have grown almost to their ultimate size, this size in effect being limited by the huge porosity (95—97%) of the freeze-dried oxide. It seems likely that the growth of the particles occurs by surface diffusion rather than by bulk diffusion since, (a) the surface to volume ratio is very large with very little interparticle contact, and (b) the absence of bulk skrinkage during sintering is compatible with surface diffusion, while bulk diffusion is not.

3. An Approach to Producing Theoretically Dense Polycrystalline Alumina

The cryochemical method as we have seen is, when applicable (and this usually means being able to get suitable aqueous solutions at the beginning of the process), an admirable way of obtaining fine uniform raw material for ceramic fabrication. It has been used by Y. S. Kim and F. R. Montforte at Bell Telephone Laboratories in an attempt to produce high density Al_2O_3. Such high density Al_2O_3 might be expected to have a very smooth surface, one of the prime objectives of the program. To obtain high densities, initial compaction is important. Probably the best compaction method is isostatic pressing in which the freeze-dried powder is carefully loaded into a shaped rubber bag; the bag is sealed and then subjected to a hydraulic force from a liquid. This even pressing results in uniform compaction which minimizes cracks, strains, and laminations in the subsequently sintered piece. Investigation has shown that this is a widely applicable process—it has been used successfully to produce shapes of garnets and ferrites for microwave purposes. It is important that the initial powder should be "dry." Binders or waxes cannot be used, at least for large pieces, as they become sealed into the body and cause cracking when they are subsequently decomposed during sintering. Kim and Montforte have produced alumina ceramic this way by isostatically pressing the powder at 20,000 psi into bars with linear dimensions of the order of inches. The bars are presintered at 1120°C for two hours in air. At this point they are in the "green" condition and may be conveniently sliced. Final sintering can be done at 1700°C for one hour in wet hydrogen gas. The slices are then ground and polished as desired.

In preparation of experimental quantities of this material one further refinement was added. In order to produce high density ceramic it is essential to eliminate the pores which inevitably result from the pressing together of the powder. These pores can be eliminated only by diffusional transport of matter to the pores from grain boundaries, and this will occur at maximum rate when the pores lie on the grain boundaries. If the grain boundaries move away from the pores, so that there are voids within a grain, then the elimination of these pores will be slower and may almost cease. One way to deal with this problem is to reduce the grain boundary mobility so that there is no breakaway of the grain boundaries from the pores. Some years ago Coble[5] showed that pore free grains of Al_2O_3 were produced in material which contained a small quantity of MgO in solid solution. It seemed likely that this came about through reduced mobility of the grain boundaries, and subsequently Jorgensen[6] showed directly that the rate of grain growth in MgO doped with Al_2O_3 was slower than in pure Al_2O_3. It is not absolutely clear why this is so, although it is probable that a second phase involving Mg concentrates at the grain boundaries and inhibits their motion.

An investigation of the optimum concentration of MgO to use resulted in the data given in Fig. 9. In this figure the calcining conditions refer to the temperature at which the freeze-dried material was fired; at 1400°C a particle diameter of a few thousand angstroms was attained. A maximum density of 3.995 gm/cc was achieved at a MgO concentration between 0.1 and 0.25 wt%. The density of sapphire is 4.00 gm/cc. Figure 10 compares the densities of a variety of high density aluminas and it can be seen that the BTL alumina most closely approaches the theoretical value. After

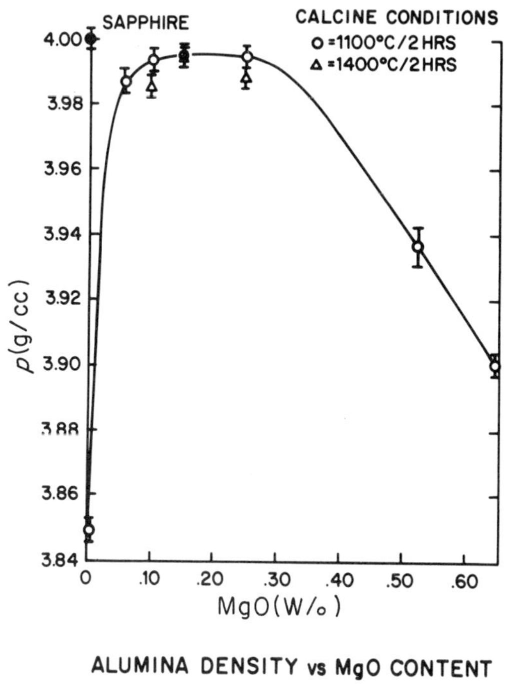

FIG. 9
The effect of the addition of MgO on the density of Al_2O_3 ceramic.

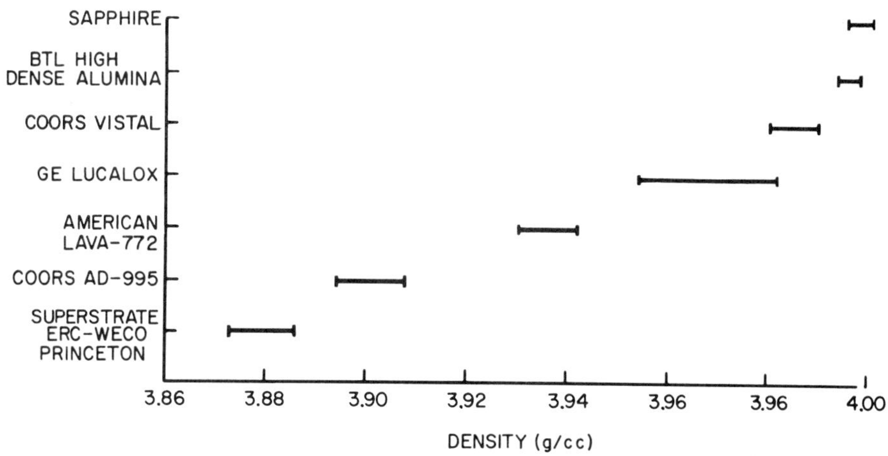

FIG. 10
The density of some alumina bodies.

polishing, a surface smoothness of about 200 Å (i.e., less than one microinch) peak to valley was obtained, considerably better than the best commercial product, and micrographs revealed essentially pore free grains, in the size range of one to six microns. However, occasional intergranular pores or irregularities, limit the usefulness of this material in its present state of development as far as capacitor substrates is concerned. It seems likely that this trouble can be traced to agglomerated oxide particles in the material prepared for sintering. Recent results in which the agglomerated particles are first removed by screening has indeed reduced the number of irregularities.

The material has a high degree of transparency, but is not as good as pure sapphire. This imperfect transparency is probably connected with the fact that since the individual grains of Al_2O_3 are doubly refracting each grain boundary is a light scattering interface. It is interesting that yttrium oxide which is cubic and so does not have light scattering grain boundaries can be fabricated into pore free ceramic which is as transparent as glass, and has the additional advantages of good transmission in the infrared and a melting point of 2400°C.

So far, freeze drying and isostatic pressing of ceramic raw material has been carried out on a laboratory scale only, but for the production of specialized ceramic pieces it is expected to be economical. J. Strasser, F. J. Schnettler, and F. R. Monforte have estimated the total cost of solution preparation, freezing and freeze drying of magnesium aluminum spinel. Excluding the cost of raw material and of calcining, and assuming a daily production of

970 pounds of the spinel, the estimated cost was found to be 64¢ per pound. While similar estimates are not yet available for the isostatic pressing process, there is no reason to suppose that exorbitant expenses will be incurred.

III. GLASSES AND TRANSMISSION AT OPTICAL FREQUENCIES

It is well known that coherent light beams generated by lasers are, in principle, capable of transmitting enormous quantities of information. It is too early to say when such an optical transmission system will become technically and economically feasible. Indeed the answer to this question will depend in part upon the availability of components for the system. Many of these components are currently thought of as being fabricated from glasses, and there is no doubt that this technology will be dependent upon the quality with which quite complex glass structures can be made. We will describe here some of the current ideas and problems in this field, for as time goes by they may become of increasing importance and may engage more and more of the attention of the materials scientist who is concerned with the noncrystalline state.

A. The Transparency of Glasses

The most obvious use of glass is as an optical transmission medium. Light can of course be transmitted through air or vacuum with low loss but even the most precisely collimated beam will gradually spread out in space due to diffraction effects and so refocusing devices such as lenses—perhaps gas lenses—will be required at quite finite intervals. In addition, light transmission in air requires very precise alignment of the optical components of the system, and this may present reliability problems with unfortunate economic consequences. One can avoid these difficulties by using flexible glass fibers as the transmission medium, but the first difficulty one now encounters is high loss in the glass.

B. Bulk Glass

Losses in bulk glass can be caused by scattering and by absorption. Providing there are no gross defects, Rayleigh scattering due to room temperature thermal fluctuations of density might be expected to dominate the scattering. The magnitude of this scattering may be calculated from a knowledge of the refractive index and the isothermal compressibility of the glass. In fact, the observed scattering from glasses is 10 to 100 times as large as the calculated values[7], and this has been explained by hypothesizing that concentration fluctuations which exist in the melt become frozen into the glass at high temperatures. Evidence in support of this comes from the fact that the scattering can depend on the cooling and annealing which a sample undergoes. Despite this extraneous scattering mechanism, Rayleigh scattering ratios as low as 1.6×10^{-6} cm^{-1} have been observed. (The Rayleigh ratio at 90° is given by

$$R_{90} = \frac{I\ r^2}{I_0\ V}$$

where I is the scattered intensity at 90° with incident unpolarized light, I_0 is the intensity of the incident beam, r is the distance from the point of observation to the scatterer, and V is the scattering volume). Such a ratio would lead to a loss of 12 db/km which is a loss rate that might be tolerable for some communication purposes.

Scattering, therefore, does not seem to be an inherent block to the use of glass as a transmission medium.

Any absorption in bulk glass at visible frequencies from causes intrinsic to the pure glass are expected to be very low. Instead absorption due to impurities has been shown to be dominant. Steel and Douglas[8] have studied the absorption due to ferrous and ferric ions in silicate and borate glasses. There can be absorption bands in the infrared and, weakly, in the visible regions of the spectrum, duc to optical transitions within the impurity atom itself modified by the crystal field, and there can be powerful absorption in the ultraviolet due to charge transfer transitions. The infrared and ultraviolet absorption peaks have tails which can produce absorption in the visible region. Based on this work, Kao and Hockham[9] have estimated that an impurity level of one part per million of Fe^{2+} can result in a loss of about 20 db/km (or an absorption coefficient of 4.6×10^{-5} cm^{-1}), still an uncomfortably high value. It seems probable from this result that the impurity level of all transition metal ion impurities will have to be reduced individually to at least one part per million and possibly lower if low loss glass is to be made. Recently Tynes, Pearson, and Bisbee[10] have examined the total extinction coefficients of some of the better optical materials and have found for Schott F2 glass, fused quartz, and specially prepared crystalline quartz, values of 1.3×10^{-3}, 7.3×10^{-4}, and 2×10^{-4} cm^{-1}, respectively. Since scattering is known to be low these losses must be produced by absorption and not by scattering. Such high absorption is quite unacceptable for optical transmission systems. Thus, much work lies ahead in the preparation of pure glass before useful bulk material is obtained. There is no reason to suppose, however, that this goal is unattainable—it should only require great attention to the purity of the raw materials, and to the details of the manufacturing processes.

C. Glass Fibers

For transmission purposes light will not be sent through large pieces of glass, but rather through channels or waveguides with dimensions which may only be of the order of 1 to 100 times that of the wavelength of the light. It is unlikely that these fibers will simply have these dimensions and be exposed to air; instead, to avoid losses which may arise from surface contacts or imperfections, it is probable that the fibers will be clad by another glass of somewhat lower refractive index, or by some means the outer layers of the guide will be fabricated as a lower index material. The outer material generally serves to isolate the cores from the outside environment and to make the fibers large enough to be conveniently handled.

The manufacturing of such clad fibers will require considerable processing, and it is possible that as a result of this additional losses may be introduced beyond those present in the bulk glass. Even though we do not yet have available bulk material of sufficient quality for long distance transmission, an effort has recently been made by Tynes, Pearson, and Bisbee[10] to see what losses may be introduced in making clad fibers. The fibers were made either from a hollow cylinder of one glass and a rod of another glass or, more conveniently, from a central rod of one glass surrounded initially by four touching rods of the surrounding different glass; when fused and drawn a diamond shaped central core is formed with an outer clover leaf shaped cladding. The core rod for the first structure was Schott SSK1 glass with a refractive index of 1.6171 and the cladding cylinder was of Schott SK14 glass with index 1.6038. In the clover leaf structure the core was of Schott F2 glass (index 1.6200), and the four cladding rods were of Schott K5 glass (index 1.5224). These glasses were chosen because of their similarity of refractive indexes and thermal expansion coefficients. The core diameters ranged from 2 to 16 μ. Measurements were made of the total light (at 6328 Å), transmitted through the fiber, and also of the light scattered from the fiber. The latter was measured by the integration of direct measurements of the scattered light at successive points along the fiber. Measurements were made in fibers 7 m long. The difference between the total losses and the losses due to scattering give the losses due to absorption. Table II indicates some results for the two types of fibers. It is apparent that at this level of

TABLE II

LOSSES IN CLAD FIBERS AT 6328 Å*

	Conventional fiber	Clover leaf fiber
α Total, cm^{-1}	4.20×10^{-3}	1.65×10^{-3}
α Scattered, cm^{-1}	0.19×10^{-3}	0.32×10^{-3}
α abs = α total - α scattered, cm^{-1}	4.01×10^{-3}	1.33×10^{-3}
α abs, measured in bulk core glass, cm^{-1}	3.1×10^{-3}	1.5×10^{-3}

*Core diameters about 8μ. Total length 7 meters.

absorption the formation of the clad fiber does not introduce significant additional absorption losses. (The differences between the α abs values for each of the fibers are within experimental error.)

Of particular interest in these experiments was a detailed examination of the scattering process. Photographs were taken of the light scattered from a clad fiber, and these revealed that scattering was produced by a large number of discrete scattering centers. Investigation showed that these centers were also present in the original bulk material. Figure 11 illustrates the nonuniform nature of the scattering; it is a recording made of the scattered light, obtained by means of a small integrating sphere which is moved along

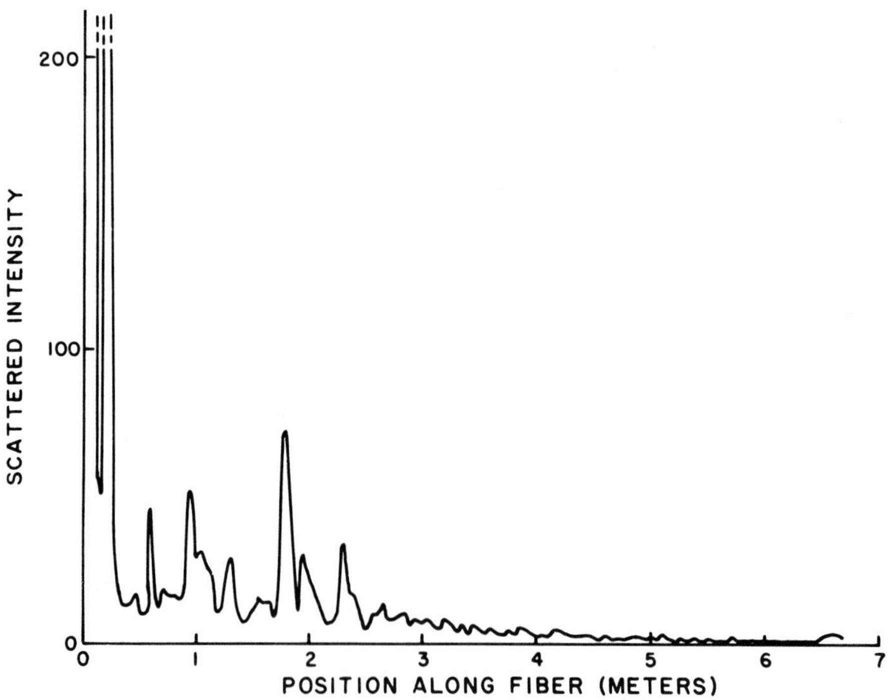

FIG. 11
The intensity of light scattered along the length of a glass fiber.

the length of the fiber. Electron microscope and X-ray investigations have revealed the presence of crystalline regions in the glass, and it is likely that these together with frozen in inhomogeneities, constitute the scattering centers. It is highly significant that in these fibers there are lengths of a few cms for which the scattering losses are very low indeed.

Again we conclude that while much work remains to be done there do not yet appear to be insuperable obstacles to overcome in the fabrication of clad fibers.

D. Glass Fibers with a Parabolic Refractive Index Gradient

So far we have had in mind a glass fiber which has a core of one index, and an abrupt change of index when the cladding commences. Fibers may also be made with a slowly changing index. An interesting case occurs when the index decreases as the square of the distance from the center. The guide now acts as a lens, so that it is possible for instance to send several spatially separate beams into the guide, and to have this separation maintained at various points along the guide. Such a guide is being considered for transmission purposes and whether this design or the one previously discussed with a rectangular index profile will ultimately prove useful will probably depend on the economics of manufacture of a low loss uniform structure.

Of more immediate interest are image transfer devices which take advantage of the lenslike property of these graded index guides. Although this principle has been understood and demonstrated for many years[11], it is only recently that concerted efforts have been made to fabricate devices from glassy materials. Uchida, et al[12] have reported such structures but have not disclosed the method of manufacture. Pearson, et al.[13] have produced a focusing rod by ion exchange. A lithium alumino silicate glass rod 1.9 mm in diameter was held in a Li, $NaNO_3$ fused salt bath at 470°C for 50 hr. Under these conditions in which the glass is above its strain point, lithium ions are replaced by sodium ions and a material of lower refractive index results. Figure 12 shows the index profile in such a rod. The solid line is that of a parabola and indicates that a reasonable approximation to the square law profile has been achieved. A rod of this material, 15 mm in length and with flat ends, was found to behave as a positive lens with an 11 mm focal length. As a lens the quality was not high since the index gradient was not precisely parabolic. Again it is clear that careful materials work will be necessary in order to exploit fully the potentialities of these devices.

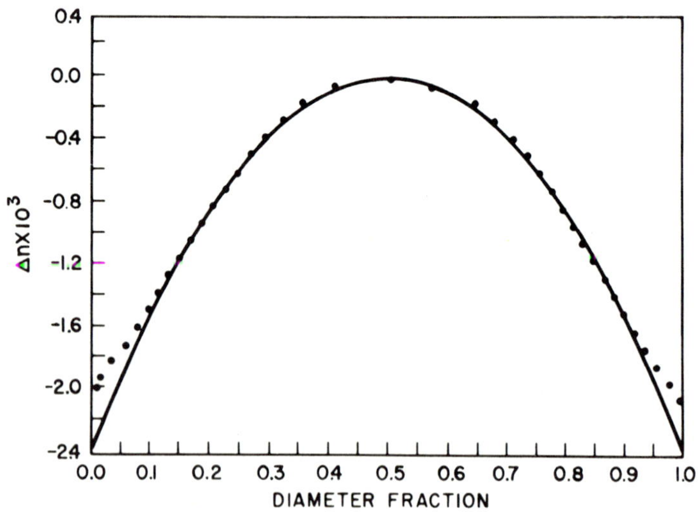

FIG. 12
The refractive index change across the diameter of a glass rod produced by ion exchange.

E. Some Proposed Components for an Optical Transmission System

So far the emphasis has been on the use of glasses as transmission media at optical frequencies. But in such an optical system components such as lasers, oscillators, modulators, and detectors will be necessary. Hitherto, comparatively large discrete building blocks have been used for these purposes. Since dimensions need really only be two to three orders of magnitude greater than the wavelength of the light used, it would be advantageous if components were developed which were analogous to the waveguides and associated devices used in the repeaters of microwave transmission systems. Recently, a series of articles have appeared in which various schemes for such devices have been described[14],[15],[16],[17]. In effect a field of "integrated optics" may be developing which will be similar to today's field of integrated electronics.

To obtain an idea of the dimensions involved one can imagine a rectangular channel of material with index n_2 and width 2a, imbedded in another medium with a lower index n_1. If a quantity Δ is defined,

$$\Delta = \frac{n_2 - n_1}{n_2}$$

then some approximate values of 2a for a guide which will support the lowest order mode, for various values of Δ are given in Table III. Estimates have also been made of allowable bending radii of such guides, and, especially for the smaller guides which require a greater index difference, sharp bends are possible so that very small circuitry becomes feasible.

The simplest form of a waveguide which might be used in such a form of optical integrated circuitry is illustrated in Fig. 13 (a). Here there is a channel with a width of a few microns consisting of

TABLE III

VALUES OF Δ FOR VARIOUS OPTICAL BEAM WIDTHS

Optical beam width, 2a	$\Delta = \frac{n_2 - n_1}{n_2}$
1 mm	10^{-6}
0.1 mm	10^{-4}
0.01 mm	10^{-2}

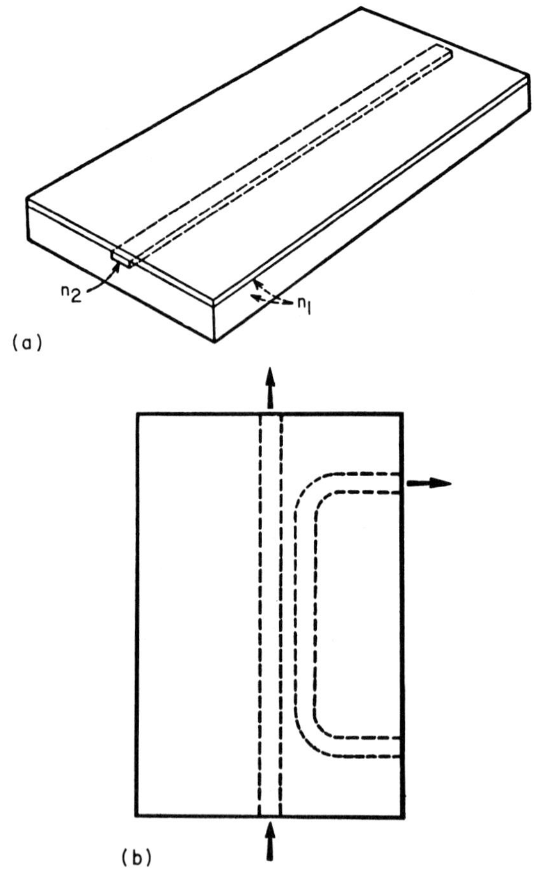

FIG. 13
(a) A schematic diagram of a planar waveguide which could be fabricated in glass using photolithographic techniques; (b) A schematic diagram of an optical microcircuit device which behaves as a directional coupler type hybrid.

material with index n_2 surrounded by a medium with index n_1. This structure might be made using photolithographic procedures similar to those currently used in making silicon integrated circuits. A window in a mask or in the photoresist could be produced on the surface of a piece of glass, and then diffusion, ion bombardment, or ionic replacement procedures could be used to change the index of the glass, thereby creating the n_2 channel imbedded in the n_1 substrate. Finally the top layer of n_1 material could be sputtered over the entire top surface. Already quite encouraging results have been obtained as to the quality of glass which has been laid down by sputtering.

It is not hard to imagine related structures which perform various circuit functions. Figure 13(b) shows a directional coupler form of hybrid. In it there are two closely spaced guides, and the exponentially decaying fields in the n_1 region provide continuous distributed coupling between the guides. In another application, since the dimensions of this type of guide are so small, quite modest voltages applied at the right points can produce very large fields. If the guides are fabricated from electrooptic materials then it is easy to see how efficient modulators can be made using techniques which incorporate very closely spaced electrodes in the circuits. Active elements may be fabricated by incorporating neodymium ions for instance, and pumping perhaps with a convenient source of light, possibly an electroluminescent device integrated into the structure. Hence, oscillators and amplifiers should become possibilities.

While work in this field is at a rudimentary level at present, it offers exciting prospects and great scope for invention in the area of materials and processing.

IV. CLOSING REMARKS

It must be clear from this talk that a rather arbitrary choice of subjects has been made. In Section II there is a description of how ceramics are being used to interconnect silicon integrated circuits and to support various passive electrical components. One hears these days a great deal about large scale integration, about crowding more and more components onto a single piece of silicon; there is no doubt that the scale of integration will steadily increase as processing techniques improve. (It is obviously of little use to increase the scale of integration if the chances of making a defect-free chip are vanishingly small.) However, interconnection patterns will always be necessary and it is likely that certain resistors and capacitors, particularly those requiring close tolerances, will always be made off the chip. So, as I mentioned, we in the Bell System expect in a few years to be using acres of high grade alumina ceramic as substrates in the integrated circuit business, and I am sure other industries will be doing likewise.

The preceding section describing some preliminary work on the components that may go into an optical communication system is of a much more nebulous character. We may be seeing the beginnings of an art which will be of immense significance, but it has some pretty fierce competition from more conventional procedures, and which system will turn out to be the most advantageous remains to be seen. But as happens so often these days success or failure of any scheme will depend in large measure on how well the materials and process scientists and engineers do their job. As a representative of this group I am at once concerned about this responsibility and worried as to how our knowledge and skills can be harnessed to these problems, yet excited by the challenge to do a job that is both intellectually satisfying and, I hope, useful to society.

ACKNOWLEDGMENTS

The author wishes to thank D. Feldman, W. B. Grupen, R. H. Meinken, S. E. Miller, A. D. Pearson, M. D. Rigterink, and others at the Bell Telephone Laboratories for many useful conversations.

REFERENCES

[1]. For a review of this subject see F. P. Gagliano, R. M. Lumley, and L. S. Watkins, Proc. IEEE, 57, 114 (1969).

[2]. See M. D. Rigterink (BTL) "The Chemical Preparation of Raw Materials for Electronic Ceramics" (Talk at the Canadian Ceramics Society Meeting in Toronto, Canada, February 11-14, 1968).

[3]. A review of this subject has been presented by J. E. Burke, Science, 161, 1205 (1968).

[4]. F. J. Schnettler, F. R. Monforte, and W. W. Rhodes, Science of Ceramics, 4, 79, (1968).

[5]. R. L. Coble, J. Amer. Ceram. Soc., 45, 123 (1961).

[6]. P. J. Jorgensen, J. Amer. Ceram. Soc., 48, 207 (1965).

[7]. R. D. Maurer, J. Chem. Phys., 25, 1206 (1956).

[8]. F. N. Steele and R. W. Douglas, Physics and Chemistry of Glasses, 6, 246 (1965).

[9]. K. C. Kao and G. A. Hockham, Proc. IEEE, 113, 1151 (1966).

[10]. Private communication from authors at Bell Telephone Laboratories, Incorporated.

[11]. R. W. Wood, Physical Optics, 3rd Ed., 1934. Macmillan, New York, p. 88.

[12]. T. Uchida, M. Furukawa, I. Kitano, K. Koizumi, and H. Matsumura, 1969 IEEE Conference on Laser Engineering and Applications, Washington, D. C.

[13]. A. D. Pearson, W. G. French, and E. G. Rawson, Appl. Phys. Letters, 15, 76 (1969).

[14]. S. E. Miller, BSTJ, 48, 2059 (1969).

[15]. E. A. J. Marcatili, BSTJ, 48, 2071 (1969).

[16]. J. E. Goell, BSTJ, 48, 2133 (1969).

[17]. E. A. J. Marcatili, BSTJ, 48, 2103 (1969).

Author Index

Numbers in brackets are reference numbers and indicate that an author's work is referred to although his name is not cited in the text. Underlined numbers show the page on which the complete reference is listed.

A

B

L

N

Q

R

T

SUBJECT INDEX